Finite Element Method with Python

An Essential Course with AI Enhanced Approaches

G.R. Liu

University of Cincinnati, USA

2026

Scientech Publisher, Mason, USA

Published by:

ScienTech Publisher, Mason, USA

Website (for ebooks): https://www.sci-en-tech.com/

Library of Congress Control Number: 2026944081

ISBN: 978-1-946018-15-1 (hardback)

ISBN: 978-1-946018-13-7 (paperback)

ISBN: 978-1-946018-14-4 (eBook)

First Edition

Printed in the United States of America

About the Author

Dr. Gui-Rong Liu received his Ph.D. from Tohoku University, Japan, in 1991. He was a Postdoctoral Fellow at Northwestern University, USA, from 1991 to 1993. He was a Professor at the National University of Singapore until 2010. He is currently a Professor at the Department of Aerospace Engineering and Engineering Mechanics, University of Cincinnati, USA. He was the Founder of the Association for Computational Mechanics (Singapore) (SACM) and served as the President of SACM until 2010. He served as the President of the Asia-Pacific Association for Computational Mechanics (APACM) (2010–2013) and an Executive Council Member of the International Association for Computational Mechanics (IACM) (2005–2010; 2020–2026). He authored a large number of journal papers and books, including two bestsellers, Mesh Free Method: Moving Beyond the Finite Element Method and Smoothed Particle Hydrodynamics: A Meshfree Particle Methods. He is the Editor-in-Chief of the International Journal of Computational Methods and served as an Associate Editor for IPSE and MANO. He is the recipient of numerous awards, including the Singapore Defence Technology Prize, NUS Outstanding University Researcher Award, NUS Best Teacher Award, APACM Computational Mechanics Award, JSME Computational Mechanics Award, ASME Ted Belytschko Applied Mechanics Award, Zienkiewicz Medal from APACM, AJCM Computational Mechanics Award, Humboldt Research Award, SACM Medal from the Association of Computational Mechanics (Singapore), and the Master Engineering Educator Award (UC, CEAS). He has been listed as one among the world's top 1% most influential scientists (Highly Cited Researchers) by Thomson Reuters for a number of years.

Contents

Chapter 1

Introduction

1.1 The Finite Element Method

The Finite Element Method (FEM) is a cornerstone of modern computational science and engineering. By simulating how physical systems behave under real-world conditions, FEM allows researchers and engineers to accurately predict deformation, vibration, heat transfer, fluid flow, and electromagnetic effects. Because of this versatility, it is an essential tool across industries like aerospace, automotive, civil infrastructure, biomechanics, energy, and advanced manufacturing.

1.1.1 Core Principles

Many fundamental problems in science, engineering, and industry are governed by **Partial Differential Equations** paired with specific **boundary conditions** (PDEs + BCs). Solving such systems to obtain reliable results presents significant numerical challenges. The primary objective of the **Finite Element Method (FEM)** is to systematically transform these continuous PDE systems into a discrete set of linear algebraic equations. These can then be efficiently solved using standard numerical routines, as illustrated in Fig. 1.1.

At its core, FEM operates by discretizing a complex physical domain into a network of smaller, interconnected sub-regions called **elements**.

Rather than attempting to solve high-order differential equations across an entire irregular geometry simultaneously, the method follows a structured three-step process:

1. **Local Formulation:** Formulates simplified governing equations for each individual element using local approximation functions.

2. **Global Assembly:** Strategically combines these local equations into a comprehensive global mathematical system (the stiffness matrix).

3. **Global Solution:** Solves the resulting system to determine the overall physical behavior, such as displacements, stresses, or temperature fields.

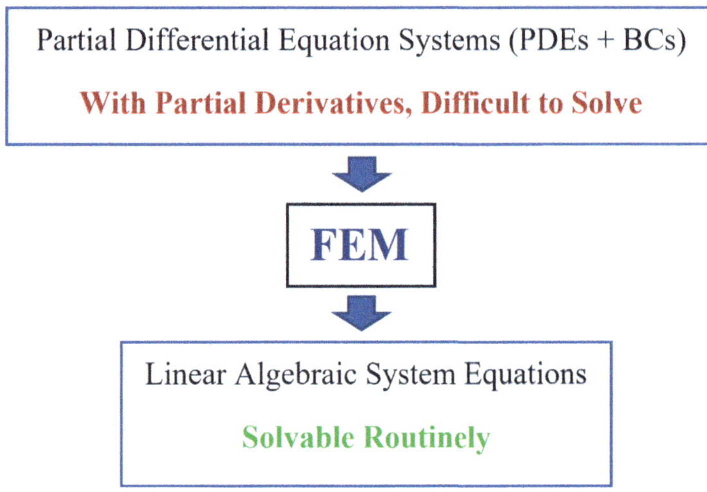

Figure 1.1: The objective of FEM: transforming continuous PDEs into discrete linear systems.

This "divide-and-conquer" strategy is what empowers engineers to analyze highly complex geometries and sophisticated nonlinear material behaviors with remarkable precision and stability.

1.1.2 Modern Impact

Today, FEM forms the computational foundation of modern Computer-Aided Engineering (CAE) software and digital simulation platforms. By allowing engineers to virtually test and optimize safer bridges, lighter aircraft, more resilient buildings, medical implants, and high-performance mechanical systems, FEM eliminates the need for costly physical prototypes. Ultimately, it has become an indispensable technology driving innovation in the modern economy.

1.2 About This Book

This book provides a comprehensive introduction to the Finite Element Method (FEM), a cornerstone of modern computational science and engineering. The text focuses on the mathematical foundations, numerical implementation, and practical coding of finite element formulations for engineering analysis.

Key topics covered include:

- Theories and principles that lead to FEM formulation
- Shape functions and interpolation theory

- Stiffness and mass matrix formulations
- Nodal Submatrix Analytical Formulations (NSAF)
- Numerical integration and Gaussian quadrature
- Boundary conditions and global assembly
- One-dimensional structural elements
- Two-dimensional solid mechanics
- hree-dimensional continuum elements
- Vectorized FEM implementations using Python

Each topic is developed systematically from first principles, combining mathematical derivations with practical computational implementation. Rather than treating FEM as a black-box numerical tool, this book emphasizes the internal structure of finite element formulations and the algorithmic logic behind professional simulation software.

1.3 Key Features of This Text

- **Integrated Approach:** Designed for classroom learning, self-study, and research-oriented exploration, the book integrates theory, numerical methods, and executable Python codes directly throughout the text. Readers are encouraged to modify, experiment with, and extend the provided implementations to develop a deeper understanding of computational mechanics and scientific computing.
- **Interactive Jupyter Notebook Format:** Presented in an interactive format, the book unifies mathematical equations, derivations, visualizations, and executable programs into a single environment. This approach allows readers to immediately test concepts, inspect numerical results, and visualize FEM behavior directly alongside the theory. Heavy symbolic manipulation and numerical computation are delegated to modern computational tools, enabling readers to focus on the underlying physical meaning and computational principles.
- **Practical Foundation:** This work draws inspiration from classical FEM literature, modern computational mechanics research, and more than four decades of the author's teaching and coding experience. Open-source scientific libraries such as **NumPy**, **SciPy**, **SymPy**, and **Matplotlib**, alongside advanced mesh generation tools, play a central role in the development of the examples and solvers. Discussions and collaborative interactions with modern AI systems, including Gemini and ChatGPT, contributed to code development, optimization, and manuscript preparation.

In summary, this book introduces FEM from first principles and gradually develops complete, working solvers through practical examples and clear mathematical derivations. Equal emphasis is placed on theory, implementation, computational efficiency, and the development of high-performance FEM codes using Python.

1.4 Why This Book?

The author, Dr. GR Liu, has been active in the field of computational methods for more than 45 years. Since developing his first nonlinear Finite Element Method (FEM) code in 1980, he has published over 600 journal papers and nearly 20 monographs, including the widely used textbook *The Finite Element Method: A Practical Course* (First Edition, 2003; Second Edition, 2013). Throughout decades of research, teaching, and software development, one conclusion has remained unchanged: FEM is one of the most powerful and practical computational tools in science, engineering, and industry.

In earlier years, FEM formulations were derived almost entirely by hand through laborious symbolic manipulations, often resulting in massive collections of handwritten manuscripts and lengthy manual calculations. With the rapid advancement of artificial intelligence and symbolic computing technologies, many of these derivations can now be performed far more efficiently and systematically. Furthermore, coding, matrix vectorization, and numerical optimization have become significantly more accessible with modern AI-assisted development tools.

More importantly, these emerging technologies enable the discovery and development of new computational formulations. For example, the Nodal Submatrix Analytical Formulations (NSAF) presented in this book were derived symbolically using **SymPy** and implemented with the assistance of AI-based coding tools such as Gemini.

This book revisits the classical subject of FEM through a modern computational lens, combining rigorous finite element theory with AI-assisted symbolic derivation, optimized numerical implementation, and high-performance Python codes. This integrated approach enables readers to study, comprehend, code, practice, and apply FEM techniques in a far more effective and streamlined manner.

1.5 Target Audience

This book is intended for beginners, university students, researchers, and engineering professionals across a wide range of STEM disciplines who wish to learn and apply the Finite Element Method (FEM) to solid mechanics problems. By combining rigorous theoretical foundations with practical implementation in Python code, the text is uniquely suited for both structured academic courses and self-directed learning.

For readers seeking additional background in continuum mechanics and structural behavior, it is highly recommended to first study *Mechanics of Materials: Formulations and Solutions with Python* [4].

Note: This book is written for human readers. It may not be used for training or testing automated systems without explicit permission from the author.

1.6 Use of the Book and Codes

Individual readers who purchase this book may contact the author at liugr100@gmail.com to request a complimentary soft copy in Jupyter Notebook format. This version includes executable code (subject to updates) and is available for academic use upon registration.

If the book is adopted as a textbook, instructors may contact the author at the same email address to request access to selected solution manuals.

Conditions of Use

The following conditions apply to the use of the book and the codes developed by the author, whether in hard copy, soft copy, or any other medium:

- **User Responsibility:** Users assume all risks associated with the application of any part of the book, codes, or techniques.
- **Code Purpose:** The provided codes are primarily intended for **conceptual demonstration** and may not be optimized for computational efficiency or industrial robustness. Most codes have not undergone third-party testing.
- **Limited Use:** The book and its accompanying codes are for the **personal use** of the individual purchaser. Redistribution or unauthorized copying is strictly prohibited without the author's prior consent.
- **No User Support:** Technical support is not provided for the codes.
- **Citation:** Proper citation and acknowledgment are mandatory when referencing the book, codes, ideas, or techniques in any subsequent work.

These codes rely on various external packages and modules. Their performance is highly dependent on the specific versions of Python and the installed libraries. If a code fails to run as expected, a version mismatch is the most likely cause. You can verify your environment versions directly within a Jupyter Notebook cell. For example, to check the current Python version, you can use:

```
!python -V                 #! is used to execute an external command
```

```
Python 3.9.16
```

```
!jupyter notebook --version
```

```
6.1.5
```

Troubleshooting Version Issues

If you encounter a version mismatch error, you can either adapt the code to match your specific environment or install the required version on your system. It is often helpful to search the web using the specific error message to find potential solutions or leads. Most conveniently, large language models can be highly effective in this regard; this approach is the author's preferred method when dealing with code execution issues.

Since our descriptions in a chapter are integrated with code cells, a specific cell often depends on the execution of those preceding it in the chapter. If an error occurs while running a cell, you should ensure that all relevant cells above it have been executed.

Learning Python

This book does not cover Python basics, as a wealth of resources is already available online. Interested readers may refer to Chapters 2 and 3 in Ref. [3] for a concise overview of using Python for scientific computations. Because Python is relatively intuitive, you can utilize the codes and examples in this book to jump-start your learning while simultaneously studying the technical subjects.

1.7 Use of External Modules and Dependencies

Module Imports in Jupyter Notebooks

Modules frequently used throughout a chapter are imported in the very first code cell, which must be executed at the beginning of your session. If you encounter a name error (such as `NameError: name 'sp' is not defined`), return to the start of the chapter and re-run that initial cell.

Editing Codes

- **Text Editors:** You can use any standard text editor to view and modify the source code.
- **Direct Editing:** Alternatively, you can copy the code into a cell within a Jupyter Notebook and edit it directly to see immediate results.

1.8 Utilizing the `help()` Function

Exploring Modules and Objects

To access detailed information about a module or an object after importing it, you can use the built-in `help()` function. This is an efficient way to explore the functionalities, methods, and attributes available within different libraries.

Code Comments

The code cells throughout the book include comments to provide additional context. These comments, denoted by a hash symbol (`#`), are typically placed on the right side of the code to offer quick references and guidance without disrupting the logical flow of the script.

Acknowledgments

The author's work has been profoundly influenced by the foundational textbook on FEM by Prof. Y.Q. Long [15]. Having studied the earliest edition in the late 1970s, the author has remained deeply engaged with this subject ever since. Furthermore, the author's research and development in the field of FEM have been greatly enriched by the seminal contributions of Prof. O.C. Zienkiewicz and Prof. R.L. Taylor [17], as well as the instructional clarity provided by the textbook of Prof. L.J. Segerlind [16].

This book was prepared with the assistance of AI language models, including Gemini, ChatGPT, and DeepSeek, which contributed to drafting text, generating code, optimization of codes, and verifying technical details. We also draw upon the author's earlier book on finite element method [12]. Part of the cover image was created using Gemini AI.

Chapter 2

Fundamentals of Solid Mechanics

Solid mechanics is one of the foundational disciplines in science and engineering. Since virtually all physical structures—from microelectromechanical systems to large aerospace structures—are composed of solid materials, the field has been extensively developed in its theoretical, mathematical, and engineering aspects.

This chapter presents a concise overview of the **governing equations** of solid mechanics that are essential for developing linear elastic finite element method (FEM) solvers. The emphasis is placed on the final formulations required for FEM implementation. Readers interested in detailed theoretical derivations are encouraged to consult foundational references such as Ref. [4].

2.1 General Formulation for 3D Solids

In this section, we establish the general three-dimensional (3D) governing equations. These formulations serve as the "parent" equations, which can subsequently be reduced to two-dimensional (2D) and one-dimensional (1D) cases through specific kinematic and kinetic assumptions.

2.1.1 Domain and Boundary

Consider a 3D solid body occupying a spatial domain Ω, enclosed by its total boundary Γ. A point within this body is defined by the Cartesian coordinate vector $\mathbf{x} = [x, y, z]^T$, or in index notation as (x_1, x_2, x_3), as illustrated in Fig. 2.1.

To ensure a well-posed problem, the boundary Γ is partitioned into two distinct regions:

1. **The Displacement (Dirichlet) Boundary (Γ_u):** Where the displacement components are prescribed (e.g., fixed supports).
2. **The Traction (Neumann) Boundary (Γ_t):** Where external surface tractions or forces are applied.

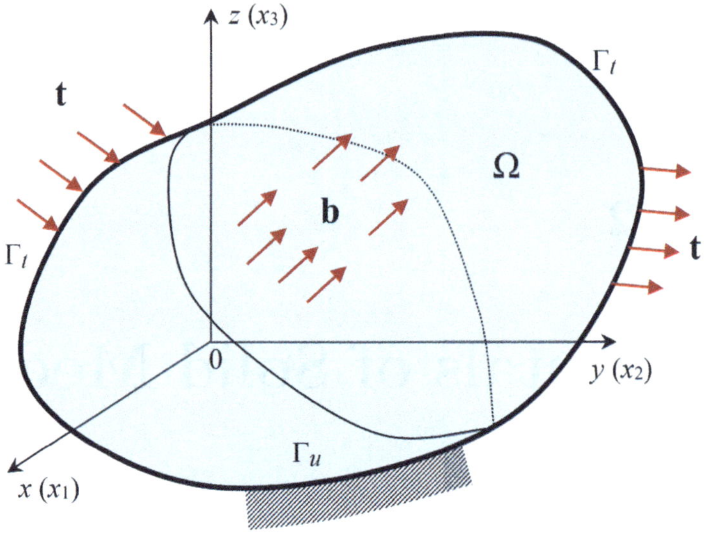

Figure 2.1: A 3D solid body in domain Ω subjected to body forces $\mathbf{b}$, prescribed displacements on Γ_u, and surface tractions on Γ_t.

These boundaries are complementary, such that $\Gamma = \Gamma_u \cup \Gamma_t$ and $\Gamma_u \cap \Gamma_t = \emptyset$.

2.1.2 Displacements

The primary unknown in a displacement-based FEM formulation is the displacement field, $\mathbf{u}$. At any point within the domain Ω, the displacement vector is defined as:

$$\mathbf{u}(x, y, z) = \begin{Bmatrix} u \\ v \\ w \end{Bmatrix}$$

where u, v, and w represent the components of displacement in the x, y, and z directions, respectively. In the context of linear elasticity, we assume these displacements (and their gradients) are small, allowing us to define the relationship between deformation and force linearly.

2.1.3 Forces or Loads

The solid is generally subjected to two types of external loads:

- **Body Forces (b):** Forces acting on the volume of the body, such as gravity or centrifugal forces, expressed as force per unit volume: $\mathbf{b} = [b_x, b_y, b_z]^T$.
- **Surface Tractions ($\bar{\mathbf{t}}$):** Forces acting on the boundary Γ_t, such as pressure or friction, expressed as force per unit area: $\bar{\mathbf{t}} = [t_x, t_y, t_z]^T$.

The internal state of the solid is described by the **stress tensor** and **strain tensor**, which must satisfy the conditions of equilibrium, compatibility, and the constitutive laws (material properties) described in the following sections.

2.1.4 Strain–Displacement Relations

To describe the deformation of the solid, we must relate the displacement field to the internal strain. This relationship is founded on the following fundamental kinematic assumption:

> **The Small Deformation Assumption:** The solid undergoes infinitesimal displacements $\mathbf{u} = (u, v, w)$. These displacements result in small strains and stresses, such that the geometry of the body and the material properties remain effectively unchanged during the loading process. Under these conditions, the kinematic behavior of any point in the solid is considered linear.

2.1.4.1 Tensor and Component Form

Under the assumption of small deformations, the strain tensor ε is defined as the symmetric part of the displacement gradient [4]:

$$\varepsilon = \frac{1}{2}\left[\nabla\mathbf{u} + (\nabla\mathbf{u})^{\top}\right] \tag{2.1}$$

In Cartesian component form (index notation), this is expressed as:

$$\varepsilon_{ij} = \frac{1}{2}\left(\frac{\partial u_i}{\partial x_j} + \frac{\partial u_j}{\partial x_i}\right), \quad i, j \in 1, 2, 3 \tag{2.2}$$

2.1.4.2 Engineering Strain and Matrix Representation

In practical engineering and FEM implementation, it is common to represent the six independent components of the symmetric strain tensor as a vector. This is known as **Voigt notation**. The explicit relations are:

$$\begin{aligned}
&\text{Normal Strains:} \quad \varepsilon_{xx} = \frac{\partial u}{\partial x}, \quad \varepsilon_{yy} = \frac{\partial v}{\partial y}, \quad \varepsilon_{zz} = \frac{\partial w}{\partial z} \\
&\text{Shear Strains:} \quad \gamma_{yz} = \frac{\partial v}{\partial z} + \frac{\partial w}{\partial y}, \quad \gamma_{zx} = \frac{\partial w}{\partial x} + \frac{\partial u}{\partial z}, \quad \gamma_{xy} = \frac{\partial u}{\partial y} + \frac{\partial v}{\partial x}
\end{aligned} \tag{2.3}$$

> **Note:** The shear components γ_{ij} are **engineering shear strains**, which represent the total change in angle between two originally perpendicular lines. They are related to the tensor shear strains by $\gamma_{ij} = 2\varepsilon_{ij}$ (for $i \neq j$).

These six relations can be written compactly in matrix-vector form as:

$$\varepsilon = \mathbf{L}_d \mathbf{u} \tag{2.4}$$

where the displacement vector $\mathbf{u}$ and the strain vector ε are:

$$\mathbf{u} = \begin{bmatrix} u \\ v \\ w \end{bmatrix}, \quad \varepsilon = \begin{bmatrix} \varepsilon_{xx} \\ \varepsilon_{yy} \\ \varepsilon_{zz} \\ \gamma_{yz} \\ \gamma_{zx} \\ \gamma_{xy} \end{bmatrix} \tag{2.5}$$

Tip: Note the ordering of the shear components. A common convention follows the "missing index" logic: σ_{yz} (missing x) appears first among the shear terms, followed by σ_{zx} (missing y) and σ_{xy} (missing z).

The differential operator matrix $\mathbf{L}_d$ is derived directly from Eq. (2.3) following the standard matrix multiplication rule:

$$\mathbf{L}_d = \begin{bmatrix} \frac{\partial}{\partial x} & 0 & 0 \\ 0 & \frac{\partial}{\partial y} & 0 \\ 0 & 0 & \frac{\partial}{\partial z} \\ 0 & \frac{\partial}{\partial z} & \frac{\partial}{\partial y} \\ \frac{\partial}{\partial z} & 0 & \frac{\partial}{\partial x} \\ \frac{\partial}{\partial y} & \frac{\partial}{\partial x} & 0 \end{bmatrix} \tag{2.6}$$

2.1.4.3 Properties of the Operator $\mathbf{L}_d$

The matrix $\mathbf{L}_d$ is a fundamental object in computational mechanics:

- **Strain Mapping:** $\mathbf{L}_d$ acts as a specialized **gradient operator**. It maps a 3D displacement field into a 6D strain space through differential operators.
- **Dual Relationship:** Interestingly, the transpose operator $\mathbf{L}_d^\top$ acts as a specialized **divergence operator**. In the following sections, we will see that $\mathbf{L}_d^\top$ maps the 6D stress vector back into a 3D internal force vector, forming the basis for the equilibrium equations.

2.1.5 Constitutive Equations

The constitutive equations, also known as the stress–strain relations, define the physical relationship between the stress tensor and the strain tensor based on the material properties. To maintain a linear framework, we introduce the second fundamental assumption:

The Linear Elastic Material Assumption: The solid is composed of a **linear elastic material**, meaning that the internal stresses depend linearly on the strains. This relationship is characterized by Generalized Hooke's Law.

2.1.5.1 Tensor Formulation

In its most general form, the relationship between the second-order stress tensor σ_{ij} and the second-order strain tensor ε_{kl} is given by [4]:

$$\sigma_{ij} = C_{ijkl}\, \varepsilon_{kl}; \quad i, j, k, l = 1, 2, 3 \tag{2.7}$$

where C_{ijkl} is a fourth-order elasticity tensor. While this tensor theoretically contains 81 components, symmetries in the stress and strain tensors, along with thermodynamic considerations, reduce the number of independent constants significantly.

For an **isotropic material**—where material properties are identical in all directions—the relationship is simplified using the Lamé constants, λ and G:

$$\sigma_{ij} = 2G\varepsilon_{ij} + \lambda\delta_{ij}\varepsilon_{kk}, \quad i, j = 1, 2, 3 \tag{2.8}$$

Here, δ_{ij} is the Kronecker delta, and ε_{kk} represents the **volumetric strain** (the trace of the strain tensor), which characterizes the change in volume per unit volume.

2.1.5.2 Matrix Formulation (Voigt Notation)

While indicial notation is concise for theory, FEM implementations rely on matrix formulations. By exploiting the symmetry of the stress tensor ($\sigma_{ij} = \sigma_{ji}$), we represent the six independent stress components as a vector:

$$\sigma = \begin{bmatrix} \sigma_{xx} & \sigma_{yy} & \sigma_{zz} & \sigma_{yz} & \sigma_{zx} & \sigma_{xy} \end{bmatrix}^\top \tag{2.9}$$

Note for the shear stress components the same "missing index" logic applies.

In this vector-matrix form, the constitutive equation is written as:

$$\sigma = \mathbf{C}\varepsilon \tag{2.10}$$

For isotropic materials, the material matrix $\mathbf{C}$ in terms of Lamé constants is:

$$\mathbf{C} = \begin{bmatrix} \lambda + 2G & \lambda & \lambda & 0 & 0 & 0 \\ \lambda & \lambda + 2G & \lambda & 0 & 0 & 0 \\ \lambda & \lambda & \lambda + 2G & 0 & 0 & 0 \\ 0 & 0 & 0 & G & 0 & 0 \\ 0 & 0 & 0 & 0 & G & 0 \\ 0 & 0 & 0 & 0 & 0 & G \end{bmatrix} \tag{2.11}$$

2.1.5.3 Engineering Constants: E and ν

In engineering practice, it is often more intuitive to define material behavior using **Young's Modulus** (E) and **Poisson's Ratio** (ν). The Lamé constants are related to these via:

$$\lambda = \frac{E\nu}{(1+\nu)(1-2\nu)}, \quad G = \frac{E}{2(1+\nu)}$$

Substituting these into Eq. (2.8), the stress components can be expressed explicitly as:

$$\begin{aligned}
\sigma_{11} &= \frac{E}{(1+\nu)(1-2\nu)}[(1-\nu)\varepsilon_{11} + \nu(\varepsilon_{22} + \varepsilon_{33})] \\
\sigma_{22} &= \frac{E}{(1+\nu)(1-2\nu)}[(1-\nu)\varepsilon_{22} + \nu(\varepsilon_{11} + \varepsilon_{33})] \\
\sigma_{33} &= \frac{E}{(1+\nu)(1-2\nu)}[(1-\nu)\varepsilon_{33} + \nu(\varepsilon_{11} + \varepsilon_{22})] \\
\sigma_{23} &= G\gamma_{23}, \quad \sigma_{13} = G\gamma_{13}, \quad \sigma_{12} = G\gamma_{12}
\end{aligned} \tag{2.12}$$

Finally, for stable materials (those that do not flow like fluids), the matrix $\mathbf{C}$ is positive definite and invertible. This allows us to calculate the strain resulting from a known state of stress:

$$\varepsilon = \mathbf{C}^{-1}\sigma \tag{2.13}$$

where $\mathbf{C}^{-1}$ is often referred to as the **compliance matrix**.

2.1.6 Equilibrium Equations

The governing equations for a solid body are derived from the principle of conservation of momentum (dynamic equilibrium). For a solid in motion, the balance of internal stresses, body forces, and inertial forces must be satisfied at every point within the domain Ω.

2.1.6.1 Matrix Formulation of Equilibrium

Using the differential operator $\mathbf{L}_d$ introduced in Section 1.2.4, the dynamic equilibrium equations can be written in a remarkably compact matrix form [12]:

$$\mathbf{L}_d^\top \sigma + \mathbf{b} = \rho\ddot{\mathbf{u}} \tag{2.14}$$

where:

- σ is the 6D stress vector.
- $\mathbf{b} = [b_x, b_y, b_z]^\top$ is the body force vector (3D) per unit volume (e.g., gravity).
- ρ is the material density (scalar).
- $\ddot{\mathbf{u}} = \frac{\partial^2\mathbf{u}}{\partial t^2}$ is the acceleration vector.

For **static equilibrium**, the acceleration term vanishes ($\rho\ddot{\mathbf{u}} = \mathbf{0}$), reducing the equation to a balance between internal stress gradients and external body forces. This book studies only static problems. However, we try to leave the dynamic terms in the equations as the by-product in the derivation process.

2.1.6.2 Displacement-Based Governing Equations

By substituting the strain–displacement relation ($\varepsilon = \mathbf{L}_d\mathbf{u}$) and the constitutive relation ($\sigma = \mathbf{C}\varepsilon$) into Eq. (2.14), we obtain the governing equations entirely in terms of the displacement field $\mathbf{u}$:

$$\mathbf{L}_d^T\mathbf{C}\mathbf{L}_d\mathbf{u} + \mathbf{b} = \rho\ddot{\mathbf{u}} \tag{2.15}$$

Note that $\mathbf{L}_d^T\mathbf{C}\mathbf{L}_d$ represents a second-order linear differential operator acting on the displacement field. Because the material constitutive matrix $\mathbf{C}$ is symmetric, this operator exhibits fundamental mathematical symmetry within the appropriate inner product space. Consequently, it is classified as a **self-adjoint** (or premium symmetric) differential operator.

As will be discussed in later chapters, when the standard **Galerkin weak form**—which utilizes the same spatial space for both the trial displacement fields and the virtual test functions—is applied to discretize the governing equations, this self-adjoint property is preserved. This mathematical inheritance is what fundamentally guarantees the symmetry of the resulting finite element system matrices, ensuring highly efficient storage schemes and the use of optimized symmetric numerical solvers.

Equation (2.15) represents a system of three coupled, second-order partial differential equations. For isotropic materials, expanding this matrix expression yields the classical **Navier Equations**:

$$\begin{aligned}
G\nabla^2 u + (G+\lambda)\frac{\partial}{\partial x}(\nabla\cdot\mathbf{u}) + b_x &= \rho\ddot{u} \\
G\nabla^2 v + (G+\lambda)\frac{\partial}{\partial y}(\nabla\cdot\mathbf{u}) + b_y &= \rho\ddot{v} \\
G\nabla^2 w + (G+\lambda)\frac{\partial}{\partial z}(\nabla\cdot\mathbf{u}) + b_z &= \rho\ddot{w}
\end{aligned} \tag{2.16}$$

where ∇^2 is the Laplacian operator and $\nabla\cdot\mathbf{u}$ is the divergence of the displacement field. Equation (2.16) can be written in the concise form of Eq.(2.15), because of the special structure of the differential operator matrix $\mathbf{L}_d$.

2.1.6.3 The "Strong Form" vs. the "Weak Form"

Equations (2.14) through (2.16) are known as the **Strong Form** of the problem because they must be satisfied exactly at every single point in the domain Ω.

Solving the Strong Form analytically is extremely difficult and is generally only possible for very simple geometries and boundary conditions [8]. This difficulty is the primary motivation for the Finite Element Method. In FEM, we do not solve these PDEs directly. Instead, we convert them into a **Weak Form** (an integral statement), which:

1. Relaxes the continuity requirements on the displacement field (reducing second-order derivatives to first-order).

2. Allows for the approximate solution of complex engineering problems by satisfying equilibrium in an average (integral) sense over discretized sub-domains (elements).

The detailed derivation of the Weak Form using the Principle of Virtual Work will be the focus of Chapter 5.

2.1.7 Summary of the 3D Solid Mechanics Framework

Before moving to 2D reductions, readers should keep this "Chain of Logic" in mind:

1. **Kinematics (strain-displacement relations):** Displacements $\mathbf{u} \xrightarrow{\mathbf{L}_d}$ Strains ε.

2. **Physics (material behavior):** Strains $\varepsilon \xrightarrow{\mathbf{C}}$ Stresses σ.

3. **Equilibrium (mechanics law):** Stresses $\sigma \xrightarrow{\mathbf{L}_d^\top}$ Forces/Balance.

This circular dependency is what defines the structural response of any solid body.

2.1.8 Boundary Conditions

To obtain a unique solution with proper causality for the governing partial differential equations, we must specify the conditions acting on the boundaries of the solid. As introduced in Section 1.2.1, the total boundary Γ is partitioned into two disjoint sets:

$$\Gamma_u \cup \Gamma_t = \Gamma, \quad \text{and} \quad \Gamma_u \cap \Gamma_t = \emptyset$$

2.1.8.1 Dirichlet (Displacement) Boundary Conditions

On the portion of the boundary denoted by Γ_u, the displacement field is explicitly prescribed. These are also known as **Dirichlet boundary conditions**, because it constraints the primary variable of the PDE:

$$\mathbf{u} = \bar{\mathbf{u}} \quad \text{on} \quad \Gamma_u \tag{2.17}$$

where $\bar{\mathbf{u}}$ is the prescribed displacement vector (e.g., $\bar{\mathbf{u}} = \mathbf{0}$ for a fixed support).

2.1.8.2 Neumann (Traction) Boundary Conditions

On the portion of the boundary denoted by Γ_t, the external surface forces or tractions are prescribed. These are known as **Neumann boundary conditions**, because it constraints the derivatives of the primary variable of the PDE. The relationship between the internal stress tensor and the external traction vector $\bar{\mathbf{t}}$ is governed by Cauchy's formula:

$$\mathbf{L}_n^\top \sigma = \bar{\mathbf{t}} \quad \text{on} \quad \Gamma_t \tag{2.18}$$

where $\mathbf{n} = [n_x, n_y, n_z]^\top$ is the outward unit normal vector to the surface. The normal mapping matrix $\mathbf{L}_n$ is defined as:

$$\mathbf{L}_n = \begin{bmatrix} n_x & 0 & 0 \\ 0 & n_y & 0 \\ 0 & 0 & n_z \\ 0 & n_z & n_y \\ n_z & 0 & n_x \\ n_y & n_x & 0 \end{bmatrix} \tag{2.19}$$

> **Insight:** Notice that the algebraic structure of $\mathbf{L}_n$ is identical to the differential operator $\mathbf{L}_d$ (replacing $\partial/\partial x_i$ with n_i). This symmetry ensures that the internal work done by stresses is consistently matched by the external work done by surface tractions.

2.1.9 Summary of the 3D Linear Elastic System

The mathematical model for a linear solid mechanics problem is now complete. It consists of a closed system of equations:

1. **Kinematics:** $\varepsilon = \mathbf{L}_d\mathbf{u}$ (Strain–Displacement)
2. **Material Law:** $\sigma = \mathbf{C}\varepsilon$ (Stress–Strain)
3. **Equilibrium:** $\mathbf{L}_d^\top\sigma + \mathbf{b} = \rho\ddot{\mathbf{u}}$ (Governing PDE)
4. **Constraints:** $\mathbf{u} = \bar{\mathbf{u}}$ on Γ_u and $\mathbf{L}_n^\top\sigma = \bar{\mathbf{t}}$ on Γ_t (BCs)

Connection to Modern Computational Methods

In the **Finite Element Method (FEM)**, these equations are transformed into a weak integral form to solve for nodal displacements. However, in the emerging field of **Physics-Informed Neural Networks (PINNs)**, these strong-form equations can also be used directly [7].

2.2 Formulation for 2D Solids

Many engineering problems can be simplified to two dimensions, when the geometry and loading exhibit specific symmetries, or when one dimension is significantly different in scale compared to the others. Consider a 2D solid body defined in the spatial domain $\Omega \subset \mathbb{R}^2$, bounded by Γ, with Cartesian coordinates (x, y) or (x_1, x_2).

As with the 3D case, we assume the material is **linear elastic** and the system undergoes **small deformations.** A significant advantage of using the matrix-vector (Voigt) representation is that the structure of the governing equations is often "dimensionality blind." Whether in 3D, 2D, or 1D, the fundamental symbolic relationships remain consistent, with only the internal dimensions of the vectors and operator matrices changing.

2.2.1 Strain–Displacement Relations in 2D

For a 2D solid, the displacement vector $\mathbf{u}$ consists of two components:

$$\mathbf{u} = \begin{bmatrix} u \\ v \end{bmatrix} \tag{2.20}$$

where u and v are displacements in the x and y directions, respectively.

The strain–displacement relation remains identical in its operator form to the 3D case:

$$\boxed{\varepsilon = \frac{1}{2}\left[\nabla \mathbf{u} + (\nabla \mathbf{u})^{\top}\right]} \tag{2.21}$$

In component form, this represents a reduction of the 3D tensor, focusing only on the in-plane deformations:

$$\begin{aligned} \text{Normal Strains:} \quad & \varepsilon_{xx} = \frac{\partial u}{\partial x}, \quad \varepsilon_{yy} = \frac{\partial v}{\partial y} \\ \text{Shear Strains:} \quad & \gamma_{xy} = \frac{\partial u}{\partial y} + \frac{\partial v}{\partial x} \end{aligned} \tag{2.22}$$

Matrix Representation

Using Voigt notation for a 2D system, we consider three independent strain components: two normal strains and one engineering shear strain. The strain-displacement relation has the same form as for 3D solid:

$$\varepsilon = \mathbf{L}_d \mathbf{u} \tag{2.23}$$

The details in the 2D differential operator matrix $\mathbf{L}_d$ becomes:

$$\mathbf{L}_d = \begin{bmatrix} \frac{\partial}{\partial x} & 0 \\ 0 & \frac{\partial}{\partial y} \\ \frac{\partial}{\partial y} & \frac{\partial}{\partial x} \end{bmatrix} \tag{2.24}$$

2.2.2 Stress and Strain Vectors

The state of stress and strain at any point in a 2D solid is represented by vectors of dimension 3:

$$\sigma = \begin{bmatrix} \sigma_{xx} \\ \sigma_{yy} \\ \sigma_{xy} \end{bmatrix}, \quad \varepsilon = \begin{bmatrix} \varepsilon_{xx} \\ \varepsilon_{yy} \\ \gamma_{xy} \end{bmatrix} \tag{2.25}$$

As previously noted, we adopt the **engineering shear strain** $\gamma_{xy} = 2\varepsilon_{xy}$. This convention ensures that the internal work density (the product of stress and strain) remains consistent with the tensor-based energy definitions without requiring additional scale factors of 2 in the matrix formulations [4].

2.2.3 Note: Dimensionality Blindness

Observe the beauty of the matrix formulation:

- **In 3D:** $\mathbf{L}_d$ is 6×3, σ and ε are 6×1.
- **In 2D:** $\mathbf{L}_d$ is 3×2, σ and ε are 3×1.

Despite these changes in size, the high-level equations for equilibrium ($\mathbf{L}_d^\top \sigma + \mathbf{b} = \mathbf{0}$) and kinematics ($\varepsilon = \mathbf{L}_d \mathbf{u}$) remain symbolically invariant. This makes the transition between different structural models (1D bars, 2D plates, 3D solids) very efficient in implementation.

2.2.4 Constitutive Equations for 2D Problems

In 2D, the relationship $\sigma = \mathbf{C}\varepsilon$ remains the central governing material law. However, the specific components of the material matrix $\mathbf{C}$ depend on the physical constraints of the problem. When reducing a 3D reality to a 2D model, we must choose between two distinct physical states: **Plane Stress** or **Plane Strain**.

2.2.4.1 Plane Stress: Thin Structures

The plane stress assumption is ideal for **thin, plate-like structures** where the thickness is significantly smaller than the other dimensions, and the loads are applied strictly within the xy-plane (see Fig. 2.2).

- **Physical Assumption:** Because the surfaces are free of external pressure, the out-of-plane stresses are assumed to be zero throughout the thickness:

$$\sigma_{zz} = \sigma_{xz} = \sigma_{yz} = 0 \tag{2.26}$$

- **Important Note:** While the stress σ_{zz} is zero, the out-of-plane strain ε_{zz} is **not** zero. The material is free to expand or contract in the thickness direction due to the Poisson effect.

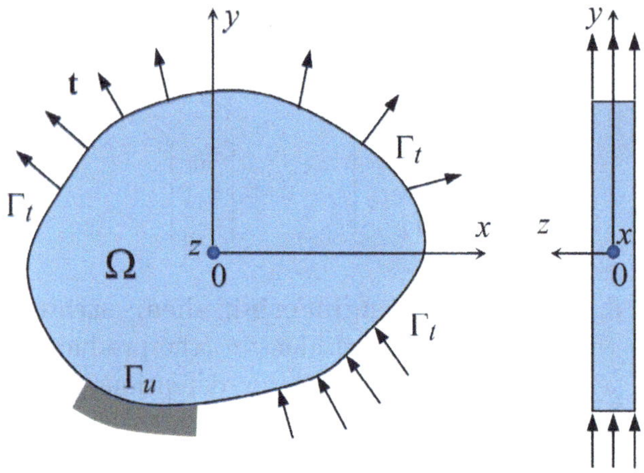

Figure 2.2: Plane stress problem: a thin plate.

The resulting constitutive matrix is [4]:

$$\mathbf{C} = \frac{E}{1-\nu^2}\begin{bmatrix} 1 & \nu & 0 \\ \nu & 1 & 0 \\ 0 & 0 & \frac{1-\nu}{2} \end{bmatrix} \tag{2.27}$$

2.2.4.2 Plane Strain: Elongated Structures

The plane strain assumption applies to **long, prismatic structures** with a constant cross-section subjected to loads that do not vary along the longitudinal (z) axis. A classic example is a gravity dam or a long buried pipe (see Fig. 2.3).

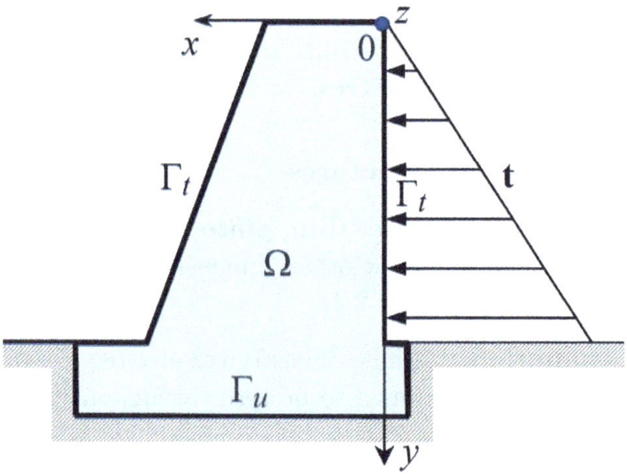

Figure 2.3: Plane strain problem: a 2D cross-section of an elongated 3D solid body.

- **Physical Assumption:** The surrounding material or supports prevent any movement in the longitudinal direction, meaning the out-of-plane strains vanish:

$$\varepsilon_{zz} = \gamma_{xz} = \gamma_{yz} = 0 \tag{2.28}$$

- **Important Note:** Because the material is prevented from expanding/contracting in the z-direction, an out-of-plane stress σ_{zz} is developed to maintain this constraint ($\sigma_{zz} = \nu(\sigma_{xx} + \sigma_{yy})$).

The constitutive matrix for plane strain is [4]:

$$\mathbf{C} = \frac{E}{(1+\nu)(1-2\nu)} \begin{bmatrix} 1-\nu & \nu & 0 \\ \nu & 1-\nu & 0 \\ 0 & 0 & \frac{1-2\nu}{2} \end{bmatrix} \tag{2.29}$$

2.2.5 Comparison of 2D Constitutive Relations

Feature	Plane Stress	Plane Strain
Typical Structure	Thin plates, membranes	Dams, tunnels, long pipes
Vanishing Component	$\sigma_{zz} = 0$	$\varepsilon_{zz} = 0$
Non-Zero Hidden Component	$\varepsilon_{zz} = -\frac{\nu}{E}(\sigma_{xx} + \sigma_{yy})$	$\sigma_{zz} = \nu(\sigma_{xx} + \sigma_{yy})$
Material Matrix C	Modified E and ν	Direct reduction of 3D **C**

2.2.6 Summary for Implementation

Regardless of the choice, the stress components are calculated as:

$$\begin{aligned} \sigma_{xx} &= C_{11}\varepsilon_{xx} + C_{12}\varepsilon_{yy} \\ \sigma_{yy} &= C_{12}\varepsilon_{xx} + C_{11}\varepsilon_{yy} \\ \sigma_{xy} &= C_{33}\gamma_{xy} \end{aligned} \tag{2.30}$$

The matrix of material constants ,**C**, can be written in a unified form of :

$$\mathbf{C} = \begin{bmatrix} C_{11} & C_{12} & 0 \\ C_{12} & C_{22} & 0 \\ 0 & 0 & C_{33} \end{bmatrix} \tag{2.31}$$

When using numerical codes (like the Python scripts discussed in previous chapters), simply swapping the definition of the **C** matrix allows the same solver to handle both Plane Stress and Plane Strain problems seamlessly.

2.2.7 Equilibrium Equations (2D)

In two dimensions, the dynamic equilibrium of a solid body is maintained by the balance of in-plane stresses and body forces, which has the same form as the 3D solid:

$$\mathbf{L}_d^\top \sigma + \mathbf{b} = \rho \ddot{\mathbf{u}} \tag{2.32}$$

where $\mathbf{b} = [b_x, b_y]^\top$ represents the body force per unit volume (which, for a 2D model of constant thickness h, is often treated as force per unit area).

Displacement-Based Governing Equation

By substituting the 2D kinematics ($\varepsilon = \mathbf{L}_d \mathbf{u}$) and the chosen 2D constitutive matrix ($\mathbf{C}$), the governing equation becomes:

$$\boxed{\mathbf{L}_d^\top \mathbf{C} \mathbf{L}_d \mathbf{u} + \mathbf{b} = \rho \ddot{\mathbf{u}}} \tag{2.33}$$

> **Key Observation:** Notice that Eq. (2.33) is symbolically identical to the 3D Eq. (2.15). This consistency ensures the high-level expression remains unchanged, while only the specific definitions of $\mathbf{L}_d$ and $\mathbf{C}$ are swapped based on the problem type (Plane Stress vs. Plane Strain).

2.2.8 Boundary Conditions (2D)

A 2D problem is only well-posed when its boundary Γ (now a closed curve rather than a surface) is fully constrained. We partition the boundary into Γ_u (displacement) and Γ_t (traction).

2.2.8.1 Dirichlet (Displacement) Boundary Conditions

On Γ_u, the displacements are fixed or prescribed:

$$\mathbf{u} = \bar{\mathbf{u}} \quad \text{on} \quad \Gamma_u \tag{2.34}$$

where $\bar{\mathbf{u}} = [\bar{u}, \bar{v}]^\top$.

2.2.8.2 Neumann (Traction) Boundary Conditions

On Γ_t, we prescribe the external forces acting on the edge of the domain. The balance between internal stress and external traction is given by:

$$\mathbf{L}_n^\top \sigma = \bar{\mathbf{t}} \quad \text{on} \quad \Gamma_t \tag{2.35}$$

where $\bar{\mathbf{t}} = [t_x, t_y]^\top$ is the prescribed traction vector. The **2D Normal Mapping Matrix $\mathbf{L}_n$** is defined using the components of the outward unit normal $\mathbf{n} = [n_x, n_y]^\top$:

$$\mathbf{L}_n = \begin{bmatrix} n_x & 0 \\ 0 & n_y \\ n_y & n_x \end{bmatrix} \tag{2.36}$$

2.2.9 Summary Remarks on 2D Problems

To summarize the transition from 3D to 2D:

- **Identical Framework:** The kinematics ($\mathbf{L}_d$) and equilibrium ($\mathbf{L}_d^\top$) operators maintain their adjoint relationship regardless of dimension.
- **The Power of C:** The choice between **Plane Stress** and **Plane Strain** is handled *exclusively* within the material matrix $\mathbf{C}$. This means a FEM code can switch between a thin plate and a massive dam simulation just by changing a few lines of material property definitions.
- **Primary vs. Secondary Variables:** In both cases, we often solve for the displacement field $\mathbf{u}(x, y)$ as the primary unknown. Once $\mathbf{u}$ is known, strains and stresses are derived through simple differentiation and multiplication.

2.3 Formulation for 1D Solids Bars)

2.3.1 Introduction

The one-dimensional (1D) formulation is the simplest case in solid mechanics, typically used to model slender structural members such as bars or rods subjected to axial loads. In this reduction, all vectors and tensors are simplified to scalar variables. Remarkably, the high-level matrix-operator form of the equations remains unchanged, proving once again that the formulation is **dimensionality blind**.

2.3.2 Constitutive Equation

For a 1D linear elastic bar, the material relationship simplifies to the classic **Hooke's Law**. The stress and strain vectors each contain only one component, and the material matrix $\mathbf{C}$ reduces to a scalar—the Young's Modulus (E):

$$\sigma = E\varepsilon \tag{2.37}$$

where:

- σ is the axial stress (σ_{xx}).
- ε is the axial strain (ε_{xx}).
- E is the Young's Modulus of the material.

2.3.3 Strain–Displacement Relations

Under the assumption of small deformations, the axial strain is defined by the gradient of the displacement u along the x-axis. The symbolic operator form $\varepsilon = \mathbf{L}_d\mathbf{u}$ is maintained:

$$\boxed{\varepsilon = \mathbf{L}_d u = \frac{du}{dx}} \tag{2.38}$$

In this 1D case, the differential operator matrix $\mathbf{L}_d$ is simply the scalar derivative operator:

$$\mathbf{L}_d = \frac{d}{dx} \tag{2.39}$$

2.3.4 Equilibrium Equation

For a bar in static equilibrium, the balance of internal axial forces and external body forces must be satisfied. The general governing equation $\mathbf{L}_d^\top \mathbf{C} \mathbf{L}_d \mathbf{u} + \mathbf{b} = \mathbf{0}$ reduces to a second-order ordinary differential equation (ODE):

$$\frac{d}{dx}\left(E\frac{du}{dx}\right) + b = 0 \tag{2.40}$$

where b is the body force per unit length (e.g., gravity or axial friction acting along the bar).

2.3.5 Boundary Conditions

In 1D, the boundary Γ consists only of the two endpoints of the bar (i.e., $x = 0$ and $x = L$). Boundary conditions must be specified at these points:

- **Dirichlet (Displacement) BC:** The displacement is fixed or prescribed at an end.

$$u = \bar{u} \quad \text{at} \quad x = 0 \text{ and/or } x = L \tag{2.41}$$

- **Neumann (Traction/Force) BC:** The axial force (traction) is prescribed at an end.

$$n_x \sigma = \bar{t} \quad \text{at} \quad x = 0 \text{ and/or } x = L \tag{2.42}$$

where n_x is the unit normal ($n_x = -1$ at the left end and $n_x = 1$ at the right end), and $\bar{t}$ is the applied stress (force divided by cross-sectional area).

2.4 Summary of Key Formulas for FEM

The following five equations constitute the "Strong Form" of a linear elastic problem. In the Finite Element Method, these are the starting points for deriving the **Weak Form** and the eventual **Global Discretized Equilibrium Equation** ($\mathbf{KD} = \mathbf{F}$).

2.4.1 Equilibrium Equation

Must be satisfied locally at every point in the solid to balance internal and external forces:

$$\boxed{\mathbf{L}_d^\top \sigma + \mathbf{b} = \mathbf{0}} \tag{2.43}$$

2.4.2 Constitutive Equation

Describes the material's physical response (Generalized Hooke's Law):

$$\boxed{\sigma = \mathbf{C}\varepsilon} \tag{2.44}$$

2.4.3 Strain–Displacement Relation

The kinematic link between the geometry (displacement) and the deformation (strain):

$$\boxed{\varepsilon = \mathbf{L}_d\mathbf{u}} \tag{2.45}$$

2.4.4 Boundary Conditions

The essential (Dirichlet) and natural (Neumann) constraints that ensure a unique solution:

$$\boxed{\mathbf{u} = \bar{\mathbf{u}} \quad \text{on} \quad \Gamma_u} \tag{2.46}$$

$$\boxed{\mathbf{L}_n^\top \sigma = \bar{\mathbf{t}} \quad \text{on} \quad \Gamma_t} \tag{2.47}$$

2.4.5 The Dimensionality Hierarchy

We have now established how the fundamental laws of mechanics are expressed across all spatial dimensions. This unified framework is the "engine" that drives other modern computational methods, such as the **Smoothed Finite Element Method (S-FEM)** [11] and **Physics-Informed Neural Networks (PINNs)** [7].

Dimension	Displacement $\mathbf{u}$	Stress σ	Operator $\mathbf{L}_d$
3D	3 components	6 components	6×3 matrix
2D	2 components	3 components	3×2 matrix
1D	1 component	1 component	Scalar (d/dx)

Basic Equation Conclusion: By adopting this matrix-operator notation, we have decoupled the **mathematical structure** of solid mechanics from the **spatial dimension** of the specific problem. In Chapter 6, we will translate these equations into Python code to build our first numerical solver.

2.5 Key Integral Identity for Matrix Operators

To transition from the governing differential equations to the numerical implementation used in the Finite Element Method (FEM) formulations [12], we require a fundamental integral identity. This identity is the matrix-operator equivalent of Green's Theorem or integration by parts in 1D. [9].

2.5.1 Identity Statement

For any arbitrary displacement-like vector $\mathbf{v}$ and a stress-like vector $\mathbf{q}$, the following identity holds across all dimensions (1D, 2D, and 3D):

$$-\int_{\Omega} \mathbf{v}^{\top} (\mathbf{L}_d^{\top} \mathbf{q})\, d\Omega = \int_{\Omega} (\mathbf{L}_d \mathbf{v})^{\top} \mathbf{q}\, d\Omega - \int_{\Gamma} \mathbf{v}^{\top} \mathbf{L}_n^{\top} \mathbf{q}\, d\Gamma \tag{2.48}$$

where for 3D cases:

- $\mathbf{v}$ is an arbitrary test function (vector of length 3 in 3D).
- $\mathbf{q}$ is a vector of length 6 (e.g., the stress vector σ).

2.5.2 Derivation (3D Case)

While this identity is general, we derive it here for the most general 3D case. Let $\mathbf{q} = [q_1, q_2, q_3, q_4, q_5, q_6]^{\top}$. Recalling the definition of the divergence-like operator $\mathbf{L}_d^{\top}$:

$$\mathbf{L}_d^{\top} \mathbf{q} = \begin{bmatrix} \frac{\partial q_1}{\partial x} + \frac{\partial q_5}{\partial z} + \frac{\partial q_6}{\partial y} \\ \frac{\partial q_2}{\partial y} + \frac{\partial q_4}{\partial z} + \frac{\partial q_6}{\partial x} \\ \frac{\partial q_3}{\partial z} + \frac{\partial q_4}{\partial y} + \frac{\partial q_5}{\partial x} \end{bmatrix} \tag{2.49}$$

Expanding the left-hand side of Eq. (2.48):

$$\begin{aligned} -\int_{\Omega} \mathbf{v}^{\top} (\mathbf{L}_d^{\top} \mathbf{q})\, d\Omega = -\int_{\Omega} \Big[& v_x \left(\frac{\partial q_1}{\partial x} + \frac{\partial q_5}{\partial z} + \frac{\partial q_6}{\partial y} \right) \\ & + v_y \left(\frac{\partial q_2}{\partial y} + \frac{\partial q_4}{\partial z} + \frac{\partial q_6}{\partial x} \right) \\ & + v_z \left(\frac{\partial q_3}{\partial z} + \frac{\partial q_4}{\partial y} + \frac{\partial q_5}{\partial x} \right) \Big] d\Omega \end{aligned}$$

Applying **integration by parts**

$$\int v \frac{\partial \phi}{\partial x} d\Omega = -\int \frac{\partial v}{\partial x} \phi d\Omega + \int v \phi n_x d\Gamma$$

to every term in Eq. (2.50) allows us to split the expression into a domain integral and a boundary integral.

Domain Contribution:

Collecting the terms involving derivatives of $\mathbf{v}$:

$$\int_{\Omega}\left[q_1\frac{\partial v_x}{\partial x}+q_2\frac{\partial v_y}{\partial y}+q_3\frac{\partial v_z}{\partial z}+q_4\left(\frac{\partial v_y}{\partial z}+\frac{\partial v_z}{\partial y}\right)+...\right]d\Omega \tag{2.50}$$

This is identically

$$\int_{\Omega}(\mathbf{L}_d\mathbf{v})^\top\mathbf{q}d\Omega$$

which is the first term in the right hand side of Eq. (2.48). It representing the **internal virtual work** if **v** is viewed as a virtual displacement.

Boundary Contribution:

Collecting the boundary terms resulting from the integration:

$$-\int_{\Gamma}\left[v_x(q_1n_x+q_5n_z+q_6n_y)+v_y(q_2n_y+q_4n_z+q_6n_x)+...\right]d\Gamma \tag{2.51}$$

By using the definition of the normal mapping matrix $\mathbf{L}_n^\top$ from Eq. (2.19), the boundary term becomes

$$-\int_{\Gamma}\mathbf{v}^\top\mathbf{L}_n^\top\mathbf{q}\,d\Gamma$$

which is the second term in the right hand side of Eq. (2.48). This completes the proof.

2.5.3 Physical Significance

This identity is the "workhorse" of computational mechanics. It allows us to:

1. **Reduce Continuity Requirements:** It moves derivatives from the stress-like vector **q** to the test function **v**, which is essential for FEM.
2. **Naturally Incorporate Boundary Conditions:** The traction vector $\mathbf{L}_n^\top\mathbf{q}$ appears explicitly as a boundary integral, allowing for the direct application of Neumann BCs.

For the 1D case, this identity simplifies back to the standard integration by parts formula used in elementary calculus [6]:

$$-\int_a^b v\frac{dq}{dx}dx=\int_a^b\frac{dv}{dx}q\,dx-\left[vn_xq\right]_a^b \tag{2.52}$$

2.6 Remarks on Strong and Weak Form Methods

The transition from a system of differential equations to a numerical solution requires an understanding of how these equations are enforced across a domain Ω. The following provides an intuitive perspective on the two primary approaches:

2.6.1 The Need for Integral Formulations

The equilibrium equation, constitutive relation, and strain–displacement relation must be satisfied locally at every point within the domain, while the boundary conditions must be satisfied along the boundary. Based on the fundamental theorem of calculus (see [6] for 1D and [9] for 2D/3D), an appropriate **integral formulation** is required to ensure that the entire system of equations is satisfied over the domain. This is particularly vital for problems defined on **irregular geometries**, where local differential schemes often struggle.

2.6.2 Limitations of Collocation Methods

In contrast, **collocation-type methods**—which enforce governing equations pointwise without an integral framework—frequently encounter stability and accuracy difficulties on irregular domains. While Taylor series–based schemes or simple finite difference methods can be effective for structured, simple geometries, their performance often deteriorates as domain complexity increases. This behavior has been documented extensively in the development of meshfree methods [2].

2.6.3 The Equivalence of Forms

In the Finite Element Method, the equilibrium equation is not solved directly in its **Strong Form** (the PDE). Instead, it serves as the conceptual starting point for deriving the **Weak Form** (the integral equation). Using the integral identity derived in Section 1.5, we move from the differential balance of forces to a global balance of energy or work.

It is important to note that these forms are often two sides of the same coin:

1. **Recovery:** The Strong Form can be recovered from a sufficiently smooth Weak Form using the fundamental lemma of the calculus of variations.
2. **Variational Origins:** The Weak Form can be derived independently without explicitly invoking the equilibrium equation—for example, through the **Principle of Minimum Total Potential Energy** [12].

In this sense, the local equilibrium of a point and the global minimization of potential energy are mathematically equivalent. The FEM leverages this equivalence to transform a continuous problem into a discrete system of algebraic equations ($\mathbf{KD} = \mathbf{F}$) that a computer can solve.

2.7 Unit Systems for Solid Mechanics

Variables in solid mechanics carry distinct physical dimensions. Analysts select a specific unit system based on geometric scale, computational efficiency, and structural convenience.

2.7.1 Consistent Metric Unit Systems

The fundamental rule in finite element analysis (FEA) is that **all variables within a given model must belong to a single, internally consistent set of units**.

Every consistent unit system satisfies Newton's second law ($F = ma$) without requiring scalar correction factors. Consequently, all derived quantities—such as stress, mass density, energy, and acceleration—remain internally consistent.

Consistency Check:

To ensure a unit system is consistent for computational mechanics (like FEM), check it against Newton's Second Law: $F = m \cdot a$. In terms of base units (Length L, Mass M, Time T):

$$[F] = \frac{[M][L]}{[T]^2}$$

Note: For computational codes, it is often safer to derive the required mass from your chosen Force, Length, and Time units to avoid hidden scaling errors.

2.7.2 Table of Some Consistent Metric Unit Systems

(Time: s for all; Density is derived by Mass/Length^3)

System Name	Length	Force	Mass	Stress (σ)	Density (ρ)	Typical Engineering Usage
Standard SI System	m	N	kg	Pa (N/m^2)	kg/m^3	General continuum mechanics, acoustics, and fluid-structure interaction
Engineering FEM System	mm	N	T (10^3 kg)	MPa (N/mm^2)	T/mm^3	Mechanical components, automotive body design, and standard CAD-integrated FEA
Structural Civil System	m	kN	T (10^3 kg)	kPa	T/m^3	Infrastructure, concrete/steel frames, and large-scale civil structures

System Name	Length	Force	Mass	Stress (σ)	Density (ρ)	Typical Engineering Usage
Millimeter-kN System	mm	kN	kT (10^6 kg)	GPa (kN/mm^2)	g/m^3	Heavy machinery, high-velocity ballistics, and high-force structural impact
Nanometer-Nano System	nm	nN	pg (10^{-15} kg)	GPa (nN/nm^2)	pg/nm^3	Molecular mechanics, nanotechnology, graphene modeling, and nanoscale solid mechanics
CGS System	cm	dyne	g	barye (dyn/cm^2)	g/cm^3	Micro-electro-mechanical systems (MEMS) and classical elasticity literature

2.8 Dimensionless Form of PDEs

In solid mechanics, material parameters can span multiple orders of magnitude depending on the chosen unit system. For instance, the Young's modulus of structural steel is approximately 200×10^9 Pa. Assembling a discretized system matrix in FEM using values this large often leads to massive numerical disparities between matrix coefficients. Because computers operate with finite floating-point precision, these large condition numbers can degrade solution accuracy and induce catastrophic round-off errors [5].

To mitigate these numerical issues, a highly effective practice is to transform the governing partial differential equations (PDEs) into a **dimensionless (non-dimensional)** form [13,14]. Non-dimensionalization can also scale the values involved in the governing equations to order unity (~ 1). This philosophy directly mirrors data normalization and standardization techniques standard in machine learning applications [3].

Though sometimes overlooked during routine engineering workflows, converting a PDE into its dimensionless counterpart represents an invaluable analytical tool. This process systematically delivers the following advantages:

- **Reveals dominant physical mechanisms** by isolating competing forces.
- **Identifies governing similarity parameters** (dimensionless scaling groups).
- **Improves numerical conditioning** of stiffness, mass, and damping matrices.
- **Simplifies scaling analysis** for scale-model physical testing.

- **Enables direct comparison** across physically disparate systems.
- **Provides mathematical clarity** regarding asymptotic and limiting behaviors.

2.8.1 General Philosophy of Non-Dimensionalization

Consider a governing physical system containing dimensional variables such as coordinates, time, displacements, stresses, strains, and body forces:

$$x_i,\ t,\ u_i,\ \sigma_{ij},\ \varepsilon_{ij},\ b_i, \ldots$$

The objective is to establish fixed **characteristic scales** for each physical dimension and define corresponding **dimensionless variables** ($\bar{q}$) that scale to order unity. Any arbitrary dimensional variable q is uniquely decoupled into:

$$q = q_c\,\bar{q}$$

where q_c denotes the constant characteristic scale carrying the physical dimensions of q, and $\bar{q}$ represents the dimensionless variable.

2.8.2 The Non-Dimensionalization Process

2.8.2.1 Step 1: Select Characteristic Scales

The analyst identifies independent physical constants native to the problem's geometry, material properties, and boundary conditions.

Physical Quantity	Symbol	Chosen Characteristic Scale
Length (Geometry)	x_i	L (e.g., total span, bar length)
Elastic Modulus (Material)	C_{ijkl}	E (e.g., Young's modulus)
Body Force (Loading)	b_i	B (e.g., peak gravitational/magnetic force)
Displacement (Unknown)	u_i	U (Derived scale)
Stress (Unknown)	σ_{ij}	Σ (Derived scale)
Strain (Unknown)	ε_{ij}	U/L (Derived scale)

2.8.2.2 Step 2: Define Dimensionless Variables

Using the chosen characteristic scales, the primary field variables are mapped into dimensionless space:

$$x_i = L\,\bar{x}_i, \quad u_i = U\,\bar{u}_i, \quad \sigma_{ij} = \Sigma\,\bar{\sigma}_{ij}, \quad b_i = B\,\bar{b}_i$$

2.8.2.3 Step 3: Transform Spatial Derivatives

Applying the multi-variable chain rule maps spatial gradients from dimensional to dimensionless coordinates:

$$\frac{\partial}{\partial x_i} = \frac{\partial \bar{x}_j}{\partial x_i}\frac{\partial}{\partial \bar{x}_j} = \frac{1}{L}\frac{\partial}{\partial \bar{x}_i}$$

This linear transformation alters the differential operators governing the continuous mechanics problem.

2.8.2.4 Step 4: Substitute into the Governing Equations

Consider the static Cauchy equation of structural equilibrium:

$$\frac{\partial \sigma_{ij}}{\partial x_j} + b_i = 0$$

Substituting the dimensionless relations from Step 2 and Step 3 yields:

$$\frac{\Sigma}{L}\frac{\partial \bar{\sigma}_{ij}}{\partial \bar{x}_j} + B\bar{b}_i = 0$$

Factoring out the leading differential coefficient $\frac{\Sigma}{L}$ produces:

$$\frac{\partial \bar{\sigma}_{ij}}{\partial \bar{x}_j} + \left(\frac{BL}{\Sigma}\right)\bar{b}_i = 0$$

The dimensionless grouping emerging from the loading terms is designated as:

$$\Pi_1 = \frac{BL}{\Sigma}$$

The non-dimensionalized Cauchy equilibrium equation is thus expressed cleanly as:

$$\frac{\partial \bar{\sigma}_{ij}}{\partial \bar{x}_j} + \Pi_1\bar{b}_i = 0$$

where all variable in the equation are dimensionless.

2.8.3 Exposing the Stress Scale via Constitutive Laws

To evaluate the dimensionless group Π_1, the unknown stress scale Σ must be determined from the constitutive relation of the material. Under linear elasticity, Hooke's law implies that stress scales proportionally with the elastic modulus and the engineering strain:

$$\sigma \sim E\varepsilon, \qquad \varepsilon \sim \frac{U}{L}$$

which gives the characteristic stress scale

$$\Sigma = \frac{EU}{L}.$$

Substituting this constitutive stress scale into the definition of Π_1 yields

$$\Pi\_1 = \frac{BL}{\left(\frac{EU}{L}\right)} \frac{BL^2}{EU}.$$

The non-dimensional equilibrium equation can therefore be written as

$$\frac{\partial \bar{\sigma}_{ij}}{\partial \bar{x}_j} + \left(\frac{BL^2}{EU}\right) \bar{b}_i = 0 \tag{2.53}$$

where all quantities inside the parentheses are known for a given problem except the displacement scale U. Choosing

$$U = \frac{BL^2}{E}$$

normalizes the coefficient to unity, and Eq.(2.53) reduces to

$$\frac{\partial \bar{\sigma}_{ij}}{\partial \bar{x}_j} + \bar{b}_i = 0. \tag{2.54}$$

This form is particularly desirable for numerical computation because all variables are dimensionless and all governing coefficients are of order unity, which generally improves numerical stability and conditioning.

2.8.4 Full Displacement Formulation

Alternatively, non-dimensionalization can be applied directly to the governing equation in terms of displacements:

$$\frac{\partial}{\partial x_j}\left(C_{ijkl}\frac{\partial u_k}{\partial x_l}\right) + b_i = 0$$

Assuming an isotropic material where $C_{ijkl} \sim E$, we substitute the spatial coordinates and displacement mappings ($u_k = U\bar{u}_k$ and $x_j = L\bar{x}_j$):

$$\frac{EU}{L^2}\frac{\partial}{\partial \bar{x}_j}\left(\bar{C}_{ijkl}\frac{\partial \bar{u}_k}{\partial \bar{x}_l}\right) + B\bar{b}_i = 0$$

Dividing the entire equation by the dominant structural stiffness scaling factor $\frac{EU}{L^2}$ yields:

$$\frac{\partial}{\partial \bar{x}_j}\left(\bar{C}_{ijkl}\frac{\partial \bar{u}_k}{\partial \bar{x}_l}\right) + \left(\frac{BL^2}{EU}\right)\bar{b}_i = 0$$

This confirms that the identical dimensionless scaling group governs both the stress and displacement formulations.

2.8.5 Dynamic Solid Mechanics (Elastodynamics)

For dynamic systems subject to inertial forces, the governing equation expands to include time-dependent momentum terms:

$$\rho \ddot{u}_i = \frac{\partial \sigma_{ij}}{\partial x_j} + b_i$$

2.8.5.1 Non-dimensionalized Elastodynamic equation

To scale the temporal domain, we introduce a characteristic time scale T such that $t = T\bar{t}$. The secondary material derivative for acceleration transforms as:

$$\ddot{u}_i = \frac{\partial^2 u_i}{\partial t^2} = \frac{U}{T^2}\frac{\partial^2 \bar{u}_i}{\partial \bar{t}^2}$$

Substituting the spatial, inertial, and constitutive scales into the dynamic equilibrium equation produces:

$$\rho\left(\frac{U}{T^2}\right)\frac{\partial^2 \bar{u}_i}{\partial \bar{t}^2} = \left(\frac{\Sigma}{L}\right)\frac{\partial \bar{\sigma}_{ij}}{\partial \bar{x}_j} + B\bar{b}_i$$

Dividing through by the spatial stress gradient scale $\frac{\Sigma}{L}$ gives:

$$\left(\frac{\rho U L}{\Sigma T^2}\right)\frac{\partial^2 \bar{u}_i}{\partial \bar{t}^2} = \frac{\partial \bar{\sigma}_{ij}}{\partial \bar{x}_j} + \left(\frac{BL}{\Sigma}\right)\bar{b}_i$$

Recalling our elasticity stress scale $\Sigma = \frac{EU}{L}$, the inertial coefficient simplifies to a new dimensionless group, designated as Π_2:

$$\Pi_2 = \frac{\rho U L}{\left(\frac{EU}{L}\right) T^2} = \frac{\rho L^2}{E T^2}$$

The fully non-dimensionalized elastodynamic governing equation is written as:

$$\Pi_2 \frac{\partial^2 \bar{u}_i}{\partial \bar{t}^2} = \frac{\partial \bar{\sigma}_{ij}}{\partial \bar{x}_j} + \Pi_1 \bar{b}_i$$

The parameter Π_2 explicitly quantifies the ratio of dynamic inertial forces to static internal elastic forces within the continuum.

2.8.5.2 The Natural Time Scale

An elegant simplification can be achieved by defining the characteristic time scale T using the structural medium's intrinsic wave speed:

$$T = L\sqrt{\frac{\rho}{E}}$$

Substituting this natural time scale directly sets $\Pi_2 = 1$, yielding the canonical dimensionless elastodynamic equation. Physically, this characteristic scale represents the exact time required for an elastic dilatation or shear wave to propagate across the characteristic domain length L.

2.8.6 Summary of Core Scaling Strategies

1. **Geometric Scaling:** Anchors transformations to global boundary dimensions ($x_i = L\bar{x}_i$). This is highly useful for structural slenderness, aspect ratio, and boundary layer analyses.

2. **Load-Based Scaling:** Defines the internal stress scale directly from applied surface tractions ($\Sigma = \|\bar{\mathbf{t}}\|$), making it ideal for force-driven simulations.

3. **Constitutive Scaling:** derives internal scales from material parameters ($\Sigma = \frac{EU}{L}$), which is the standard protocol for elasticity and stiffness-dominated systems.

4. **Energy Scaling:** Normalizes variables via total strain energy density ($W \sim E\varepsilon^2$), commonly employed in variational formulations, weakened-weak (W^2) methods, and fracture mechanics.

2.8.7 Common Dimensionless Groups in Solid Mechanics

Dimensionless Group	Algebraic Expression	Underlying Physical Interpretation
Force Ratio (Π_1)	$\frac{BL^2}{EU}$	Ratio of external body forces to internal structural restoring forces.
Inertia Ratio (Π_2)	$\frac{\rho L^2}{ET^2}$	Ratio of localized transient inertial forces to continuous elastic forces.
Normalized Stress	$\frac{\sigma}{E}$	Continuum stress magnitude scaled against the baseline material stiffness.
Geometric Strain	$\frac{u}{L}$	Global normalized displacement indicating structural deformation scale.
Slenderness Ratio	$\frac{h}{L}$	Aspect ratio comparing structural thickness (h) to primary span length (L).

2.8.8 Example: 1D Axial Bar Equilibrium

2.8.8.1 Problem Formulation

Consider the static, one-dimensional deformation of a continuous structural bar:

$$\frac{d}{dx}\left(EA\frac{du}{dx}\right) + b(x) = 0$$

To isolate the core mechanics, we assume a homogeneous material with a constant cross-sectional area:

- Constant Young's modulus: E
- Constant cross-sectional area: A
- Distributed axial body force per unit length: $b(x)$
- Total structural length: L

The governing ODE simplifies to:

$$EA\frac{d^2u}{dx^2} + b(x) = 0 \implies \frac{d^2u}{dx^2} + \frac{b(x)}{EA} = 0$$

The critical task is defining an appropriate characteristic displacement scale U using only our known parameters (E, b, L, A).

2.8.8.2 Dimensional Analysis & Balancing

We evaluate the fundamental units of our known variables:

- $[E] = \text{Force} \cdot \text{Length}^{-2}$
- $[b] = \text{Force} \cdot \text{Length}^{-1}$
- $[A] = \text{Length}^2$
- $[L] = \text{Length}$

Approximating the spatial derivative magnitudes within the differential equation yields:

$$\frac{d^2u}{dx^2} \sim \frac{U}{L^2} \implies EA\left(\frac{U}{L^2}\right) \sim b$$

Solving explicitly for our unknown displacement scale U establishes the natural characteristic scale for a force-driven bar:

$$U = \frac{bL^2}{EA}$$

2.8.8.3 Ideal Characteristic Quantities

Physical Parameter	Ideal Scale Definition	Physical Derivation
Length Scale	L	Total physical span of the bar
Displacement Scale	$U = \dfrac{bL^2}{EA}$	Peak expected static deflection
Strain Scale	$\dfrac{U}{L} = \dfrac{bL}{EA}$	Normalized internal kinematic elongation
Stress Scale	$E\frac{U}{L} = \dfrac{bL}{A}$	Average internal normal stress axial intensity

2.8.8.4 Transformation to Parameter-Free Form

We introduce our dimensionless coordinate and field variables:

$$x = L\bar{x}, \quad u = U\bar{u}, \quad \text{where } U = \frac{bL^2}{EA} \tag{2.55}$$

Transforming the second-order spatial derivative gives:

$$\frac{d^2u}{dx^2} = \frac{U}{L^2}\frac{d^2\bar{u}}{d\bar{x}^2}$$

Substituting this structural scale back into the original 1D ODE yields:

$$EA\left(\frac{bL^2}{EAL^2}\right)\frac{d^2\bar{u}}{d\bar{x}^2}+b=0 \implies b\frac{d^2\bar{u}}{d\bar{x}^2}+b=0$$

Dividing through by the common loading factor b isolates the final dimensionless system:

$$\frac{d^2\bar{u}}{d\bar{x}^2}+1=0 \qquad \text{for } \bar{x}\in(0,1) \tag{2.56}$$

This represents the ideal outcome of non-dimensionalization: all physical units are successfully eliminated, the remaining coefficients resolve to unity, and the governing system becomes completely parameter-free. This reveals that the chosen scale perfectly captures the underlying physics of the problem.

2.8.8.5 Solution Procedure

For this 1D bar problem, we only need to solve Eq.(2.56) over the normalized 1D domain $(0,1)$ (with $\bar{L}=1$), for a body force $\bar{b}=1$, material parameter $E=1$, area $A=1$, and the fixed displacement boundary condition $\bar{u}=0$ at $\bar{x}=0$.

Once the dimensionless solution $\bar{u}$ is obtained, the solution u to the original problem can be recovered using Eq.(2.55).

This dimensionless formulation will be adopted throughout the following chapters, so that our focus will be on how to develop FEM models to solve the problem without distraction from the units. For simplicity, we will remove the bar for all the dimensionless variables.

2.8.9 Alternative Boundary-Driven Scaling Choices

Depending on boundary conditions, alternative characteristic scales may be more appropriate:

2.8.9.1 Prescribed End Displacement

If the bar is subject to a fixed target displacement $u(L)=u_0$ rather than a distributed force, the analyst defines the displacement scale as $U=u_0$. Applying this transformation changes the dimensionless ODE to:

$$\frac{d^2\bar{u}}{d\bar{x}^2}+\left(\frac{bL^2}{EAu_0}\right)=0 \implies \frac{d^2\bar{u}}{d\bar{x}^2}+\Pi_{\text{disp}}=0$$

Here, the dimensionless parameter $\Pi_{\text{disp}}=\frac{bL^2}{EAu_0}$ measures the relative importance of internal body-force loading compared to the imposed geometric boundary constraint.

2.8.9.2 Boundary Traction-Based Scaling

If the system is driven by a discrete point traction P applied at the free end ($x = L$) with zero distributed body force ($b = 0$), the governing equation is $EA\frac{d^2u}{dx^2} = 0$, bound by the boundary condition $EAu'(L) = P$. Under these conditions, the natural displacement scale shifts to:

$$U = \frac{PL}{EA}$$

2.9 Grant Summary

We have now established the full mathematical "Strong Form" of solid mechanics, the hierarchy of dimensions (3D, 2D, 1D), and the integral tools needed to bridge the gap to numerical methods. The dimensionless form of the equations are also derived.

In the following chapters, we will use these foundations to build practical solvers.

2.10 Exercises

1. The Dimensionality-Blind Principle Explain what is meant by the "dimensionality-blind" nature of matrix-vector formulations in solid mechanics. How does this simplify the development of finite element software?

2. Generalized Hooke's Law In the 3D constitutive equation $\sigma_{ij} = C_{ijkl}\varepsilon_{kl}$, the fourth-order tensor C theoretically has 81 components. For an isotropic material, how many independent material constants remain, and what are their common names in engineering practice?

3. Volumetric Strain and the Kronecker Delta Given the isotropic relation $\sigma_{ij} = 2G\varepsilon_{ij} + \lambda\delta_{ij}\varepsilon_{kk}$, define the physical meaning of the term ε_{kk}. What role does the Kronecker delta δ_{ij} play in this equation?

4. Voigt Notation Why do we use engineering shear strain ($\gamma = 2\varepsilon$) instead of tensorial shear strain when representing the strain field as a vector ε?

5. Plane Stress vs. Plane Strain Compare and contrast "Plane Stress" and "Plane Strain." Provide one real-world example for each and identify which component of the stress or strain tensor is assumed to be zero in each case.

6. The Normal Mapping Matrix The boundary condition $\mathbf{L}_n^\top\sigma = \bar{\mathbf{t}}$ relates internal stress to external traction. Describe the relationship between the differential operator matrix $\mathbf{L}_d$ and the normal mapping matrix $\mathbf{L}_n$.

7. Static vs. Dynamic Equilibrium Starting from the general equilibrium equation $\mathbf{L}_d^\top\sigma + \mathbf{b} = \rho\ddot{\mathbf{u}}$, identify the term that must be neglected to arrive at a static solution. What physical property does ρ represent in this context?

8. The Strong Form What is the "Strong Form" of a solid mechanics problem, and why is it generally difficult to solve analytically for practical engineering applications?

9. The Integral Identity The fundamental integral identity moves the operator $\mathbf{L}_d$ from the stress vector $\mathbf{q}$ to the test vector $\mathbf{v}$. What is the mathematical name for this process in 1D, and why is it a requirement for formulating the Finite Element Method?

10. Essential vs. Natural Boundary Conditions Distinguish between Dirichlet and Neumann boundary conditions. In the context of a 1D bar fixed at one end and pulled at the other, which end represents which type of boundary condition?

11. Unit Consistency and Matrix Conditioning

A structural analyst is modeling a micro-electro-mechanical system (MEMS) cantilever actuator. The device has a characteristic length of 250 μm, a material Young's modulus of $E = 160$ GPa, and a mass density of $\rho = 2330$ kg/m^3.

1. **Identify the Failure:** Explain what happens to the numerical condition number of the global FEM stiffness matrix if the analyst inputs these exact numerical values directly into a standard commercial solver configured for the metric **SI System** (m, N, kg, s).

2. **Select a Consistent System:** Using the table provided in this section, select a consistent metric unit system that natively scales the length to order unity (~ 1 to 100) while maintaining a consistent mass and time profile.

3. **Convert the Properties:** Convert the given Young's modulus (E) and mass density (ρ) into the numerical values required by your selected consistent unit system.

12. Non-Dimensionalization of a 1D Heat Conduction PDE

The 1D transient heat conduction equation with internal heat generation in a structural component is governed by:

$$\rho c_p \frac{\partial T}{\partial t} = k \frac{\partial^2 T}{\partial x^2} + q_g$$

where T is temperature, x is position, t is time, ρ is density, c_p is specific heat capacity, k is thermal conductivity, and q_g is a constant uniform volumetric heat source. The domain spans a length $x \in [0, L]$.

1. **Define the Variables:** Introduce the standard dimensionless spatial scale $\bar{x} = x/L$ and a general dimensionless temperature scale $\bar{T} = (T - T_0)/\Delta T_c$, where T_0 is the initial temperature and ΔT_c is an unknown characteristic temperature difference.

2. **Derive the Parameter-Free Forms:**

- Show that if the system is **transient-driven** with zero internal heat generation ($q_g = 0$), choosing the characteristic time scale $T_{\text{thermal}} = \frac{\rho c_p L^2}{k}$ reduces the PDE to a parameter-free form ($\frac{\partial T}{\partial t} = \frac{\partial^2 T}{\partial x^2}$). Physically interpret this time scale.
- Show that if the system reaches a **steady-state equilibrium** ($\frac{\partial T}{\partial t} = 0$), balancing the differential operators yields a natural characteristic temperature scale of $\Delta T_c = \frac{q_g L^2}{k}$.

13. Scaling Strategy for a Thin Elastic Plate (Kirchhoff-Love Theory) (Optional)

The static out-of-plane deflection $w(x, y)$ of a thin, isotropic elastic plate subjected to a uniform transverse pressure load p_0 is governed by the fourth-order PDE:

$$D\left(\frac{\partial^4 w}{\partial x^4} + 2\frac{\partial^4 w}{\partial x^2 \partial y^2} + \frac{\partial^4 w}{\partial y^4}\right) = p_0$$

where $D = \frac{Eh^3}{12(1-\nu^2)}$ represents the flexural rigidity (bending stiffness) of the plate, h is the plate thickness, E is Young's modulus, and ν is Poisson's ratio. Let the lateral dimensions of the plate be governed by a characteristic length scale L (where $x = L\bar{x}$ and $y = L\bar{y}$).

1. **Balance the Operators:** Apply the multi-variable chain rule to scale the fourth-order spatial derivatives.
2. **Derive the Displacement Scale:** By balancing the bending stiffness operator with the transverse pressure load, derive the natural characteristic out-of-plane displacement scale W (where $w = W\bar{w}$) such that the resulting dimensionless PDE becomes entirely parameter-free.
3. **Analyze the Physics:** Based on your derived scale W, what happens to the peak structural deflection if the plate thickness (h) is cut in half while keeping the pressure load and lateral dimensions identical?

Pro-Tip for Study

If you can derive the 1D equilibrium ODE $\frac{d}{dx}(E\frac{du}{dx}) + b = 0$ directly from the 3D matrix form $\mathbf{L}_d^\top \mathbf{C} \mathbf{L}_d \mathbf{u} + \mathbf{b} = \mathbf{0}$, you have successfully mastered the hierarchical logic of this chapter!

Chapter 3

Domain Discretization

```
import importlib
from sympy import init_printing
import numpy as np
import matplotlib.pyplot as plt                                         # For plots
init_printing(use_unicode=True)     # For latex-like quality printing
np.set_printoptions(precision=4,suppress=True) # Digits in print-out
```

Domain discretization is one of the most fundamental and practical concepts in the Finite Element Method (FEM) for solving solid mechanics problems involving complex geometries and boundary conditions. The core principle is to subdivide a continuous physical domain into smaller, simpler subdomains so that the governing equations can be systematically formulated and solved numerically.

A continuum body possesses infinitely many degrees of freedom because the displacement field is defined at every point in the domain. Consequently, direct analytical solutions are only feasible for relatively simple problems. The FEM overcomes this challenge by replacing the continuum with a finite-dimensional approximation. Through discretization, the original continuum problem is transformed into a system of algebraic equations that can be solved efficiently using numerical methods.

In FEM, unknown field variables—such as displacement—are approximated locally over each small subdomain, known as an **element**. The collection of all elements forms a **mesh** that approximates the geometry of the original domain. The quality of this discretization significantly influences the accuracy, convergence, and computational efficiency of the numerical solution.

3.1 Partitioning of the Domain

Consider a solid occupying a domain Ω. In the FEM, this domain is subdivided into a finite number of small, non-overlapping subdomains called *elements*. This process is commonly referred to as *mesh generation* or *meshing*. The assembly of all elements is defined as the *finite element mesh*.

Each element is connected to its neighbors through shared *nodes*. These nodes serve as the primary locations where the unknown degrees of freedom are defined. In one-dimensional (1D) problems, elements are bounded by point, so mesh generation is trivial. In two-dimensional (2D) problems, elements are bounded by edges, while in three-dimensional (3D) problems, they are bounded by faces. Meshing 2D, especially 3D domain can be a challenge.

A variety of element shapes may be utilized depending on the geometry

and specific application, including:

- Triangular elements (2D)
- Quadrilateral elements (2D)
- Tetrahedral elements (3D)
- Hexahedral elements (3D)

For large-scale engineering applications, mesh generation is typically performed using professional preprocessing software, such as:

- Altair HyperMesh
- ANSYS Workbench
- Abaqus/CAE

In this book, however, we will also develop Python-based mesh generation procedures. This allows all finite element studies to be conducted directly within a Jupyter notebook environment, maintaining a cohesive workflow without the need for external commercial tools.

Mathematically, the discretized domain is represented as:

$$\Omega \approx \Omega^h = \bigcup_{i=1}^{N_e} \Omega_e^{(i)}, \qquad \Omega_e^{(i)} \cap \Omega_e^{(j)} = \emptyset \quad (i \neq j) \tag{3.1}$$

Where:

- $\Omega_e^{(i)}$ denotes the i-th element domain.
- N_e is the total number of elements.
- Ω^h denotes the discretized approximation of the original domain Ω.
- h represents a characteristic element size.

Equation (3.1) establishes that elements do not overlap and that their union forms an approximation of the original continuum. In a properly constructed mesh, neighboring elements share common nodes and boundaries, ensuring the compatibility of the approximation across the entire domain.

The parameter h plays a critical role in finite element analysis. It represents a characteristic measure of the mesh density. As the mesh is refined and the characteristic element size decreases:

$$h \to 0 \tag{3.2}$$

the finite element approximation generally becomes more accurate, converging toward the exact continuum solution—provided the formulation satisfies the necessary conditions of consistency, compatibility, and stability.

The parameter h must be representative of the entire mesh. Therefore, the limiting process $h \to 0$ implies that the sizes of all elements in the mesh decrease simultaneously. Convergence is associated with global refinement of the discretization rather than the refinement of a few isolated elements.

For a uniform mesh, h is simply the characteristic size of an individual element. For irregular or non-uniform meshes, h is more broadly defined as a representative measure of global mesh density, such as:

- The maximum element diameter.
- The average element size.
- An equivalent global measure of element dimensions.

More formally, for an element Ω_e, the local element size may be defined as:

$$h_e = \mathrm{diam}(\Omega_e) \tag{3.3}$$

where $\mathrm{diam}(\Omega_e)$ denotes the diameter of the element (the maximum distance between any two points within the element). The global mesh parameter is then defined as:

$$h = \max_e h_e \tag{3.4}$$

With this definition, the condition $h \to 0$ ensures that every element in the mesh becomes sufficiently small during refinement, which is essential for establishing rigorous convergence properties.

Other practical choices of h are given in detail in Section 4.10.7 in Ref. [11].

The process of discretization provides the essential bridge between the mathematical continuum model and its numerical approximation. Once the domain is discretized, approximation functions can be introduced over each element, leading to the assembly of the global finite element equations.

3.2 Code for Generating a Triangular Mesh

The following Python code demonstrates a straightforward procedure for generating and visualizing a triangular finite element mesh independently of commercial software. This example is primarily educational, designed to facilitate the creation and inspection of meshes directly within a Jupyter notebook. By exploring this implementation, readers can gain a deeper understanding of the underlying data structures and connectivity that define finite element discretizations.

The code includes detailed documentation and is designed to be self-explanatory. While the implementation is comprehensive, readers are encouraged to study it carefully to build a solid foundation in the mechanics of mesh generation.

```
# ===================================================================
# Triangular Mesh Generation for a Plate with an Elliptical Hole
# ===================================================================
#
# This example demonstrates how to:
#   1. Generate a 2D triangular finite element mesh
#   2. Create an internal elliptical hole
#   3. Visualize the mesh
#   4. Display node and element numbering
# The mesh consists of linear triangular elements (T3 elements),
# where each triangle contains 3 nodes.
# ===================================================================

import meshpy.triangle as triangle

# Mesh Generation Function
def generate_hole_mesh(nx_boundary=100, max_area=0.001,
        refine_factor=0.3, a=0.2, b=0.1, h=0.5, k=0.5):
    """
    Generate a triangular mesh for a square plate
    containing an elliptical hole.
    Parameters:
    nx_boundary : int
        Number of nodes used to discretize the ellipse boundary.
    max_area : float
        Maximum allowable triangle area.
        Smaller values generate finer meshes.
    refine_factor : float
        Reserved for future mesh refinement control.
        (Currently not used directly.)
    a : float, Semi-major axis of ellipse.
    b : float, Semi-minor axis of ellipse.
```

```
    h : float, x-coordinate of ellipse center.
    k : float, y-coordinate of ellipse center.

    Returns:
    nodes :     ndarray of shape (n_nodes, 2)
                Coordinates of mesh nodes.
    elements : ndarray of shape (n_elements, 3)
                Connectivity matrix for triangular elements.
                Each row stores the 3 node numbers of one triangle.
    """
    # 1. Define the outer boundary of the square plate
    # Node numbering:
    #   (3) -------- (2)
    #    |            |
    #    |            |
    #   (0) -------- (1)

    outer_pts = [(0.0, 0.0),    # node 0
                 (1.0, 0.0),    # node 1
                 (1.0, 1.0),    # node 2
                 (0.0, 1.0) ]   # node 3

    # Boundary edges of the square
    outer_facets = [(0, 1),     # edge formed by nodes 0 and 1
                    (1, 2),
                    (2, 3),
                    (3, 0) ]
    # 2. Define the elliptical hole boundary
    # Parametric equation of ellipse:
    #   x = h + a cos(theta)
    #   y = k + b sin(theta)
    # where: (h,k) = center of ellipse
    #         a     = horizontal radius
    #         b     = vertical radius

    theta = np.linspace(0, 2*np.pi, nx_boundary, endpoint=False)
    ellipse_pts = [(h + a*np.cos(t), k + b*np.sin(t))
                   for t in theta]

    # Create boundary edges for ellipse:
    # Since ellipse nodes are appended after outer nodes,
    # their numbering starts from:
    #     offset = number of outer nodes
```

```
    offset = len(outer_pts)
    ellipse_facets = [(offset + i,
        offset + (i + 1) symmetric positive definite nx_boundary)
        for i in range(nx_boundary) ]

    # 3. Create mesh information object
    mesh_info = triangle.MeshInfo()

    # Add all mesh nodes
    mesh_info.set_points(outer_pts + ellipse_pts)

    # Add all boundary edges
    mesh_info.set_facets(outer_facets + ellipse_facets)

    # Define the hole region
    # Any point inside the ellipse can be used.
    # MeshPy removes elements surrounding this point.
    mesh_info.set_holes([(h, k)])

    # 4. Generate triangular mesh
    # max_volume controls triangle size
    # min_angle improves triangle quality
    mesh=triangle.build(mesh_info, max_volume=max_area, min_angle=30)

    # Convert MeshPy output to NumPy arrays
    nodes = np.array(mesh.points)
    elements = np.array(mesh.elements)
    return nodes, elements

# Mesh Visualization Function
def plot_mesh_T3(pts, tris, a=0.2, b=0.1, h=0.5, k=0.5,
        show_node_ids=True, show_elem_ids=True):
    """
    Plot triangular mesh with optional node and element numbering.
    Parameters:
    pts : ndarray, Node coordinates.
    tris : ndarray, Element connectivity matrix.
    a, b : float, Ellipse radii.
    h, k : float, Ellipse center coordinates.
    show_node_ids : bool, Show node numbering if True.
    show_elem_ids : bool, Show element numbering if True.
    """
    # Create figure
    plt.figure(figsize=(6, 6), dpi=300)
```

```
    # Plot triangular elements
    for elem_id, tri in enumerate(tris):
        # Coordinates of triangle vertices
        poly = pts[tri]
        # Close polygon by repeating first point
        poly_closed = np.vstack([poly, poly[0]])
        # Draw triangle edges
        plt.plot(poly_closed[:, 0], poly_closed[:, 1],
                 color='black', linewidth=0.6)

        # Plot element number at centroid
        if show_elem_ids:
            centroid = np.mean(poly, axis=0)
            plt.text(centroid[0], centroid[1], str(elem_id),
                fontsize=8, color='blue', ha='center', va='center')

    # Plot nodes and node numbers
    if show_node_ids:
        for node_id, (x, y) in enumerate(pts):
            # Draw node
            plt.plot(x, y, 'o', color='black', markersize=4 )
            # Draw node number
            plt.text(x + 0.002, y + 0.002, str(node_id),
                fontsize=7,color='red',ha='left',va='bottom' )

    # Plot exact ellipse boundary
    theta = np.linspace(0, 2*np.pi, 300)
    x_ellipse = h + a*np.cos(theta)
    y_ellipse = k + b*np.sin(theta)
    plt.plot(x_ellipse, y_ellipse, color='black', linewidth=2 )

    # Figure formatting
    plt.gca().set_aspect('equal')
    plt.xlabel("x")
    plt.ylabel("y")
    plt.title("Triangular Mesh with Node and Element Numbering" )
    plt.tight_layout()
    plt.show()
```

```
# Example Usage: a T3 mesh for a plate with an elliptic hole.
# Ellipse geometry
a = 0.2         # Horizontal radius
b = 0.1         # Vertical radius
```

```
cx = 0.5        # Ellipse center x-coordinate
cy = 0.5        # Ellipse center y-coordinate
# Generate mesh
nodes, elements = generate_hole_mesh(
    # Number of nodes on ellipse boundary
    nx_boundary=10,
    # Maximum triangle area
    max_area=0.1,
    # Reserved refinement parameter
    refine_factor=0.5,
    # Ellipse parameters
    a=a, b=b, h=cx, k=cy )

print("Number of nodes    :", nodes.shape[0])
print("Number of elements :", elements.shape[0])
plot_mesh_T3(nodes, elements, a=a, b=b, h=cx, k=cy )
```

```
Number of nodes    : 32;    Number of elements : 42
```

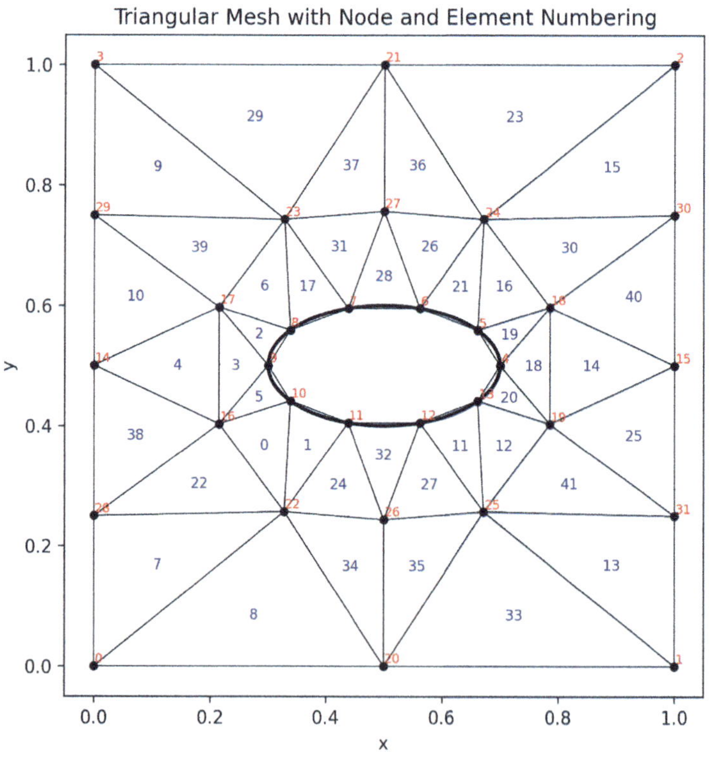

We note the following observations regarding the generated mesh:

1. The implementation generates a mesh composed of triangular elements, each defined by three edges and three nodes. Within this text, such an element is

referred to as a **T3 element**.

2. Each node (labeled in red) is shared by the neighboring edges and elements connected to it. In this specific example, the discretized domain consists of a total of 32 nodes.

3. Because all element edges are represented by straight line segments defined by two nodes, adjacent elements interface exactly along their common boundaries. As a result, a properly constructed mesh contains no gaps or overlaps between neighboring elements.

4. Along the boundary of the elliptical hole, the true curved geometry is approximated by a series of straight line segments. This introduces a **geometric approximation error** in the discretized model. It is important to recognize that the finite element solution is calculated for the discretized mesh rather than the exact original geometry; therefore, any assessment of the FEM results must account for this geometric discrepancy.

5. If the initial geometric approximation is not sufficiently accurate,

   the discretization can be improved by:

   - **Refining the mesh** using a higher density of smaller elements.
   - **Employing higher-order elements** with curved edges (isoparametric elements) that can represent the geometry more precisely.

The following example illustrates a refined mesh, providing a more accurate approximation of the elliptical boundary. It uses the same `generate_hole_mesh` function with parameters tuned for refinement near the hole.

```
# Example: refinement near the hole:
nodes, elements = generate_hole_mesh(
    nx_boundary=50,                    # number of node on the hole edge
    max_area=0.1,                      # control element size
    refine_factor=0.8,                 # control the mesh gradient
    a=a, b=b, h=cx, k=cy )

print("Nodes:", nodes.shape[0])
print("Elements:", elements.shape[0])
plot_mesh_T3(nodes, elements,a=a, b=b, h=cx, k=cy, \
             show_node_ids=False, show_elem_ids=False)
```

```
Nodes: 258       Elements: 446
```

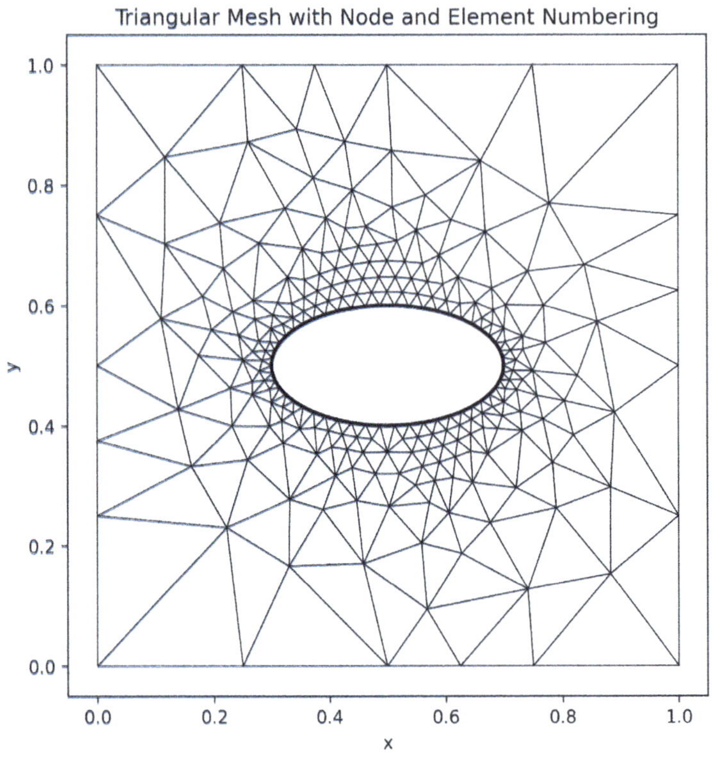

As observed, the refined mesh provides a significantly better approximation of the curved boundary of the elliptical hole. By utilizing smaller elements along the boundary, the discretized geometry more closely resembles the true physical domain. In general, mesh refinement improves both the geometric fidelity of the model and the numerical accuracy of the finite element solution.

3.3 Role of Nodes and Connectivity

Through domain discretization, the continuous displacement field of the solid is replaced by a discrete approximation. Instead of solving for the displacement at every point in the continuum, the FEM approximates the field using values defined specifically at the mesh nodes. These unknowns are referred to as **nodal displacements**.

Using appropriate finite element formulations—which will be developed in the following chapters—the original continuum problem is transformed into a system of algebraic equations where the nodal displacements are the primary unknowns. The central objective of the FEM procedure is to solve for these values.

The mesh data structure primarily consists of the following components:

- **Nodal Coordinates:** Nodes are defined by their spatial coordinates within the problem domain. These coordinates establish the geometry of the discretized structure and mark the precise locations where the unknown degrees of freedom are located.

- **Connectivity:** Connectivity defines how nodes are grouped to form individual elements. In practice, this is managed by an indexing system that specifies which nodes belong to each element. Connectivity is a critical component of the FEM because it establishes the relationships between neighboring elements. Since adjacent elements share common nodes, deformation compatibility is maintained throughout the mesh, ensuring that no gaps or overlaps occur during deformation.

- **Nodal Displacements:** The nodes carry the unknown displacement values in the finite element model. Once these values are solved, the displacement field, strains, and stresses within each element can be computed through interpolation procedures.

The following example prints a subset of the nodes and element connectivity arrays generated by the same `generate_hole_mesh` function to illustrate how this data is structured.

```
# Key output from a mesh generator:
print(f"Coordinates of the first 10 nodes: \n{nodes[:6]}")
print(f"The connectivity matrix with first 10 elements "
      f"(defined by nodes):\n{elements[:6]}")
```

```
Coordinates of the first 10 nodes:
[[0.     0.    ]
 [1.     0.    ]
 [1.     1.    ]
 [0.     1.    ]
 [0.7    0.5   ]
 [0.6984 0.5125]]
The connectivity matrix with first 10 elements (defined by nodes):
[[ 33  62  76]
 [ 34  33  76]
 [201  93  73]
 [ 31  63  68]
 [ 59   4  53]
 [ 63  30  56]]
```

The connectivity matrix defines the nodal composition of each element. For example, in the matrix above, Element 1 is formed by nodes 33, 62, and 76. These node indices are also used during the assembly process to map element-level contributions into the global system matrices.

3.4 Types of Finite Elements

A wide variety of finite elements has been developed within the FEM community to model distinct physical behaviors, complex geometries, and diverse engineering applications. Selecting the appropriate element type is one of the most critical decisions in finite element modeling, as it directly influences:

- Solution accuracy
- Computational efficiency
- Numerical stability
- Convergence behavior
- Overall reliability of the simulation

Different elements are tailored for specific classes of problems, such as truss structures, beams, thin shells, two-dimensional continua, or three-dimensional solids.

Figure 3.1 illustrates several commonly used one-dimensional and two-dimensional finite elements.

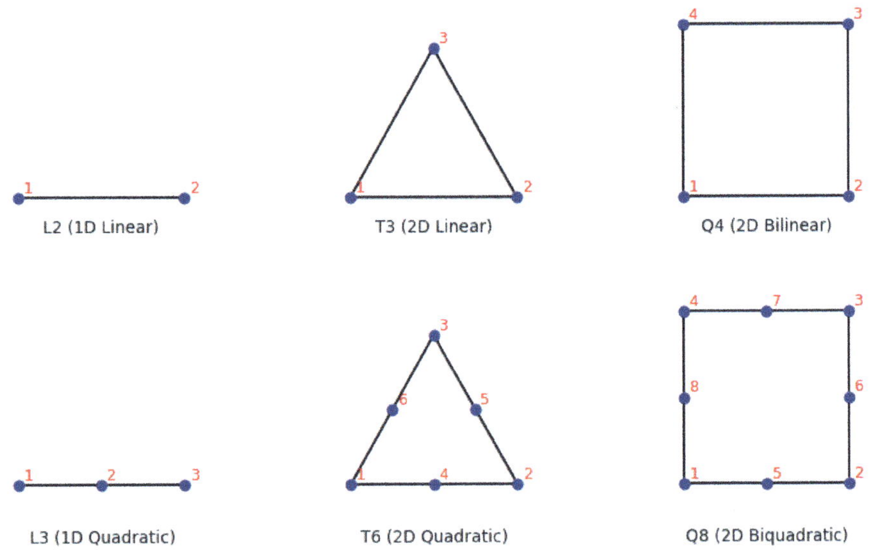

Figure 3.1: Commonly used one-dimensional and two-dimensional finite elements.

3.4.1 One-Dimensional Elements

One-dimensional elements are primarily used to model slender structural members where the length is significantly greater than the cross-sectional dimensions.

Element Type	Typical Nodes	DOFs per Node	Primary Behavior	Advantages	Limitations	Typical Applications
Truss (Bar)	2 (or higher)	1 axial displacement	Axial deformation only	Highly efficient	Cannot model bending	Trusses and links
Beam	2–3	2–6	Bending and shear	Efficient for slender structures	Limited by beam theory	Beams and frames
Frame	2–3	3–6	Axial, bending, torsion	General structural modeling	1D assumptions	Structural systems

While truss elements resist only axial forces, beam and frame elements are capable of modeling bending effects and rotational degrees of freedom.

3.4.2 Two-Dimensional Elements

Two-dimensional elements are extensively used for plane stress, plane strain, plate, and shell problems.

Element Type	Nodes	Order	Typical Formulation	Advantages	Limitations	Typical Applications
Triangle (T3)	3	Linear	Plane stress/strain	Simple automatic meshing	Constant strain	Complex 2D geometries
Triangle (T6)	6	Quadratic	Plane stress/strain	Improved accuracy	Higher computational cost	General 2D analysis
Quad (Q4)	4	Linear	Plane stress/strain	Efficient, widely used	Sensitive to distortion	Structured meshes
Quad (Q8)	8	Quadratic	Plane stress/strain	High accuracy	More difficult to mesh	High-fidelity simulations

Element Type	Nodes	Order	Typical Formulation	Advantages	Limitations	Typical Applications
Shell	4–8	Mixed	Membrane + bending	Efficient for thin structures	Thickness assumptions	Plates and shells
Axi-symmetric	3–8	Mixed	Axi-symmetric	Reduces 3D to 2D	Requires rotational symmetry	Pressure vessels

Triangular elements are particularly effective for complex geometries due to their ability to fill irregular domains. Quadrilateral elements, however, often provide superior accuracy per degree of freedom when well-shaped meshes are utilized.

3.4.3 Three-Dimensional Solid Elements

Figure 3.2 shows several standard three-dimensional solid elements.

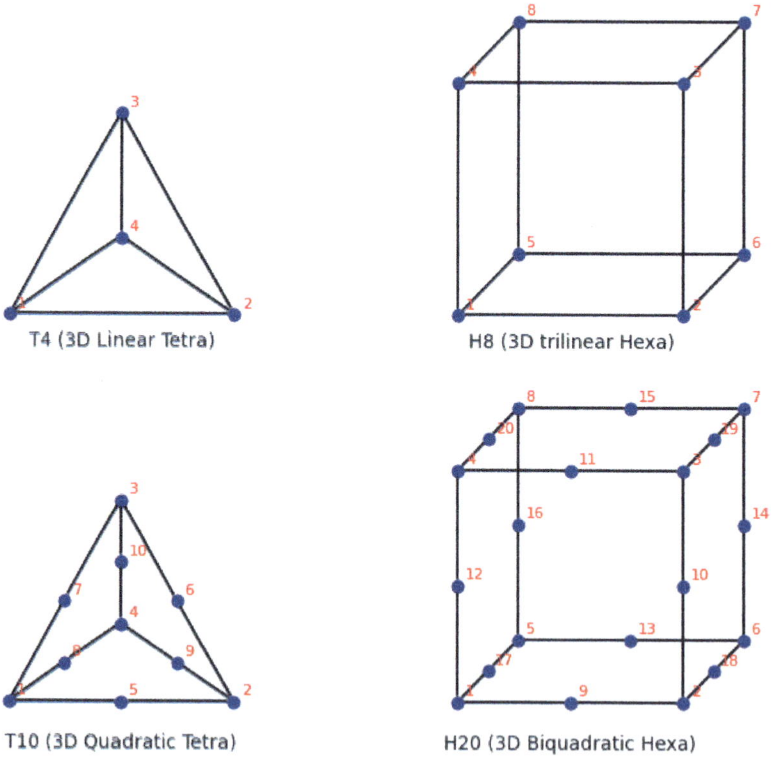

Figure 3.2: Commonly used three-dimensional finite elements.

The following sections provide a compact summary of these finite element types and their typical characteristics.

Three-dimensional solid elements are required when the full triaxial stress and deformation state must be captured.

Element Type	Nodes	Order	Advantages	Limitations	Typical Applications
Tetrahedral (T4)	4	Linear	Easy automatic meshing	Often overly stiff (locking)	Complex geometries
Tetrahedral (T10)	10	Quadratic	Good accuracy	Higher cost	General 3D analysis
Hexahedral (H8)	8	Linear	Computationally efficient	Reduced bending performance	Structured meshes
Hexahedral (H20)	20	Quadratic	Very accurate	Difficult mesh generation	High-fidelity models
Wedge (Prism)	6–15	Mixed	Ideal for layered domains	Less flexible geometry handling	Boundary layers
Pyramid	5–13	Mixed	Mesh transition	Rarely used alone	Hex–Tet transitions

Hexahedral (brick) elements generally offer superior numerical performance. The tetrahedral elements are, however, significantly easier to generate automatically for intricate geometries.

3.4.4 Element Order and Approximation Behavior

Finite elements are further classified by the mathematical order of their interpolation functions.

Category	Description	Advantages	Limitations	Typical Usage
Linear	Corner nodes only	Simple and efficient	Lower stress accuracy	Coarse/initial meshes
Quadratic	Includes mid-side nodes	Accurate curvature representation	Higher computational cost	Preferred for most analysis

Category	Description	Advantages	Limitations	Typical Usage
Higher-order	High-degree polynomials	Very high accuracy	Expensive and complex	Specialized research

Higher-order elements represent curved boundaries and steep solution gradients more effectively than linear elements.

3.4.5 Special Element Formulations

Specialized formulations are often employed to improve numerical performance in specific scenarios.

Element Type	Purpose	Main Advantage	Main Limitation	Typical Applications
Reduced Integration	Fewer integration points	Lower computational cost	Hourglass instability	Large-scale models
Full Integration	Standard integration	Stable behavior	Higher cost; potential locking	General-purpose analysis
Hybrid (Mixed)	Combined variables	Avoids volumetric locking	Complex formulation	Incompressible materials
Contact Elements	Surface interaction	Realistic interaction	Strong nonlinearity	Contact mechanics

3.4.6 General Guidelines for Element Selection

The choice of element type depends on the geometry, the dominant physical behavior, and the available computational resources.

Modeling Scenario	Recommended Element Type
Complex Geometries	Tetrahedral elements
High-Accuracy Analysis	Hexahedral elements
Thin/Slender Structures	Shell, beam, or frame elements

In general, linear elements are robust for solutions with less regularity (low smoothness). However, quadratic elements are preferred when higher accuracy is required—

particularly in bending-dominated problems—provided the solution exhibits sufficient regularity.

The choice of element significantly affects not only the accuracy and efficiency of the computation but also the convergence of the numerical solution. A deeper understanding of these factors will emerge as we study the finite element formulations in the following chapters.

3.5 Mesh Quality

Mesh quality is a decisive factor in the accuracy, stability, and convergence of finite element solutions. Even with a mathematically rigorous formulation, a poor-quality mesh can lead to inaccurate stress distributions, slow convergence, or entirely invalid numerical results.

An ideal mesh must approximate the geometry accurately while maintaining well-shaped elements throughout the domain to ensure the resulting system of equations is numerically well-conditioned.

3.5.1 Types of Element Distortion

In FEM formulations, each element is a "mapped" version of an ideal, perfect shape—known as the reference configuration—over which the function approximation is performed.

An element is considered **distorted** when its physical shape deviates excessively from this ideal reference. Common examples include:

- **Highly skewed quadrilaterals:** Deviating significantly from an ideal square.
- **Stretched triangles:** Deviating from an ideal equilateral triangle (high aspect ratio).
- **Warped hexahedra:** In 3D, when the faces of a brick element are no longer planar.
- **Collapsed elements:** When nodes overlap, causing "degenerated" edges or surfaces.

Key Takeaway: Mapping allows for irregular geometries, but extreme irregularity can produce poorly conditioned stiffness matrices. High-quality results necessitate high-quality elements; therefore, we must ensure the mesh is as regular and undistorted as possible.

3.5.2 Common Mesh Quality Measures

Indicators typically used to assess element quality include:

- **Aspect Ratio:** The ratio of the largest to the smallest dimension; a value of 1 is ideal.

- **Skewness:** The deviation from the ideal angle (90° for quads/hexes; 60° for triangles/tets).

Controlling these factors ensures that elements maintain a preferred shape. At the global level, a high-quality mesh should:

- Feature smoothly varying element sizes.
- Avoid abrupt density transitions.
- Align with critical geometric features and expected loading gradients.

3.5.3 Structured vs. Unstructured Meshes

Finite element meshes are generally classified into two categories.

3.5.3.1 Structured Meshes

If a domain is geometrically regular, structured meshes are preferred. They possess regular connectivity patterns and offer several advantages:

- Predictable node numbering.
- Grid-like arrangements that simplify data handling.
- Efficient memory access and faster computation.
- Quadrilateral (2D) and hexahedral (3D) elements are most common here.

3.5.3.2 Unstructured Meshes

Since most engineering structures possess complex, irregular geometries, unstructured meshes are often a necessity. Triangular and tetrahedral elements dominate this category because they can adapt to nearly any arbitrary geometry with ease.

3.5.4 Mesh Refinement Strategies

Refinement is the process of improving the accuracy of the finite element approximation. Three major strategies are commonly employed:

3.5.4.1 h-Refinement

The characteristic element size h is reduced by using a larger number of smaller elements:

$$h \to 0$$

This is the most common strategy. Its advantages include simple implementation and effectiveness near singularities. However, it can rapidly increase the total degrees of freedom (DOFs), leading to higher computational costs.

3.5.4.2 p-Refinement

The polynomial order of the interpolation functions is increased while the mesh geometry remains unchanged (e.g., transitioning from T3 to T6 or Q4 to Q8). This is excellent for achieving high accuracy in smooth solution fields but increases the computational cost per element.

3.5.4.3 hp-Refinement

By combining both strategies—reducing element size (h) and increasing polynomial order (p) simultaneously—extremely rapid convergence rates can be achieved. This is considered one of the most powerful refinement approaches in advanced FEM.

3.5.5 Compatibility Between Elements

In standard FEM, the assumed displacement function must be continuous over the entire domain. This is known as **compatibility**. At the element level, the displacement field must remain continuous across element interfaces.

Continuity is naturally enforced in **conforming meshes**, where neighboring elements share the same nodes and possess the same interpolation order. In contrast, **non-conforming meshes** contain mismatched interfaces, such as "hanging nodes" or mismatched approximation orders. These require specialized numerical treatments like Multi-Point Constraints (MPC) [12], which can be complex to implement. Best practice dictates using consistent element orders to maintain compatibility.

3.5.6 Boundary Representation and Geometry Error

Solid mechanics problems are defined by a domain (Ω) and its boundaries (Γ). Boundary conditions and external loads are applied through the mesh boundaries. Consequently, it is best practice to define specific boundary sets during mesh generation, such as:

- **Node sets** for point loads or supports.
- **Edge/Face sets** for distributed tractions or contact modeling.

Geometry Approximation Error

Discretization introduces two errors: **solution error** and **geometry error**. The latter occurs when curved boundaries are approximated by straight-edged elements. While h-refinement reduces this error, an alternative is to use **higher-order elements** with isoparametric mapping. In the physical domain, these elements can have curved edges that fit the actual solid geometry much more precisely.

3.5.7 Automatic Mesh Generation

Modern FEM relies heavily on automated algorithms to transition from CAD models to simulation-ready meshes. Key algorithms include:

- **Delaunay Triangulation:** Favored for its robust automation in generating triangular and tetrahedral meshes. It is particularly effective for methods like the **Smoothed Finite Element Method (S-FEM)** [11].

- **Advancing Front Method:** Constructs the mesh from the boundaries inward, offering excellent boundary conformity and control over mesh growth.
- **Mapped Meshing:** Used to create high-quality structured meshes for regular geometries.

While these technologies are foundational for large-scale engineering, the specific algorithmic details of their implementation fall outside the scope of this book.

3.6 Local and Global Numbering

Finite element implementation utilizes both local and global numbering systems. This dual-system approach provides a systematic framework for managing complex geometries and large-scale computations.

3.6.1 Local Numbering

Each individual element possesses its own local node numbering system. This is used to formulate equations at the element level—generating element stiffness matrices and force vectors—independent of the overall structure of the mesh.

3.6.2 Global Numbering

The entire mesh is governed by a global node numbering system generated during the meshing process. A **connectivity matrix** (or mapping array) relates local node indices to their corresponding global indices. This mapping is essential for the **assembly process**, where individual element matrices are strategically "summed" into the global system matrix.

3.7 Topological vs. Physical Dimension

Because FEM subdivides a domain into a large number of elements, the resulting system of equations can become computationally massive. To enhance efficiency, solid mechanics formulations often use theoretical assumptions to reduce the dimensionality of a problem based on its geometric features:

3.7.1 Plate and Shell Structures

If a 3D solid is significantly thinner in one direction than the others, it can be modeled as a plate or shell. In these cases, the structure is **topologically two-dimensional** but exists in three-dimensional space. Mathematically, the 3D problem is reduced to a 2D formulation. While this reduces the number of elements, the governing partial differential equations (PDEs) typically become higher-order. Correspondingly, plate and shell elements are formulated with both translational and rotational degrees of freedom (DOFs).

3.7.2 Beam and Frame Structures

If a 3D solid is slender (small dimensions in two directions), it is modeled as a beam or frame. These structures are **topologically one-dimensional** but exist in 3D space. This reduces the 3D problem to a 1D mathematical model. Like shells, beam elements incorporate both translational and rotational DOFs to capture bending, torsion, and axial effects.

Using these geometrically reduced elements is an effective strategy for minimizing model size. While highly important in structural engineering, the detailed formulation of these elements falls outside the scope of this book; interested readers may refer to Ref. [12].

3.8 Mesh Convergence Studies

Because the FEM provides, in general, an approximate solution, a single simulation result should never be assumed accurate by default. Verification is a critical phase of the engineering process, primarily conducted through a **mesh convergence study**. The standard procedure includes:

1. Obtaining a solution using an initial **coarse mesh**.
2. **Refining the mesh** (increasing the number of elements).
3. **Comparing critical quantities** between the different mesh densities, such as:
    - Maximum displacement
    - Peak von Mises stress
    - Total strain energy
    - Reaction forces at supports

This process is repeated until the change in results between successive refinements becomes negligible (stabilized). This study confirms if the mesh is sufficiently refined, ensures the reliability of the stress data, and helps identify potential numerical singularities.

3.9 The Shape Functions

After the domain is discretized into elements connected by nodes, the next phase of the FEM is to approximate the unknown field variables (such as displacement) within each element.

This is achieved using interpolation functions, commonly referred to as **shape functions**. For a given element, the approximation is expressed as:

$$u^h(\mathbf{x}) = \sum_{i=1}^{n} N_i(\mathbf{x}) d_i \tag{3.5}$$

where:

- $N_i(\mathbf{x})$ are the **shape functions**, which are predetermined for a specific element type.
- d_i are the **nodal unknowns** (degrees of freedom).

Equation (3.5) is the most fundamental expression in FEM approximation enabled by domain discretization. It effectively converts an unknown continuous function into a discrete set of nodal values. This mathematical bridge is at the heart of the discretization process. The systematic derivation of these shape functions will be the focus of the next chapter.

3.10 Exercises

1. **The Principle of Discretization** Explain why domain discretization is necessary for solving solid mechanics problems in the Finite Element Method. How does this process transform a continuum problem with infinitely many degrees of freedom into a solvable numerical system?

2. **The Role of the Mesh Parameter (h)** Define the characteristic element size h for both uniform and non-uniform meshes. Mathematically and physically, what does the condition $h \to 0$ signify in the context of a mesh convergence study?

3. **Connectivity and Compatibility** What is the "connectivity matrix," and why is it considered a critical component of the mesh data structure? Explain how shared nodes between neighboring elements ensure deformation compatibility in a conforming mesh.

4. **Element Selection and Geometry** Compare the advantages and limitations of triangular (T3/T6) versus quadrilateral (Q4/Q8) elements for 2D problems. In what scenarios would a researcher choose tetrahedral elements over hexahedral elements for a 3D simulation?

5. **Mesh Quality and Distortion** List and describe three common types of element distortion. Why is it important to maintain an aspect ratio and skewness close to ideal values during the mesh generation process?

6. **Approximation through Shape Functions** Based on the fundamental approximation equation $u^h(\mathbf{x}) = \sum_{i=1}^{n} N_i(\mathbf{x}) d_i$, describe the relationship between the continuous displacement field, the shape functions, and the nodal unknowns.

Chapter 4

Shape Function Construction

```
# Place cursor here, and press Ctrl+Enter to import dependencies.
import importlib
from sympy import init_printing
import numpy as np
import sympy as sp
import matplotlib.pyplot as plt                              # For plots
from IPython.display import display, Math
from sympy import symbols, lambdify, latex, Matrix, Function

init_printing(use_unicode=True)     # For latex-like quality printing
np.set_printoptions(precision=4,suppress=True) # Digits in print-out
```

Once the problem domain has been discretized into a set of elements with established connectivity, the next stage of the FEM procedure is to approximate the unknown continuous displacement field within each element. This approximation is achieved using **shape functions** as interpolators. These functions, in conjunction with the nodal displacements, define the spatial variation of the displacement field across the element. The construction of shape functions is the cornerstone of nearly all FEM formulations, as it determines how the governing equations are transformed into a discrete system.

4.1 Shape Functions for 1D Bar Elements

We will utilize the simplest case to explicitly illustrate the construction process. Below is a step-by-step derivation of the shape functions for a **1D linear finite element**, commonly referred to as a **two-node bar element**.

4.1.1 Problem Setup

Consider a 1D element defined on a straight line segment of length l_e. We introduce a local coordinate system oriented along the axial direction of the bar. The two nodes are positioned as follows:

- **Node 1** at $x = x_1$
- **Node 2** at $x = x_2$

The configuration is illustrated in Fig. 4.1.

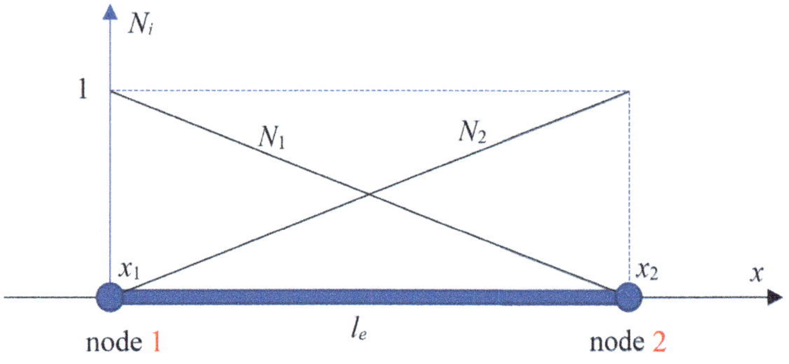

Figure 4.1: Shape functions of a 1D bar element: node 1 is associated with shape function $N_1(x)$ and node 2 with $N_2(x)$. Both are linear functions.

Let $u(x)$ denote a field variable (such as axial displacement). Within the element, it is approximated as:

$$u^h(x) = N_1(x)u_1 + N_2(x)u_2 \tag{4.1}$$

where:

- The superscript h denotes the finite element (approximate) solution.
- $N_1(x)$ and $N_2(x)$ are the shape functions associated with the respective nodes.
- u_1 and u_2 are the nodal values of the field variable.

Since the element is defined by only two nodes, the corresponding shape functions must be **linear** to uniquely interpolate the field between the nodal points.

4.1.2 Requirements for Shape Functions

4.1.2.1 Kronecker Delta Property

To ensure that the approximation reproduces the nodal values exactly, we require that the interpolated field satisfies the nodal conditions at both ends of the element.

At **Node 1**:

$$u^h(x_1) = N_1(x_1)u_1 + N_2(x_1)u_2 = u_1 \tag{4.2}$$

For this equality to hold for arbitrary values of u_2, it is necessary that $N_2(x_1) = 0$. Consequently, to satisfy the equation, we must also have $N_1(x_1) = 1$. Therefore:

$$N_1(x_1) = 1, \quad N_2(x_1) = 0 \tag{4.3}$$

Similarly, at **Node 2**:

$$u^h(x_2) = N_1(x_2)u_1 + N_2(x_2)u_2 = u_2 \tag{4.4}$$

By the same reasoning, this implies:

$$N_1(x_2) = 0, \quad N_2(x_2) = 1 \tag{4.5}$$

Combining Eqs. (4.3) and (4.5), we obtain the **Kronecker delta property**:

$$\boxed{N_i(x_j) = \delta_{ij}} \tag{4.6}$$

where δ_{ij} denotes the **Kronecker delta**. Conceptually:

> Each shape function takes the value 1 at its own (home) node and 0 at all other (remote) nodes.

4.1.2.2 Remark: Linear Independence

The Kronecker delta property leads directly to an important conclusion regarding the mathematical nature of the approximation:

> **Independence Remark:** Shape functions are linearly independent [5]. That is, no shape function can be expressed as a linear combination of the others. Equivalently, if a linear combination of all shape functions vanishes identically, then all coefficients must be zero.

For the 1D bar element, this means that $N_1(x)$ cannot be represented as a scalar multiple of $N_2(x)$, and vice versa. More generally, suppose:

$$\alpha_1 N_1(x) + \alpha_2 N_2(x) = 0 \quad \text{for all } x. \tag{4.7}$$

Using the Kronecker delta property and evaluating at the specific nodal locations:

- At $x = x_1$: $\alpha_1 N_1(x_1) + \alpha_2 N_2(x_1) = \alpha_1(1) + \alpha_2(0) = \alpha_1 = 0$
- At $x = x_2$: $\alpha_1 N_1(x_2) + \alpha_2 N_2(x_2) = \alpha_1(0) + \alpha_2(1) = \alpha_2 = 0$

Thus, $\alpha_1 = \alpha_2 = 0$, proving linear independence.

This simple 1D example illustrates a fundamental property—**linear independence**—that holds for standard finite element shape functions across all element types. It is an essential requirement for constructing a well-posed finite element approximation.

This property is inherently tied to the spatial discretization: because nodes are defined at distinct physical locations, each node represents a unique discrete degree of freedom. Properly constructed shape functions inherit this distinction, guaranteeing their linear independence. This is also why valid FEM models do not permit duplicated nodes at the same coordinate.

> **Basis Remark:** Linear independence ensures that the shape functions form a valid set of basis functions, thereby defining the associated finite element function space.

4.1.2.3 Partition of Unity

The approximation given in Eq. (4.1) must be capable of reproducing a constant field exactly. Let $u^h(x) = c$, where c is a non-zero constant. Substituting this into the interpolation formula:

$$c = N_1(x)c + N_2(x)c \tag{4.8}$$

which implies:

$$N_1(x) + N_2(x) = 1 \tag{4.9}$$

This is known as the **partition of unity property**:

> Each shape function is a part of the unity.

It is critical for ensuring **rigid-body consistency**, meaning the element can undergo a constant displacement (rigid translation) without generating internal strains.

4.1.2.4 Linearity

Since the element has only two nodes, it can represent at most a linear variation of the field variable. Therefore, the shape functions are assumed to take the following linear form:

$$N_1(x) = a_1 + b_1 x, \quad N_2(x) = a_2 + b_2 x \tag{4.10}$$

4.1.3 Derivation in Physical Coordinate x

We first derive $N_1(x)$ by applying the Kronecker delta property to the linear form $N_1(x) = a_1 + b_1 x$.

At $x = x_1$:

$$1 = a_1 + b_1 x_1 \tag{4.11}$$

At $x = x_2$:

$$0 = a_1 + b_1 x_2 \tag{4.12}$$

Solving this system of equations yields:

$$N_1(x) = \frac{x_2 - x}{x_2 - x_1} \tag{4.13}$$

Following the same procedure for $N_2(x)$:

$$N_2(x) = \frac{x - x_1}{x_2 - x_1} \tag{4.14}$$

These functions are linear and **dimensionless**. Their spatial behavior is characterized by:

- $N_1(x)$ decreasing linearly from 1 to 0 across the element.
- $N_2(x)$ increasing linearly from 0 to 1 across the element.

This behavior is illustrated in Fig. 4.1. Notably, Eqs. (4.13) and (4.14) are identical to first-order **Lagrange interpolation functions** [5]. Recognizing this connection allows us to directly define shape functions for higher-order 1D elements using Lagrange polynomials without repeating the algebraic derivation.

4.1.4 Linearity, Alternative Expression

Let us now examine this

$$\sum_{i=1}^{2} N_i(x)x_i = \frac{x_2 - x}{x_2 - x_1}x_1 + \frac{x - x_1}{x_2 - x_1}x_2 = x \tag{4.15}$$

which is an alternative expression of the linearity Eq.(4.10), which is used more often in studying shape function properties.

4.1.5 Standard Coordinate Transformation

In finite element implementation, it is often more efficient to map the physical element to a **natural (dimensionless) coordinate system**:

$$\xi \in [-1, 1] \tag{4.16}$$

The mapping between the physical coordinate x and the natural coordinate ξ is defined as:

$$x(\xi) = \frac{1 - \xi}{2}x_1 + \frac{1 + \xi}{2}x_2 \tag{4.17}$$

4.1.6 Shape Functions in Natural Coordinates

Expressed in terms of ξ, the shape functions take a universal form:

$$N_1(\xi) = \frac{1 - \xi}{2}, \quad N_2(\xi) = \frac{1 + \xi}{2} \tag{4.18}$$

These forms are widely used in FEM software because they are simple, computationally efficient, and independent of the element's physical length. Using these expressions, the coordinate mapping can be rewritten as:

$$x(\xi) = N_1(\xi)x_1 + N_2(\xi)x_2 \tag{4.19}$$

This representation shares the exact same form as the displacement interpolation in Eq. (4.1). Elements that use the same shape functions to approximate both the unknown field variables and the geometric coordinates are known as **isoparametric elements**.

4.1.7 Summary of Key Properties of Shape Functions

The shape functions satisfy several fundamental properties essential for the validity of the FEM:

4.1.7.1 Kronecker Delta Property

$$\boxed{N_i(\xi_j) = \delta_{ij}}, \quad i, j = 1, \dots, n_e \tag{4.20}$$

This ensures that the approximated field exactly matches the nodal values at the node locations. While standard in FEM, it is worth noting that this condition is sometimes relaxed in meshfree methods [2].

4.1.7.2 Partition of Unity

$$\boxed{\sum_{i=1}^{n_e} N_i = 1} \tag{4.21}$$

This property guarantees the exact reproduction of a constant field (rigid-body translation). A direct consequence of this is:

$$\boxed{\sum_{i} \frac{dN_i}{d\xi} = 0} \tag{4.22}$$

This ensures that a uniform displacement field results in zero strain, satisfying a core requirement of mechanics.

4.1.7.3 Linear Field Reproduction

$$\boxed{\sum_{i=1}^{n_e} N_i x_i = x} \tag{4.23}$$

This ensures that the element can represent a linear displacement gradient exactly, which is necessary for convergence.

4.1.8 Derivatives of Shape Functions

To compute strains, we must calculate the derivatives of the shape functions with respect to the physical coordinate:

$$\frac{dN_1}{dx} = -\frac{1}{l_e}, \quad \frac{dN_2}{dx} = \frac{1}{l_e} \tag{4.24}$$

where $l_e = x_2 - x_1$ is the element length. These derivatives are **constant**, implying that for a linear element:

- The strain is constant within the element.
- The stress is constant (assuming a linear elastic material).

4.1.9 Remarks

Shape functions serve as the mathematical bridge that defines how a field variable varies within an element based on its nodal values. By analyzing the simplest 1D two-node bar element, we have established the essential framework and properties—such as the Kronecker delta and partition of unity—that govern shape function construction for all higher-order and multi-dimensional elements in the finite element method.

4.2 Shape Functions for Quadratic 1D Elements

We now utilize Lagrange interpolation polynomials [5] to construct higher-order shape functions for 1D bar elements. This systematic approach generalizes effectively to any polynomial order.

4.2.1 General Lagrange Shape Function Formula

For a 1D element containing n nodes at positions ξ_i, the Lagrange shape function for node i is defined as:

$$N_i(\xi) = \prod_{\substack{j=1 \\ j \neq i}}^{n} \frac{\xi - \xi_j}{\xi_i - \xi_j} \tag{4.25}$$

By construction, this formula automatically satisfies the Kronecker delta property:

$$N_i(\xi_j) = \delta_{ij} \tag{4.26}$$

4.2.2 Quadratic 1D Element (3-Node, L3)

Consider the L3 element with three nodes as illustrated in Fig. 3.1. In the **natural coordinate system**, the node locations are typically selected as:

$$\xi_1 = -1, \quad \xi_2 = 0, \quad \xi_3 = 1 \tag{4.27}$$

In this configuration, the second node is located precisely at the geometric center of the element.

Construction of Shape Functions

Applying the Lagrange formula, the set of three quadratic shape functions is derived as follows:

$$N_1(\xi) = \frac{(\xi - 0)(\xi - 1)}{(-1 - 0)(-1 - 1)} = \frac{1}{2}\xi(\xi - 1) \tag{4.28}$$

$$N_2(\xi) = \frac{(\xi+1)(\xi-1)}{(0+1)(0-1)} = 1-\xi^2 \tag{4.29}$$

$$N_3(\xi) = \frac{(\xi+1)(\xi-0)}{(1+1)(1-0)} = \frac{1}{2}\xi(\xi+1) \tag{4.30}$$

In summary, the quadratic shape functions are:

$$N_1(\xi) = \frac{1}{2}\xi(\xi-1), \quad N_2(\xi) = 1-\xi^2, \quad N_3(\xi) = \frac{1}{2}\xi(\xi+1) \tag{4.31}$$

Their corresponding derivatives with respect to the natural coordinate are:

$$\frac{dN_1}{d\xi} = \xi - \frac{1}{2}, \quad \frac{dN_2}{d\xi} = -2\xi, \quad \frac{dN_3}{d\xi} = \xi + \frac{1}{2} \tag{4.32}$$

It is straightforward to verify that these shape functions satisfy the essential properties of partition of unity and linear independence mentioned earlier. The linearity in the natural coordinate can also be shown with ease:

$$\sum_{i=1}^{n_e} N_i\xi_i = \frac{1}{2}\xi(\xi-1)(-1) + (1-\xi^2)(0) + \frac{1}{2}\xi(\xi+1)(1) = \xi \tag{4.33}$$

This ensures that the element can represent a linear displacement gradient exactly, which is necessary for convergence.

The following Python code visualizes these shape functions and their derivatives to illustrate their spatial behavior.

```
# 1. SETUP: Natural Coordinate System
# We map physical elements to a "natural" space where xi in [-1, 1].
# This makes integration and derivation universal for all elements.
xi = np.linspace(-1, 1, 400)

# 2. DEFINE QUADRATIC SHAPE FUNCTIONS (L3 Element)
# N1, N2, N3 define how displacement is interpolated between nodes.
# Node 1 is at -1, Node 2 is at 0 (center), Node 3 is at +1.
N1 = 0.5 * xi * (xi - 1)    # Equal to 1 at xi =-1, 0 elsewhere
N2 = 1 - xi**2              # Equal to 1 at xi = 0, 0 elsewhere
N3 = 0.5 * xi * (xi + 1)    # Equal to 1 at xi = 1, 0 elsewhere

# Check Partition of Unity: The sum of shape functions must be 1.0.
# This ensures that rigid-body translation is captured correctly.
N_total_sum = N1 + N2 + N3
```

```
# 3. DEFINE DERIVATIVES (Strain Basis)
# Strains are calculated using the derivatives of shape functions.
# The shape functions are quadratic, their derivatives are linear.
dN1 = xi - 0.5
dN2 = -2 * xi
dN3 = xi + 0.5

# Check Derivative Consistency: The sum of derivatives must be 0.
# This ensures that a constant displacement results in zero strain.
dN_total_sum = dN1 + dN2 + dN3

# 4. VISUALIZATION
fig, axes = plt.subplots(1, 2, figsize=(10, 4), dpi=300)

# --- Left Plot: Shape Functions ---
ax0 = axes[0]
ax0.plot(xi, N1, label=r'$N_1$', linewidth=2, color='tab:blue')
ax0.plot(xi, N2, label=r'$N_2$', linewidth=2, color='tab:orange')
ax0.plot(xi, N3, label=r'$N_3$', linewidth=2, color='tab:green')

# Highlight Partition of Unity
ax0.plot(xi, N_total_sum, '--', color='black', alpha=0.5,\
                                        label='Sum (Unity)')

# Mark nodal points: the Delta property (Value=1 at home node)
nodes_xi = [-1, 0, 1]
ax0.scatter(nodes_xi, [1, 1, 1], color='red', marker='o', zorder=5,\
                                           label='Nodal Values')
ax0.set_title('Quadratic Shape Functions (L3)')
ax0.set_xlabel(r'Natural Coordinate ($\xi$)')
ax0.set_ylabel(r'Interpolation Value ($N$)')
ax0.axhline(0, color='black', linewidth=0.8)
ax0.grid(True, linestyle=':', alpha=0.6)
ax0.legend(loc='best')

# --- Right Plot: Derivatives ---
ax1 = axes[1]
ax1.plot(xi, dN1, label=r'$dN_1/d\xi$', linewidth=2, color='blue')
ax1.plot(xi, dN2, label=r'$dN_2/d\xi$', linewidth=2, color='orange')
ax1.plot(xi, dN3, label=r'$dN_3/d\xi$', linewidth=2, color='green')

# Highlight Derivative Consistency
ax1.plot(xi, dN_total_sum, '--', color='black', alpha=0.5, label=\
                                                  'Sum (Zero)')
```

```
ax1.set_title('Derivatives (Strain Distribution)')
ax1.set_xlabel(r'Natural Coordinate ($\xi$)')
ax1.set_ylabel(r'Rate of Change ($dN/d\xi$)')
ax1.axhline(0, color='black', linewidth=0.8)
ax1.grid(True, linestyle=':', alpha=0.6)
ax1.legend(loc='best')

plt.tight_layout()
plt.show()
```

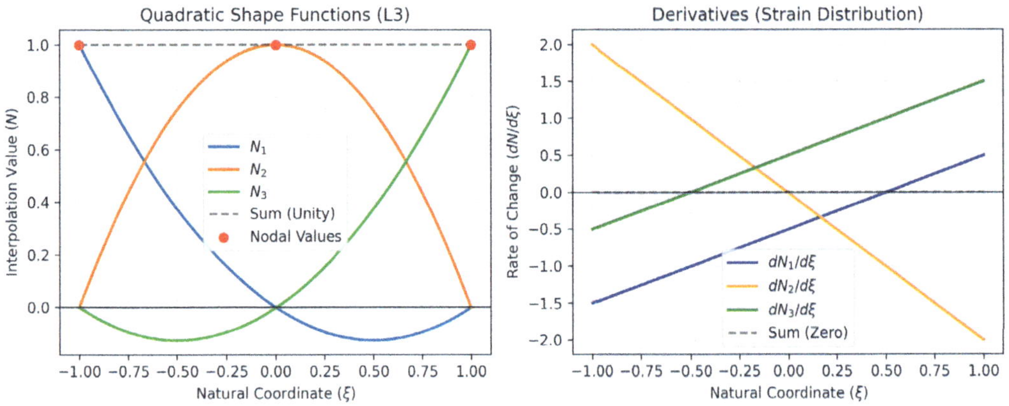

The visualization confirms the following essential properties:

- **Kronecker Delta Property:** $N_i(\xi_j) = \delta_{ij}$, where $N_i = 1$ at its own node and 0 at all other nodes.
- **Partition of Unity:** $\sum N_i = 1$, which ensures that rigid-body motion is represented exactly.
- **Derivative Consistency:** $\sum \frac{dN_i}{d\xi} = 0$, which guarantees that constant displacement produces zero strain.

For quadratic shape functions, the derivatives are linear, implying that both strain and stress vary linearly within the element. By applying Lagrange interpolation polynomials once more, we can extend this framework to even higher-order elements.

4.3 Shape Functions for Cubic 1D Elements

Consider a 1D cubic element (**L4**) with four nodes. To maintain symmetry, the two internal nodes are evenly spaced within the element. In the **natural coordinate system**, the node locations are:

$$\xi_1 = -1, \quad \xi_2 = -\frac{1}{3}, \quad \xi_3 = \frac{1}{3}, \quad \xi_4 = 1 \tag{4.34}$$

In this configuration, the origin is located at the geometric center of the element. The set of **cubic shape functions** is obtained directly using the Lagrange formula:

$$N_1(\xi) = \frac{(\xi+\frac{1}{3})(\xi-\frac{1}{3})(\xi-1)}{(-1+\frac{1}{3})(-1-\frac{1}{3})(-1-1)} = -\frac{9}{16}(\xi+\tfrac{1}{3})(\xi-\tfrac{1}{3})(\xi-1) \tag{4.35}$$

$$N_2(\xi) = \frac{(\xi+1)(\xi-\frac{1}{3})(\xi-1)}{(-\frac{1}{3}+1)(-\frac{1}{3}-\frac{1}{3})(-\frac{1}{3}-1)} = \frac{27}{16}(\xi+1)(\xi-\tfrac{1}{3})(\xi-1) \tag{4.36}$$

$$N_3(\xi) = \frac{(\xi+1)(\xi+\frac{1}{3})(\xi-1)}{(\frac{1}{3}+1)(\frac{1}{3}+\frac{1}{3})(\frac{1}{3}-1)} = -\frac{27}{16}(\xi+1)(\xi+\tfrac{1}{3})(\xi-1) \tag{4.37}$$

$$N_4(\xi) = \frac{(\xi+1)(\xi+\frac{1}{3})(\xi-\frac{1}{3})}{(1+1)(1+\frac{1}{3})(1-\frac{1}{3})} = \frac{9}{16}(\xi+1)(\xi+\tfrac{1}{3})(\xi-\tfrac{1}{3}) \tag{4.38}$$

Summary of Cubic Shape Functions:

$$\begin{aligned}
N_1(\xi) &= -\frac{9}{16}(\xi^2-\tfrac{1}{9})(\xi-1)\\
N_2(\xi) &= \frac{27}{16}(\xi+1)(\xi-\tfrac{1}{3})(\xi-1)\\
N_3(\xi) &= -\frac{27}{16}(\xi+1)(\xi+\tfrac{1}{3})(\xi-1)\\
N_4(\xi) &= \frac{9}{16}(\xi+1)(\xi^2-\tfrac{1}{9})
\end{aligned} \tag{4.39}$$

These cubic shape functions satisfy all the fundamental properties mentioned previously. Their derivatives with respect to the natural coordinate are expressed in matrix form as:

$$\frac{d\mathbf{N}}{d\xi} = \begin{bmatrix} -\frac{27}{16}\xi^2+\frac{9}{8}\xi+\frac{1}{16} \\ \frac{81}{16}\xi^2-\frac{9}{8}\xi-\frac{27}{16} \\ -\frac{81}{16}\xi^2-\frac{9}{8}\xi+\frac{27}{16} \\ \frac{27}{16}\xi^2+\frac{9}{8}\xi-\frac{1}{16} \end{bmatrix} \tag{4.40}$$

A quick inspection confirms the identity $\sum_i \frac{dN_i}{d\xi} = 0$. The following code snippet plots these cubic shape functions and their derivatives for a four-node 1D bar element.

```
# Natural coordinate
xi = np.linspace(-1, 1, 500)

# Cubic shape functions (L4)
```

```
N1 = -9/16*(xi + 1/3)*(xi - 1/3)*(xi -  1)
N2 = 27/16*(xi + 1  )*(xi - 1/3)*(xi -  1)
N3 =-27/16*(xi + 1  )*(xi + 1/3)*(xi -  1)
N4 =  9/16*(xi + 1  )*(xi + 1/3)*(xi -1/3)

# Partition of unity
Nsum = N1 + N2 + N3 + N4

# Derivatives (analytical)
dN1 = -9/16*((xi-1/3)*(xi-1) + (xi+1/3)*(xi-1) + (xi+1/3)*(xi-1/3))
dN2 = 27/16*((xi-1/3)*(xi-1) + (xi + 1)*(xi-1) + (xi + 1)*(xi-1/3))
dN3 =-27/16*((xi+1/3)*(xi-1) + (xi + 1)*(xi-1) + (xi + 1)*(xi+1/3))
dN4 =  9/16*((xi+1/3)*(xi-1/3)+(xi+1)*(xi-1/3) + (xi + 1)*(xi+1/3))

# Derivative consistency
dNsum = dN1 + dN2 + dN3 + dN4

# Plotting
fig, axes = plt.subplots(1, 2, figsize=(12,5), dpi=300)

# ---- Left: Shape functions ----
ax = axes[0]
ax.plot(xi, N1, label='N1', linewidth=2)
ax.plot(xi, N2, label='N2', linewidth=2)
ax.plot(xi, N3, label='N3', linewidth=2)
ax.plot(xi, N4, label='N4', linewidth=2)

# Partition of unity
ax.plot(xi, Nsum, '--', linewidth=2, label='Sum = 1')
ax.axhline(1, linestyle=':', linewidth=1)

# Node locations
nodes = [-1, -1/3, 1/3, 1]
ax.scatter(nodes, [1,1,1,1], color='black', zorder=5)

ax.set_title('Cubic Shape Functions (L4)')
ax.set_xlabel(r'$\xi$')
ax.set_ylabel('N')
ax.axhline(0, linewidth=1)
ax.axvline(0, linewidth=1)
ax.grid(True)
ax.legend()

# ---- Right: Derivatives ----
```

```
ax = axes[1]
ax.plot(xi, dN1, label='dN1/d ', linewidth=2)
ax.plot(xi, dN2, label='dN2/d ', linewidth=2)
ax.plot(xi, dN3, label='dN3/d ', linewidth=2)
ax.plot(xi, dN4, label='dN4/d ', linewidth=2)

# Derivative consistency
ax.plot(xi, dNsum, '--', linewidth=2, label='Sum = 0')
ax.axhline(0, linestyle=':', linewidth=1)

ax.set_title('Derivatives of Shape Functions')
ax.set_xlabel(r'$\xi$')
ax.set_ylabel('dN/d ')
ax.axhline(0, linewidth=1)
ax.axvline(0, linewidth=1)
ax.grid(True)
ax.legend()
plt.tight_layout()
plt.show()
```

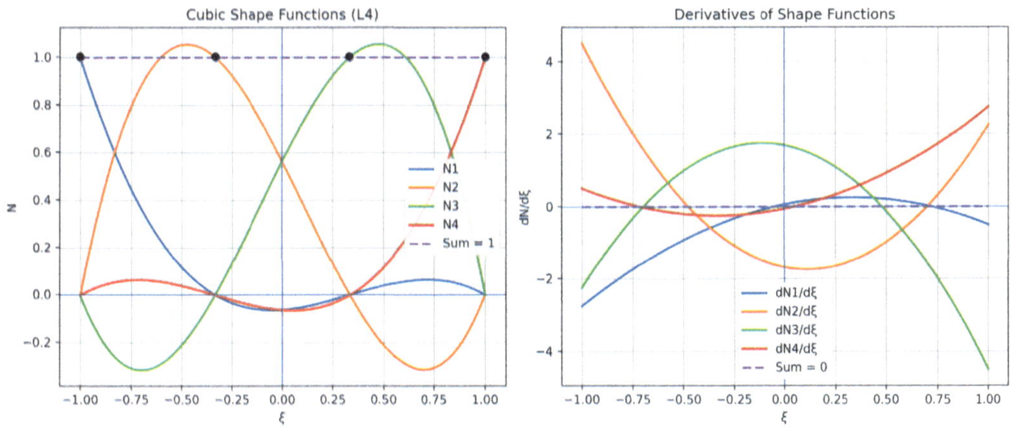

As shown, this cubic element features four evenly distributed nodes (indicated by black dots). The plots confirm that these shape functions satisfy all the fundamental properties established in the preceding sections.

4.4 Shape Functions for T3 Elements

When extending finite element analysis to a two-dimensional (2D) domain, an element must possess **at least three nodes** to define a non-zero area. Using the governing properties of shape functions, we now derive the formulation for a **linear triangular element (T3)**.

4.4.1 Problem Setup for the T3 Element

Consider a triangular element with three nodes (T3) in a 2D Cartesian space, as illustrated in Fig. 3.1.

4.4.2 Local Node Numbering

Local node numbering at element level follows the following counterclockwise rule:

The three nodes (1, 2, and 3) are ordered counterclockwise, which ensures a positive element area (see Section 5.5.5) [6].

The nodal coordinates and their respective degrees of freedom are:

- **Node 1:** (x_1, y_1) with nodal displacements (u_1, v_1)
- **Node 2:** (x_2, y_2) with nodal displacements (u_2, v_2)
- **Node 3:** (x_3, y_3) with nodal displacements (u_3, v_3)

For a valid element, these three nodes must not be collinear; otherwise, the element area would vanish, which is no longer a T3 element. When properly defined, they form a triangle with a non-zero area A_e, as shown in Fig. 4.2.

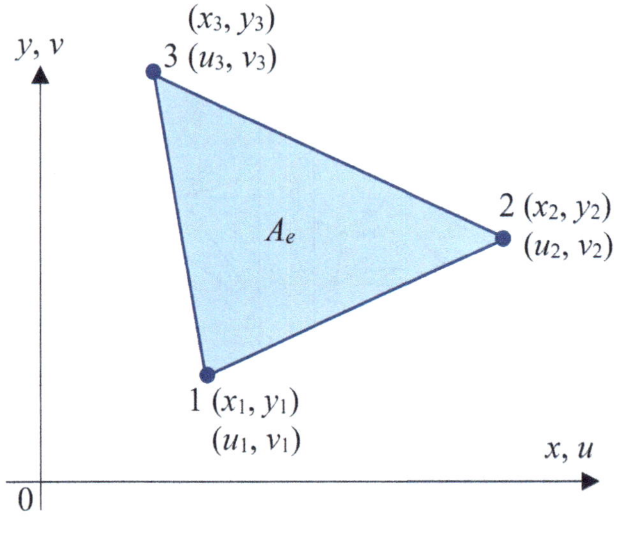

Figure 4.2: A three-node triangular (T3) element in two-dimensional space.

Let $u(x, y)$ represent a field variable (such as a displacement component). Its finite element approximation is expressed as:

$$u^h(x, y) = N_1(x, y)u_1 + N_2(x, y)u_2 + N_3(x, y)u_3 \tag{4.41}$$

where u^h denotes the numerical approximation and $N_i(x, y)$ are the shape functions associated with each node. Since there are **three nodes** defining a plane in a 2D domain, each shape function is modeled as a **linear polynomial** in x and y.

4.4.3 Assumed Form of Shape Functions

Each shape function is assumed to take the following linear form:

$$N_i(x, y) = a_i + b_i x + c_i y \tag{4.42}$$

This introduces three unknown coefficients (a_i, b_i, c_i) for each shape function, which are uniquely determined by the three nodal conditions of the triangle.

4.4.4 Kronecker Delta Property

Consistent with the 1D case, we enforce the **Kronecker delta property**:

$$N_i(x_j, y_j) = \delta_{ij} \tag{4.43}$$

For the first shape function N_1, this requirement results in three simultaneous equations:

$$\begin{aligned} N_1(x_1, y_1) &= 1 \\ N_1(x_2, y_2) &= 0 \\ N_1(x_3, y_3) &= 0 \end{aligned} \tag{4.44}$$

This system can be written in matrix form as:

$$\underbrace{\begin{bmatrix} 1 & x_1 & y_1 \\ 1 & x_2 & y_2 \\ 1 & x_3 & y_3 \end{bmatrix}}_{\mathbf{P}\text{ (Moment Matrix)}} \begin{bmatrix} a_1 \\ b_1 \\ c_1 \end{bmatrix} = \begin{bmatrix} 1 \\ 0 \\ 0 \end{bmatrix} \tag{4.45}$$

The matrix **P** is known as the **moment matrix**. The area of the triangular element, A_e, is directly related to the determinant of this matrix [10]:

$$2A_e = \det(\mathbf{P}) = \begin{vmatrix} 1 & x_1 & y_1 \\ 1 & x_2 & y_2 \\ 1 & x_3 & y_3 \end{vmatrix} \tag{4.46}$$

4.4.5 Solution Using Area Coordinates

As long as these three nodes are not inline, the area is nonzero, and hence the moment matrix is invertible. Hence, Eq.(4.45) can be solved for unknowns a_1, b_1, c_1. The Sympy code to get this done is as follows:

```
# Define symbols for coordinates
x1, y1 = sp.symbols('x1 y1')
x2, y2 = sp.symbols('x2 y2')
x3, y3 = sp.symbols('x3 y3')

# Define unknown coefficients
a1, b1, c1 = sp.symbols('a1 b1 c1')

# Construct the moment matrix P
P = sp.Matrix([[1, x1, y1],
               [1, x2, y2],
               [1, x3, y3] ])

# Right-hand side vector
rhs = sp.Matrix([1, 0, 0])

# Solve the linear system
solution = P.LUsolve(rhs)

# Display results
a1_sol, b1_sol, c1_sol = solution
display(Math(f" \\text{{$a_1$=}} {latex(sp.simplify(a1_sol))}"))
display(Math(f" \\text{{$b_1$=}} {latex(sp.simplify(b1_sol))}"))
display(Math(f" \\text{{$c_1$=}} {latex(sp.simplify(c1_sol))}"))
```

$$a_1 = \frac{x_2y_3 - x_3y_2}{x_1y_2 - x_1y_3 - x_2y_1 + x_2y_3 + x_3y_1 - x_3y_2}$$

$$b_1 = \frac{y_2 - y_3}{x_1y_2 - x_1y_3 - x_2y_1 + x_2y_3 + x_3y_1 - x_3y_2}$$

$$c_1 = \frac{-x_2 + x_3}{(x_1 - x_2)(y_1 - y_3) - (x_1 - x_3)(y_1 - y_2)}$$

The denominator of our coefficients is the determinant of the moment matrix, which is geometrically equal to $2A_e$ [10]. Consequently, the shape function N_1 is expressed as:

$$N_1(x, y) = a_1 + b_1x + c_1y \tag{4.47}$$

where the specific coefficients are:

$$a_1 = \frac{x_2y_3 - x_3y_2}{2A_e}, \quad b_1 = \frac{y_2 - y_3}{2A_e}, \quad c_1 = \frac{x_3 - x_2}{2A_e} \tag{4.48}$$

By applying a **cyclic permutation** of indices $(1 \rightarrow 2 \rightarrow 3 \rightarrow 1)$, we directly obtain the coefficients for the remaining shape functions:

$$\begin{aligned} a_2 &= \frac{x_3y_1 - x_1y_3}{2A_e}, & b_2 &= \frac{y_3 - y_1}{2A_e}, & c_2 &= \frac{x_1 - x_3}{2A_e} \\ a_3 &= \frac{x_1y_2 - x_2y_1}{2A_e}, & b_3 &= \frac{y_1 - y_2}{2A_e}, & c_3 &= \frac{x_2 - x_1}{2A_e} \end{aligned} \tag{4.49}$$

These coefficients remain well-defined provided $A_e \neq 0$. If $A_e = 0$, the nodes are collinear and the element degenerates. Furthermore, if A_e is very small, the shape functions become highly sensitive to nodal variations. This highlights why maintaining a **high-quality mesh** (avoiding heavily distorted elements) is critical in FEM.

4.4.6 Final Shape Functions

The three shape functions for a T3 element can be summarized in the following compact form:

$$\boxed{N_i(x, y) = a_i + b_i x + c_i y, \quad i = 1, 2, 3} \tag{4.50}$$

4.4.7 Area Coordinates

An alternative, more geometric approach to deriving T3 shape functions is through **area (or barycentric) coordinates**. These are defined by the ratios of sub-triangle areas, as illustrated in Fig. 4.3.

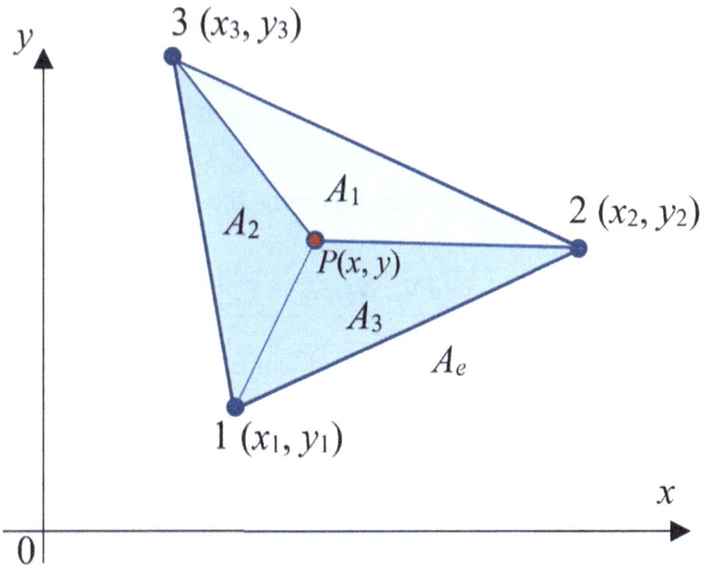

Figure 4.3: A triangular element subdivided into three sub-triangles by an arbitrary internal point $P(x, y)$.

The area coordinates are defined as:

$$L_i(x, y) = \frac{A_i(x, y)}{A_e} \tag{4.51}$$

where A_e is the total element area and $A_i(x, y)$ is the area of the sub-triangle opposite node i. Since $A_1 + A_2 + A_3 = A_e$, the sum of the coordinates is always unity ($L_1 + L_2 + L_3 = 1$). Because they satisfy both the Kronecker delta and partition of unity properties, area coordinates are identical to the shape functions: $L_i = N_i$.

4.4.8 Derivatives and Linear Completeness

The derivatives of the T3 shape functions are:

$$\frac{\partial N_i}{\partial x} = b_i, \quad \frac{\partial N_i}{\partial y} = c_i \tag{4.52}$$

Because these derivatives are **constants**, the T3 element is known as a **Constant Strain Triangle (CST)**. This implies that:

- Strains and stresses are uniform throughout the element.
- The element satisfies **linear completeness**, meaning it can exactly reproduce any linear field $u(x, y) = a_0 + a_1 x + a_2 y$.

4.4.9 Proof of Linear Completeness

Let $\mathbb{P}_1$ denote the space of polynomials in (x, y) of degree ≤ 1, where $\dim(\mathbb{P}_1) = 3$.

1. **Polynomial Degree:** Each N_i is linear, so $N_i \in \mathbb{P}_1$.
2. **Linear Independence:** The Kronecker delta property ensures that if $\sum \alpha_i N_i = 0$, then each α_i must be zero.

Since we have three linearly independent functions in a three-dimensional space:

$$\boxed{\text{span} N_1, N_2, N_3 = \mathbb{P}_1} \tag{4.53}$$

Thus, the T3 element is linearly complete and can exactly represent any constant or linear variation. The logic is straightforward: linearly independent basis $\Rightarrow$ dimensionality match $\Rightarrow$ the same space. $\square$

The following snippet plots these three T3 shape functions:

```
import matplotlib.tri as tri
from mpl_toolkits.mplot3d import Axes3D

# Equilateral triangle (T3 reference element)
```

```
x1, y1 = 0.0, 0.0
x2, y2 = 1.0, 0.0
x3, y3 = 0.2, np.sqrt(3)/2

nodes_x = [x1, x2, x3]
nodes_y = [y1, y2, y3]

edge_x = [x1, x2, x3, x1]
edge_y = [y1, y2, y3, y1]
edge_z = [0, 0, 0, 0]

# Generate barycentric grid
n = 60
L1_list, L2_list, L3_list = [], [], []
x_list, y_list = [], []

for i in range(n + 1):
    for j in range(n + 1 - i):
        L1 = i / n
        L2 = j / n
        L3 = 1.0 - L1 - L2

        x = L1*x1 + L2*x2 + L3*x3
        y = L1*y1 + L2*y2 + L3*y3

        L1_list.append(L1)
        L2_list.append(L2)
        L3_list.append(L3)
        x_list.append(x)
        y_list.append(y)

L1 = np.array(L1_list)
L2 = np.array(L2_list)
L3 = np.array(L3_list)
x = np.array(x_list)
y = np.array(y_list)

# T3 shape functions
N1 = L1; N2 = L2; N3 = L3
N_all = [N1, N2, N3]
titles = ["N1", "N2", "N3"]
triang = tri.Triangulation(x, y)

# Plot (1 × 3 layout)
```

```python
fig = plt.figure(figsize=(15, 4), dpi=300)
for i in range(3):
    ax = fig.add_subplot(1, 3, i+1, projection='3d')
    # surface
    ax.plot_trisurf(triang, N_all[i], zorder=20)
    # reference triangle
    ax.plot(edge_x, edge_y, edge_z, 'k-', linewidth=1, zorder=2)
    ax.scatter(nodes_x, nodes_y, [0, 0, 0], color='b', s=50)
    # node labels
    for k, (xx, yy) in enumerate(zip(nodes_x, nodes_y), start=1):
        ax.text(xx, yy, 0, f"{k}", color='red')

    # dashed "home node" line for each shape function
    for k in range(3):
        Nk_at_node = 1.0 if k == i else 0.0  # Kronecker delta
        ax.plot([nodes_x[k], nodes_x[k]],
                [nodes_y[k], nodes_y[k]],
                [0, Nk_at_node], 'k--', linewidth=1.5 )

    ax.set_title(titles[i])
    ax.set_xlabel('x')
    ax.set_ylabel('y')
    ax.set_zlabel('N')
    ax.set_box_aspect([1, 1, 0.6])

plt.tight_layout()
plt.show()
```

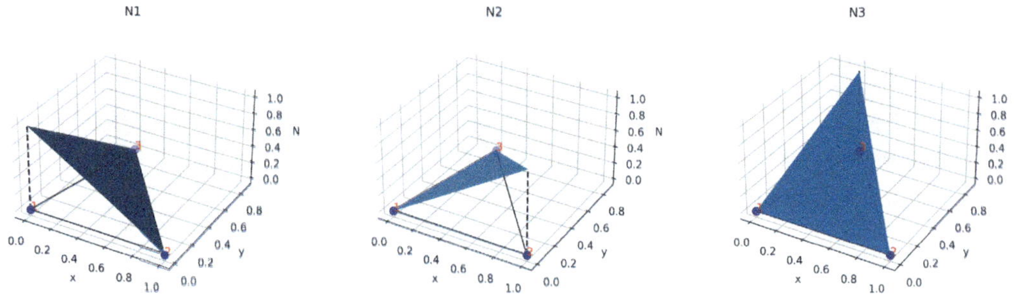

Visual analysis of the T3 element shows that each shape function defines a plane, with a value of 1 at its home node and 0 at all remote nodes.

4.4.10 Patch test

In the finite element method, the **patch test** is a consistency check used to verify whether an FEM formulation can exactly reproduce a prescribed polynomial displacement field over a small group of connected elements (a *patch*).

For solid mechanics, the most common form is the **constant-strain patch test**, in which nodal displacements corresponding to an exact linear displacement field are imposed. An element passes the patch test if the numerical solution reproduces:

- the exact displacement field,
- the exact constant strains,
- and the exact constant stresses,

independent of mesh distortion and element connectivity.

Passing the patch test is a necessary condition for convergence and indicates that the element satisfies the required completeness and inter-element compatibility properties.

It follows that, to pass the patch test, an FEM formulation must at least be able to reproduce a linear displacement field exactly. Since the T3 element shape functions are linearly complete, the T3 element is capable of passing the patch test.

4.5 Shape Functions for T6 Elements

The **quadratic shape functions** for a six-node triangular element (T6) can be derived by leveraging the fundamental properties of shape functions in conjunction with **area coordinates**.

4.5.1 Problem Setup for the T6 Element

Consider a triangular element in a 2D domain, as shown in Fig. 3.1. The T6 element consists of six nodes:

- **Corner nodes:** 1, 2, 3
- **Mid-edge nodes:** 4, 5, 6

By standard convention, the nodes are arranged such that:

- Node 4 is the midpoint of edge (1–2)
- Node 5 is the midpoint of edge (2–3)
- Node 6 is the midpoint of edge (3–1)

The corner nodes must not be collinear to ensure the element possesses a well-defined area A_e. Assuming the edges are straight, the configuration is illustrated in Fig.4.4.

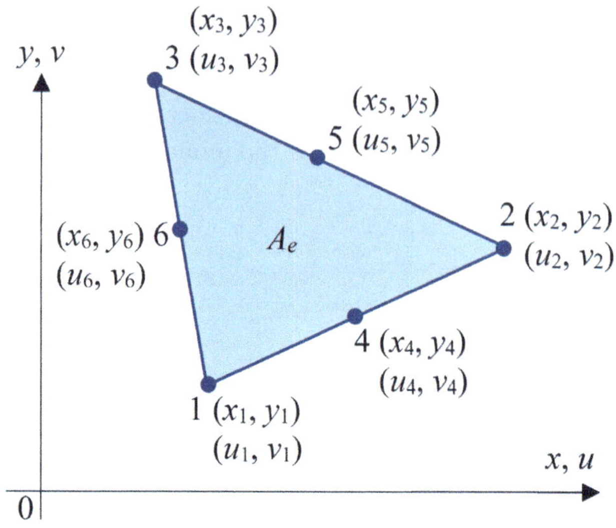

Figure 4.4: A six-node triangular (T6) element in two-dimensional space.

The finite element approximation for a field variable $u(x, y)$ is given by:

$$u^h(x, y) = \sum_{i=1}^{6} N_i(x, y) u_i \tag{4.54}$$

Since this is a quadratic element, each N_i must be a **quadratic polynomial** in x and y.

4.5.2 Use of Area Coordinates

Using the area coordinates L_1, L_2, L_3defined in Eq.(4.51), their values at the six nodal positions are summarized in the table below:

Node	Description	(L_1, L_2, L_3)
1	Corner	$(1, 0, 0)$
2	Corner	$(0, 1, 0)$
3	Corner	$(0, 0, 1)$
4	Mid-edge (1–2)	$(\frac{1}{2}, \frac{1}{2}, 0)$
5	Mid-edge (2–3)	$(0, \frac{1}{2}, \frac{1}{2})$
6	Mid-edge (3–1)	$(\frac{1}{2}, 0, \frac{1}{2})$

4.5.3 Construction of Shape Functions

The geometric interpretation of the area coordinates L_i is illustrated in Fig. 4.5. It allows us to construct the shape functions by simple (but careful) inspection of nodal constraints.

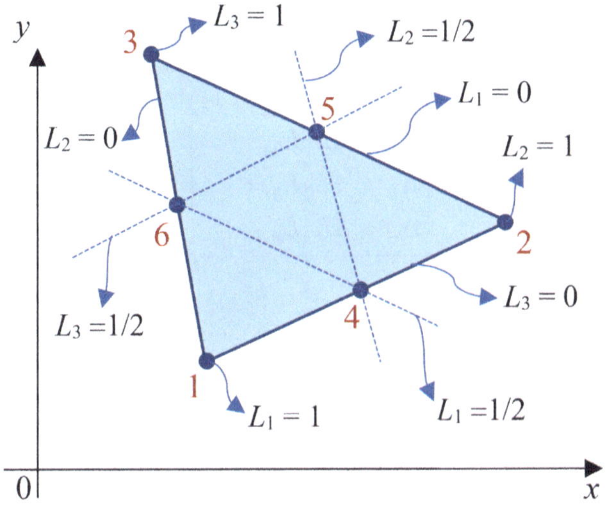

Figure 4.5: The area coordinates for a six-node triangular (T6) element in two-dimensional space.

4.5.3.1 Corner Nodes

For node 1, we require $N_1 = 1$ at $(1, 0, 0)$ and $N_1 = 0$ at all other nodes.

- $L_1 = 0$ along edge (2–3), which contains nodes 2, 3, and 5.
- $L_1 = \frac{1}{2}$ along the line passing through mid-edge nodes 4 and 6.

To satisfy these zeros, we include the factors L_1 and $(L_1 - \frac{1}{2})$. The resulting form is:

$$N_1 = CL_1(L_1 - \tfrac{1}{2}) \tag{4.55}$$

Setting $N_1 = 1$ at $L_1 = 1$ yields $1 = C(1)(1 - \frac{1}{2})$, so $C = 2$. Thus, for the corner nodes:

$$N_1 = L_1(2L_1 - 1), \quad N_2 = L_2(2L_2 - 1), \quad N_3 = L_3(2L_3 - 1) \tag{4.56}$$

4.5.3.2 Mid-Edge Nodes

For node 4 (midpoint of 1–2), we require $N_4 = 1$ at $(\frac{1}{2}, \frac{1}{2}, 0)$ and $N_4 = 0$ elsewhere.

- $L_1 = 0$ along edge (2–3), passing through nodes 2, 3, and 5.
- $L_2 = 0$ along edge (1–3), passing through nodes 1, 3, and 6.

The form $N_4 = CL_1L_2$ satisfies these zeros. Setting $N_4 = 1$ at $L_1 = L_2 = \frac{1}{2}$ yields $1 = C(\frac{1}{4})$, so $C = 4$. Thus, for the mid-edge nodes:

$$N_4 = 4L_1L_2, \quad N_5 = 4L_2L_3, \quad N_6 = 4L_3L_1 \tag{4.57}$$

4.5.4 Final Shape Functions for the T6 Element

$$\boxed{\begin{aligned} N_1 &= L_1(2L_1 - 1) \\ N_2 &= L_2(2L_2 - 1) \\ N_3 &= L_3(2L_3 - 1) \\ N_4 &= 4L_1L_2 \\ N_5 &= 4L_2L_3 \\ N_6 &= 4L_3L_1 \end{aligned}} \tag{4.58}$$

4.5.5 Verification

The derivation of the T6 shape functions confirms the following critical properties:

- **Kronecker Delta Property:** Each $N_i = 1$ at its associated node and zero at all other nodes.
- **Partition of Unity:**

$$\sum_{i=1}^{6} N_i = 1 \tag{4.59}$$

- **Quadratic Completeness:** The interpolation exactly reproduces any quadratic field:

$$\sum_{i=1}^{6} N_i p(\mathbf{x}_i) = p(\mathbf{x}) \tag{4.60}$$

  for any quadratic polynomial $p(x, y)$ (e.g., $1, x, y, x^2, xy, y^2$).

4.5.5.1 Proof of Quadratic Completeness

Our proof uses the same logic: linearly independent basis $\Rightarrow$ dimensionality match $\Rightarrow$ the same space.

Let $\mathbb{P}_2$ denote the space of polynomials in (x, y) of degree ≤ 2, with $\dim(\mathbb{P}_2) = 6$.

1. **Polynomial Degree:** Since L_i are linear in (x, y), each N_i (being a product of L_i) is quadratic; hence $N_i \in \mathbb{P}_2$.
2. **Linear Independence:** If $\sum_{i=1}^{6} \alpha_i N_i = 0$, evaluating at node j gives $\alpha_j = 0$ due to the Kronecker delta property. Thus, all $\alpha_i = 0$.

Because we have six linearly independent functions and $\dim(\mathbb{P}_2) = 6$, they form a basis for the space:

$$\boxed{\text{span} N_1, \ldots, N_6 = \mathbb{P}_2} \tag{4.61}$$

This ensures that T6 shape functions can exactly reproduce any quadratic function:

$$u(x, y) = a_0 + a_1 x + a_2 y + a_3 xy + a_4 x^2 + a_5 y^2 \tag{4.62}$$

Insight: This proof scales effortlessly. For any P_k triangle, the exact same logic applies where $\dim(\mathbb{P}_k) = \frac{(k+1)(k+2)}{2}$. Nodal shape functions are a specialized polynomial basis that facilitates effective FEM modeling for complex domains in higher dimensions [5].

4.5.6 Derivatives of T6 Shape Functions

To compute strains, we derive the derivatives of the T6 shape functions using the chain rule via area coordinates:

$$\frac{\partial N}{\partial x} = \sum_{k=1}^{3} \frac{\partial N}{\partial L_k} \frac{\partial L_k}{\partial x} \tag{4.63}$$

4.5.6.1 Derivatives with respect to Area Coordinates

- **Corner nodes ($N_i = 2L_i^2 - L_i$):**

$$\frac{\partial N_i}{\partial L_i} = 4L_i - 1, \quad \frac{\partial N_i}{\partial L_j} = 0 \quad (j \neq i) \tag{4.64}$$

- **Mid-edge nodes ($N_{ij} = 4L_iL_j$):**

$$\frac{\partial N_{ij}}{\partial L_i} = 4L_j, \quad \frac{\partial N_{ij}}{\partial L_j} = 4L_i \tag{4.65}$$

4.5.6.2 Derivatives of Area Coordinates

Since $L_i(x, y) = a_i + b_i x + c_i y$, their derivatives are the constant coefficients:

$$\frac{\partial L_i}{\partial x} = b_i, \quad \frac{\partial L_i}{\partial y} = c_i \tag{4.66}$$

For example, for L_1:

$$\frac{\partial L_1}{\partial x} = \frac{y_2 - y_3}{2A_e}, \quad \frac{\partial L_1}{\partial y} = \frac{x_3 - x_2}{2A_e} \tag{4.67}$$

4.5.6.3 Final Explicit Expressions

Using the gradient operator $\nabla = [\partial/\partial x, \partial/\partial y]^T$, the derivatives are:

- **Corner Nodes:**

$$\nabla N_i = (4L_i - 1)\nabla L_i \tag{4.68}$$

- **Mid-edge Nodes:**

$$\nabla N_{ij} = 4(L_i \nabla L_j + L_j \nabla L_i) \tag{4.69}$$

Key Observations:

- ∇L_i are **constants** determined by the element geometry.
- ∇N_i are **linear functions** of the coordinates, meaning the T6 element produces **linear strain fields**.
- This formulation is highly efficient for implementation as only the L_i values vary spatially.

The following Python snippet visualizes the spatial distribution of these quadratic shape functions.

```
import matplotlib.tri as tri
from mpl_toolkits.mplot3d import Axes3D

# Equilateral triangle vertices
x1, y1 = 0.0, 0.0
x2, y2 = 1.0, 0.0
x3, y3 = 0.5, np.sqrt(3)/2

# mid-edge nodes
x4, y4 = 0.5*(x1+x2), 0.5*(y1+y2)
x5, y5 = 0.5*(x2+x3), 0.5*(y2+y3)
x6, y6 = 0.5*(x3+x1), 0.5*(y3+y1)

# Full T6 node coordinates
nodes_x = [x1, x2, x3, x4, x5, x6]
nodes_y = [y1, y2, y3, y4, y5, y6]

# Triangle outline
edge_x = [x1, x4, x2, x5, x3, x6, x1]
edge_y = [y1, y4, y2, y5, y3, y6, y1]
edge_z = [0]*len(edge_x)

# Generate barycentric grid
n = 50
L1_list, L2_list, L3_list = [], [], []
x_list, y_list = [], []

for i in range(n + 1):
    for j in range(n + 1 - i):
        L1 = i / n
        L2 = j / n
        L3 = 1.0 - L1 - L2

        x = L1*x1 + L2*x2 + L3*x3
```

```
        y = L1*y1 + L2*y2 + L3*y3

        L1_list.append(L1)
        L2_list.append(L2)
        L3_list.append(L3)
        x_list.append(x)
        y_list.append(y)

L1 = np.array(L1_list)
L2 = np.array(L2_list)
L3 = np.array(L3_list)
x = np.array(x_list)
y = np.array(y_list)

# T6 shape functions
N1 = L1*(2*L1 - 1)
N2 = L2*(2*L2 - 1)
N3 = L3*(2*L3 - 1)
N4 = 4*L1*L2
N5 = 4*L2*L3
N6 = 4*L3*L1

N_all = [N1, N2, N3, N4, N5, N6]
titles = ["N1", "N2", "N3", "N4", "N5", "N6"]

triang = tri.Triangulation(x, y)

# Plot
fig = plt.figure(figsize=(14, 8), dpi=300)

for i in range(6):
    ax = fig.add_subplot(2, 3, i+1, projection='3d')
    # surface
    ax.plot_trisurf(triang, N_all[i], alpha=0.9)

    # reference element (edges + nodes)
    ax.plot(edge_x, edge_y, edge_z, 'k-', linewidth=1)
    ax.scatter(nodes_x, nodes_y, [0]*6, color='b', s=40)

    # node labels
    for k, (xx, yy) in enumerate(zip(nodes_x, nodes_y), start=1):
        ax.text(xx, yy, 0, f"{k}", color='red')

    # dashed vertical "home node" lines (the delta visualization)
```

```
    for k in range(6):
        Nk_at_node = 1.0 if k == i else 0.0
        ax.plot([nodes_x[k], nodes_x[k]],
                [nodes_y[k], nodes_y[k]],
                [0, Nk_at_node], 'k--', linewidth=1 )

    ax.set_title(titles[i])
    ax.set_xlabel('x')
    ax.set_ylabel('y')
    ax.set_zlabel('N')
    ax.set_box_aspect([1, 1, 0.6])

plt.tight_layout()
plt.show()
```

The plots are shown in the following page. It is observed that

- Each of the shape functions has 1 at its home node, and 0 at all the remote nodes
- Corner functions: "peaked" at vertices
- mid-edge functions: "bubble-like" along edges

Since the T6 element is quadratically complete, it is capable of passing the patch test, which requires only linear complete.

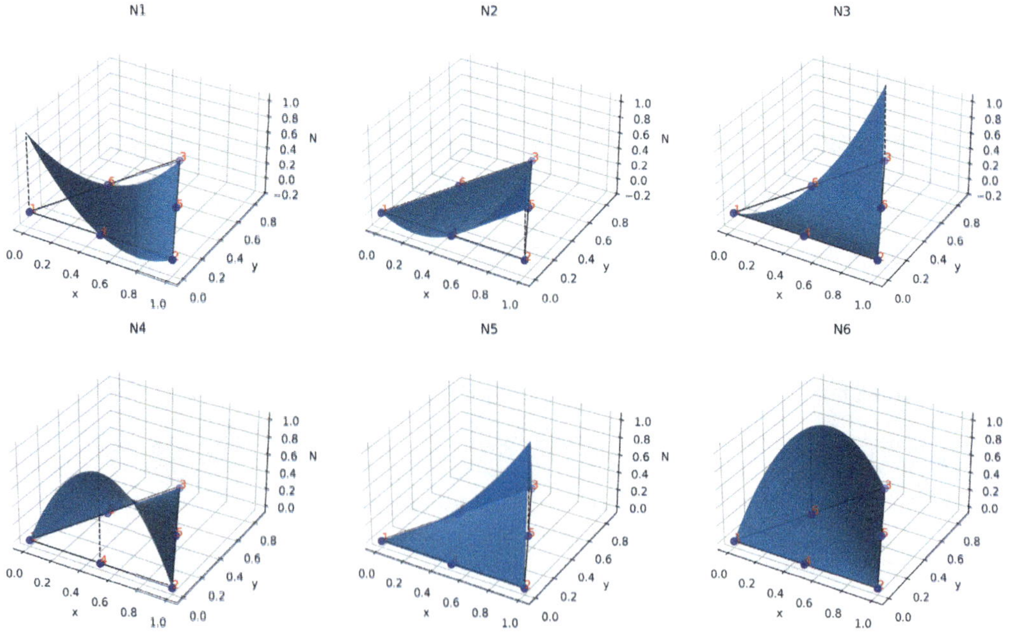

4.6 Shape Functions for Q4 Elements

When a quadrilateral element with four nodes (Q4) is used, the shape functions are most conveniently derived in a natural coordinate system. It is also a **necessity**, because it serves two **critical purposes**:

1. It ensures continuity of the field approximation along the edges shared by adjacent elements.

2. It allows the element to be represented as a simple square in the natural coordinate system, which greatly simplifies the construction of shape functions.

4.6.1 Element Description

Consider a square element with four edges and four nodes in the natural coordinate system, as shown in Fig.4.6.

Local Node Numbering

Local node numbering (1→2→3 →4) at element level follows the following counter-clockwise rule, which ensures a positive element area when standard formula is used to compute it (see Section 5.5.4) [6].

We introduce the **natural coordinates** (ξ, η) and define the Q4 element over the domain

$$\Omega_e = [-1, 1] \times [-1, 1] \tag{4.70}$$

The coordinates of the four nodes in the (ξ, η) system are:

$$(1); (-1, -1), \quad (2); (1, -1), \quad (3); (1, 1), \quad (4); (-1, 1) \tag{4.71}$$

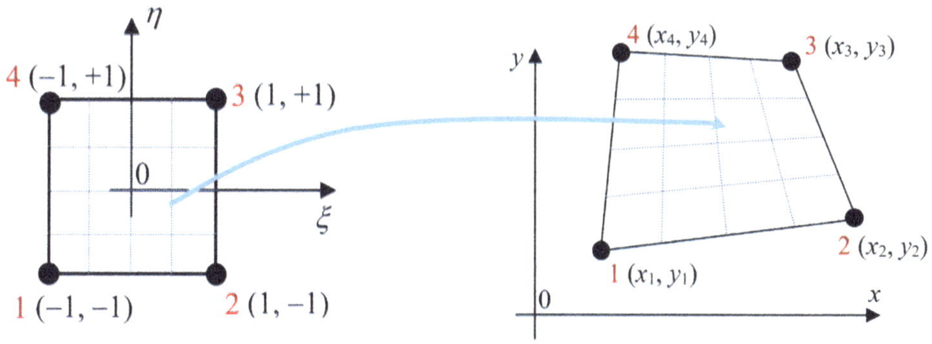

Figure 4.6: A quadrilateral element with four edges and four nodes in two-dimensional space. The shape functions, and hence the approximated field, vary linearly along the dashed lines.

Let $u(x, y)$ denote a field variable (e.g., a displacement component). Its finite element approximation is given by:

$$u^h(\xi, \eta) = \sum_{i=1}^{4} N_i(\xi, \eta)\, u_i \tag{4.72}$$

The goal is to construct four shape functions $N_i(\xi, \eta)$, each associated with one node.

4.6.2 Shape Function Construction

Because the Q4 element is a square in the natural coordinate system (orthogonal), the shape functions can be constructed as products of one-dimensional linear Lagrange polynomials.

The 1D Lagrange polynomials in ξ are:

$$N^-(\xi) = \frac{1-\xi}{2}, \quad N^+(\xi) = \frac{1+\xi}{2} \tag{4.73}$$

Similarly, in η:

$$N^-(\eta) = \frac{1-\eta}{2}, \quad N^+(\eta) = \frac{1+\eta}{2} \tag{4.74}$$

Each 2D shape function is obtained by combining the appropriate 1D factors.

4.6.3 Shape Functions at Each Node

For node 1 at $(-1, -1)$:

$$N_1(\xi, \eta) = N^-(\xi), N^-(\eta) = \frac{1}{4}(1-\xi)(1-\eta) \tag{4.75}$$

For node 2 at $(1, -1)$:

$$N_2(\xi, \eta) = N^+(\xi), N^-(\eta) = \frac{1}{4}(1+\xi)(1-\eta) \tag{4.76}$$

For node 3 at $(1, 1)$:

$$N_3(\xi, \eta) = N^+(\xi), N^+(\eta) = \frac{1}{4}(1+\xi)(1+\eta) \tag{4.77}$$

For node 4 at $(-1, 1)$:

$$N_4(\xi, \eta) = N^-(\xi), N^+(\eta) = \frac{1}{4}(1-\xi)(1+\eta) \tag{4.78}$$

4.6.4 Final Q4 Shape Functions

$$\boxed{\begin{aligned} N_1 &= \frac{1}{4}(1-\xi)(1-\eta) \\ N_2 &= \frac{1}{4}(1+\xi)(1-\eta) \\ N_3 &= \frac{1}{4}(1+\xi)(1+\eta) \\ N_4 &= \frac{1}{4}(1-\xi)(1+\eta) \end{aligned}} \tag{4.79}$$

4.6.5 Derivatives of the Shape Functions

$$\begin{aligned} \frac{\partial N_1}{\partial \xi} &= -\frac{1}{4}(1-\eta), & \frac{\partial N_1}{\partial \eta} &= -\frac{1}{4}(1-\xi) \\ \frac{\partial N_2}{\partial \xi} &= \frac{1}{4}(1-\eta), & \frac{\partial N_2}{\partial \eta} &= -\frac{1}{4}(1+\xi) \\ \frac{\partial N_3}{\partial \xi} &= \frac{1}{4}(1+\eta), & \frac{\partial N_3}{\partial \eta} &= \frac{1}{4}(1+\xi) \\ \frac{\partial N_4}{\partial \xi} &= -\frac{1}{4}(1+\eta), & \frac{\partial N_4}{\partial \eta} &= \frac{1}{4}(1-\xi) \end{aligned} \tag{4.80}$$

It can be observed that each of the four shape functions is linear in ξ for any fixed η, and linear in η for any fixed ξ. In other words, each function is linear along two independent directions: the ξ-direction and the η-direction. Consequently, the shape functions vary linearly along the dashed lines shown in Fig.4.6, which correspond to lines of constant ξ or η.

This linearity along the dashed lines is effectively exploited in the development of smoothed finite element models (S-FEM), where it is not even necessary to explicitly construct the shape functions; only their values at critical locations are required [11].

Note that along any other direction, the variation of these shape functions is generally nonlinear (quadratic). For this reason, the Q4 shape functions defined in Eq.(4.79) are referred to as **bilinear elements**.

Note also that a Q4 element in physical space is not restricted to a square; it is generally a quadrilateral with four straight edges that are not necessarily parallel or perpendicular. Because the shape functions are constructed in natural coordinates, we must establish a mapping from $(\xi, \eta) \in [-1, 1] \times [-1, 1]$ to the physical coordinates $(x, y) \in \Omega_e$. As illustrated in Fig. 4.6, this mapping is achieved using the same bilinear shape functions.

4.6.6 Mapping from Natural to Physical Coordinates

4.6.6.1 Node Definition in Physical Space

The four nodes of the quadrilateral in physical space are defined as:

$$(x_1, y_1), \quad (x_2, y_2), \quad (x_3, y_3), \quad (x_4, y_4) \tag{4.81}$$

These correspond to the natural coordinate nodes:

$$(-1, -1), \quad (1, -1), \quad (1, 1), \quad (-1, 1) \tag{4.82}$$

4.6.6.2 Isoparametric Mapping

The coordinate transformation from the natural system (ξ, η) to the physical system (x, y) is defined by:

$$x(\xi, \eta) = \sum_{i=1}^{4} N_i(\xi, \eta) x_i, \quad y(\xi, \eta) = \sum_{i=1}^{4} N_i(\xi, \eta) y_i \tag{4.83}$$

Since the **same set of shape functions** is used for both the geometry mapping and the field variable interpolation (Eq. 4.72), this approach is termed **isoparametric mapping**.

Equation (4.83) can be expanded explicitly:

$$\begin{aligned} x(\xi, \eta) &= \frac{1}{4}\Big[(1-\xi)(1-\eta)x_1 + (1+\xi)(1-\eta)x_2 \\ &\quad +(1+\xi)(1+\eta)x_3 + (1-\xi)(1+\eta)x_4\Big] \\ y(\xi, \eta) &= \frac{1}{4}\Big[(1-\xi)(1-\eta)y_1 + (1+\xi)(1-\eta)y_2 \\ &\quad +(1+\xi)(1+\eta)y_3 + (1-\xi)(1+\eta)y_4\Big] \end{aligned} \tag{4.84}$$

This transformation is expressed compactly in vector form as:

$$\boxed{\mathbf{x}(\xi, \eta) = \sum_{i=1}^{4} N_i(\xi, \eta) \mathbf{x}_i} \quad \text{where } \mathbf{x}_i = (x_i, y_i) \tag{4.85}$$

We note the following:

Linear Edges: Despite the presence of the $\xi\eta$ product, the mapping ensures that the edges of the element remain straight in physical space. For example, along the edge $\eta = -1$, the mapping becomes a linear function of ξ.

4.6.7 Verification of Key Properties

4.6.7.1 Kronecker Delta Property

The construction of the Q4 shape functions ensures that each function attains a value of unity at its associated node and zero at all other nodes:

$$N_i(\xi_j, \eta_j) = \delta_{ij} \tag{4.86}$$

For instance, at **Node 1** where $(\xi, \eta) = (-1, -1)$:

- $N_1 = \frac{1}{4}(1 - (-1))(1 - (-1)) = 1$
- $N_2 = N_3 = N_4 = 0$

4.6.7.2 Partition of Unity

The sum of the shape functions at any point within the element is identically equal to one:

$$\sum_{i=1}^{4} N_i = 1 \tag{4.87}$$

Proof:

$$\frac{1}{4}\Big[(1-\xi)(1-\eta) + (1+\xi)(1-\eta) + (1+\xi)(1+\eta) + (1-\xi)(1+\eta)\Big] \tag{4.88}$$

Factoring $(1-\eta)$ from the first two terms and $(1+\eta)$ from the last two yields $\frac{1}{4}[2(1-\eta) + 2(1+\eta)] = 1$.

4.6.7.3 Bilinear Completeness in the Natural Coordinates

The Q4 element employs a **bilinear** interpolation scheme with the functional basis $1, \xi, \eta, \xi\eta$. Consequently, it can exactly reproduce any field variable of the form

$$u(\xi, \eta) = a_0 + a_1\xi + a_2\eta + a_3\xi\eta \tag{4.89}$$

Thus, the Q4 element satisfies the completeness requirements for convergence in the natural coordinate system.

4.6.7.4 Linear Completeness in the Physical Coordinates

When the isoparametric mapping given in Eq.(4.83) is used, the Q4 shape functions can exactly reproduce any linear field in the physical coordinates.

To show this, consider an arbitrary linear field in the physical coordinates:

$$u(x, y) = a + bx + cy \tag{4.90}$$

The nodal values at the four nodes of a Q4 element are therefore

$$u_i = a + bx_i + cy_i \tag{4.91}$$

Applying the finite element interpolation with the Q4 shape functions gives

$$\begin{aligned} u^h(x,y) &= \sum_{i=1}^{4} N_i u_i \\ &= \sum_{i=1}^{4} N_i (a + bx_i + cy_i). \end{aligned} \tag{4.92}$$

Rearranging the summation yields

$$\begin{aligned} u^h(x,y) &= a\sum_{i=1}^{4} N_i + b\sum_{i=1}^{4} N_i x_i + c\sum_{i=1}^{4} N_i y_i \\ &= a + bx + cy. \end{aligned} \tag{4.93}$$

Here, the partition of unity property of the shape functions and the isoparametric mapping relations in Eq.(4.83) have been used. Therefore,

$$u^h(x,y) = u(x,y), \tag{4.94}$$

which proves that the Q4 element exactly reproduces any linear field in the physical coordinates, if the isoparametric mapping is implied.

This linear completeness ensures that Q4 elements will pass the patch test.

4.6.7.5 Bilinear Incompleteness in the Physical Coordinates

The Q4 element cannot, in general, exactly reproduce bilinear fields containing xy terms in the physical coordinates. Reproduction of such terms requires quadratic contributions in ξ and η (see Section 3.10.6 of Ref. [6]).

4.6.8 Key Properties of Isoparametric Mapping

The isoparametric mapping for a Q4 element allows for general distortion, including skewing and stretching, as illustrated in Fig. 4.6. Its fundamental properties are summarized below:

1. **Exact Nodal Mapping:** At each node i, the mapping satisfies $(\xi, \eta)_i \Rightarrow (x, y)_i$. Every node in the natural coordinate system is mapped exactly to its corresponding physical location.

2. **Bilinear Geometry and Edge Correspondence:** The mapping is **bilinear**, meaning x and y vary linearly with ξ (for fixed η) and linearly with η (for fixed ξ). Consequently, any line of constant ξ or η in the natural domain maps to a straight line in the physical domain, illustrated in Fig. 4.6 as the dished lines.

3. **Edge Continuity:** Since element edges remain straight and nodes are shared, adjacent Q4 elements match exactly along common boundaries, preventing gaps or overlaps across the mesh. This is said that the function has a C^0 **continuity** over the meshed domain.

4. **Area Correspondence:** For a **convex** Q4 element, there is a consistent one-to-one mapping between the standard square and the physical quadrilateral area.

5. **Bijectivity:** For convex elements, the mapping is bijective (one-to-one). If an element becomes concave or excessively distorted, this correspondence is lost, which may lead to numerical instability.

4.6.9 Jacobian Matrix

To perform integration and differentiation in the physical domain, we utilize the Jacobian matrix $\mathbf{J}$, which relates gradients in natural coordinates to gradients in physical coordinates:

$$\mathbf{J} = \begin{bmatrix} \frac{\partial x}{\partial \xi} & \frac{\partial x}{\partial \eta} \\ \frac{\partial y}{\partial \xi} & \frac{\partial y}{\partial \eta} \end{bmatrix} \tag{4.95}$$

Using the bilinear shape functions and the coordinate interpolation $\mathbf{x}(\xi, \eta) = \sum N_i \mathbf{x}_i$, the Jacobian matrix is expressed in terms of nodal coordinates as:

$$\mathbf{J} = \begin{bmatrix} \sum_{i=1}^4 x_i \frac{\partial N_i}{\partial \xi} & \sum_{i=1}^4 x_i \frac{\partial N_i}{\partial \eta} \\ \sum_{i=1}^4 y_i \frac{\partial N_i}{\partial \xi} & \sum_{i=1}^4 y_i \frac{\partial N_i}{\partial \eta} \end{bmatrix} \tag{4.96}$$

Therefore, the Jacobian matrix is known for a given element.

4.6.9.1 Explicit Jacobian Entries

By differentiating the shape functions, we obtain the following explicit terms:

First column ($\partial/\partial\xi$):

$$\frac{\partial x}{\partial \xi} = \frac{1}{4}\Big[-(1-\eta)x_1 + (1-\eta)x_2 + (1+\eta)x_3 - (1+\eta)x_4\Big] \tag{4.97}$$

$$\frac{\partial y}{\partial \xi} = \frac{1}{4}\Big[-(1-\eta)y_1 + (1-\eta)y_2 + (1+\eta)y_3 - (1+\eta)y_4\Big] \tag{4.98}$$

Second column ($\partial/\partial\eta$):

$$\frac{\partial x}{\partial \eta} = \frac{1}{4}\Big[-(1-\xi)x_1 - (1+\xi)x_2 + (1+\xi)x_3 + (1-\xi)x_4\Big] \tag{4.99}$$

$$\frac{\partial y}{\partial \eta} = \frac{1}{4}\Big[-(1-\xi)y_1 - (1+\xi)y_2 + (1+\xi)y_3 + (1-\xi)y_4\Big] \tag{4.100}$$

4.6.9.2 Key Observations

- **Non-Constant Mapping:** Each Jacobian entry is **linear in ξ and η**. Because $\mathbf{J}$ is not constant over the element, strain fields derived from the displacement interpolation are generally non-uniform inside element.

- **Geometric Determinant:** The determinant of the Jacobian, $|\mathbf{J}|$, relates the infinitesimal area elements: $dA = dxdy = |\mathbf{J}|d\xi d\eta$.

- **Uniqueness and Convexity:** For a valid, one-to-one mapping, $|\mathbf{J}|$ must be positive everywhere within the element. This mathematically requires the quadrilateral to be convex (no interior angle exceeds 180°), consistent with the angular distortion limits discussed in Chapter 3.

4.6.10 Plot Q4 Displacements in Physical Domain

Computing a shape function or displacement field over a Q4 element in the natural coordinate system is straightforward once the nodal values are provided: we simply evaluate the interpolation given in Eq. (4.72). However, performing the same computation in the physical domain is more subtle. This is because the bilinear mapping defined in Eq. (4.84) is, in general, not invertible uniquely in a simple closed form. As a result, the field cannot usually be expressed as an explicit function $u(x, y)$.

The standard approach is to use a **parametric representation**: we map points from the natural coordinate system (ξ, η) to the physical domain (x, y) using the bilinear shape functions, and then visualize the field accordingly using the bijectivity property.

4.6.10.1 Procedure

1. **Parametric Evaluation:** Evaluate the quantities over the natural domain:

$$x(\xi, \eta), \quad y(\xi, \eta), \quad u(\xi, \eta) \tag{4.101}$$

2. **Surface Visualization:** Visualize the field as a surface in physical space by plotting the coordinates:

$$(x(\xi, \eta), y(\xi, \eta), u(\xi, \eta)) \tag{4.102}$$

This method is the standard and robust approach used in the finite element method for handling distorted elements.

4.6.10.2 Implementation Steps

1. **Define Inputs:** Specify the nodal coordinates (x_i, y_i) and the nodal values u_i (representing either displacements or specific shape function values).

2. **Interpolate:** Use the Q4 shape functions to compute the mapping and the field simultaneously:

$$u(\xi, \eta) = \sum_{i=1}^{4} N_i(\xi, \eta) u_i \tag{4.103}$$

$$x(\xi, \eta) = \sum_{i=1}^{4} N_i(\xi, \eta) x_i, \quad y(\xi, \eta) = \sum_{i=1}^{4} N_i(\xi, \eta) y_i \tag{4.104}$$

3. **Sample:** Generate a grid of points over the square natural domain $(\xi, \eta) \in [-1, 1]^2$.

4. **Map and Plot:** Transform these points to the (x, y, u) space to generate the final visualization.

The following code snippet demonstrates this procedure for a distorted quadrilateral element.

```
# Q4 element geometry (general distorted quad)
x = np.array([0.0, 2.0, 2.5, -0.5])
y = np.array([0.0, 0.5, 1.5, 1.5])

# Example: Nodal values to extract N1
u = np.array([0.0, 1.0, 0.0, 0.0])
# Natural grid
xi = np.linspace(-1, 1, 50)
eta = np.linspace(-1, 1, 50)
XI, ETA = np.meshgrid(xi, eta)
# Shape functions
N1 = 0.25*(1-XI)*(1-ETA)
N2 = 0.25*(1+XI)*(1-ETA)
N3 = 0.25*(1+XI)*(1+ETA)
N4 = 0.25*(1-XI)*(1+ETA)

# Mapping to physical coordinates
X = N1*x[0] + N2*x[1] + N3*x[2] + N4*x[3]
Y = N1*y[0] + N2*y[1] + N3*y[2] + N4*y[3]

# N1 field
# U = N2  # same as interpolation with u = [0,1,0,0]
U = N1*u[0] + N2*u[1] + N3*u[2] + N4*u[3]

# Plot
fig = plt.figure(figsize=(6,6), dpi=300)
ax = fig.add_subplot(111, projection='3d')
```

```
# Surface of N1
ax.plot_surface(X, Y, U, cmap='viridis', alpha=0.9)

# Overlay original Q4 element (z = 0 plane)
# Close the loop
x_closed = np.append(x, x[0])
y_closed = np.append(y, y[0])
z_closed = np.zeros_like(x_closed)

# Draw edges
ax.plot(x_closed, y_closed, z_closed, 'k-', linewidth=1)

# Add projection of contours onto the element
ax.contour(X, Y, U, zdir='z', offset=0)
# Draw nodes
ax.scatter(x, y, np.zeros_like(x), c='r', s=20)
# Label nodes
for i in range(4):
    ax.text(x[i], y[i], 0, f'{i+1}', color='red')

ax.set_xlabel('x')
ax.set_ylabel('y')
plt.show()
```

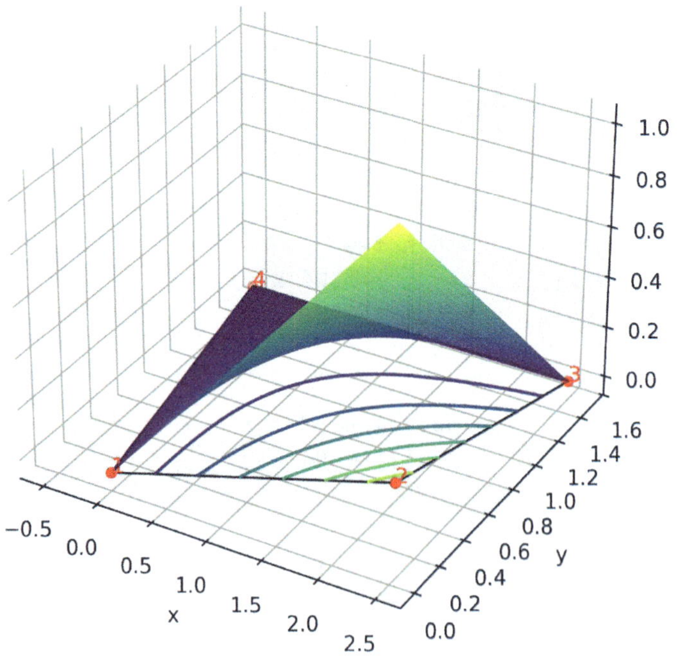

As illustrated in the foregoing plot, the shape functions exhibit a more complex variation in the physical domain. They are strictly linear only along specific directions—namely, the images of the constant-( ) and constant-( ) lines under the isoparametric mapping (shown as dashed blue lines in Fig.4.6). Along any other direction, however, the variation is generally quadratic.

If the Q4 element is rectangular in the physical domain (i.e., the mapping is affine), an explicit closed-form expression for the interpolated field (u(x,y)) can be derived. In this special case, (u(x,y)) remains bilinear in the physical coordinates, and the Jacobian matrix is constant throughout the element, with its determinant proportional to the element area.

4.7 Shape Functions for Q8 Elements

When an eight-node quadrilateral element (Q8) is used, the shape functions are constructed in the natural coordinate system. This formulation provides:

1. **Continuity** of the field approximation along element edges, including mid-edge nodes.
2. A **simple reference geometry** (a square), which facilitates the systematic construction of higher-order shape functions.

4.7.1 Element Description

Consider a square element with eight nodes in the natural coordinate system, as shown in Fig.4.7.

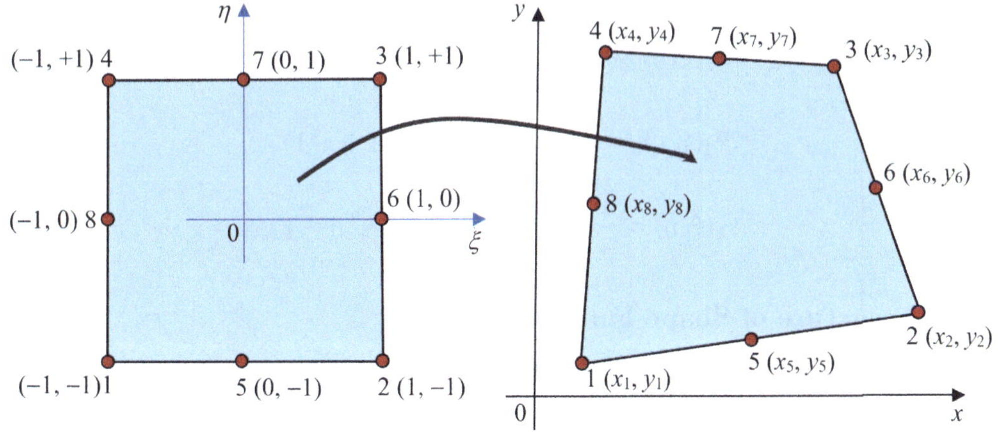

Figure 4.7: A quadrilateral element with four edges and eight nodes in two-dimensional space. The shape functions, and hence the approximated field, vary quadratically along the edges and biquadratically within the element interior.

The domain is defined as $\Omega_e = [-1,1] \times [-1,1]$. The coordinates of the nodes in both physical and natural coordinate systems are shown in Fig.4.7.

Let $u(x,y)$ be a field variable. Its finite element approximation is defined in natural coordinates as:

$$u^h(\xi,\eta) = \sum_{i=1}^{8} N_i(\xi,\eta)u_i \tag{4.105}$$

Our task is to construct the eight shape functions $N_i(\xi,\eta)$.

4.7.2 Shape Function Construction

A Q8 element exhibits quadratic variation along all edges and do not contain an interior node. It belongs to the **serendipity family**, meaning that the shape functions are constructed by enforcing interpolation conditions and edge consistency, rather than using a full tensor-product polynomial basis.

The Q8 shape functions are constructed to satisfy:

- The **Kronecker delta property** at all 8 nodes.
- **Quadratic variation** along each edge.
- **Compatibility** with adjacent elements.

4.7.3 Corner Node Shape Functions

For the four corner nodes, the shape functions take the following form:

$$N_1(\xi,\eta) = \frac{1}{4}(1-\xi)(1-\eta)(-\xi-\eta-1) \tag{4.106}$$

$$N_2(\xi,\eta) = \frac{1}{4}(1+\xi)(1-\eta)(\xi-\eta-1) \tag{4.107}$$

$$N_3(\xi,\eta) = \frac{1}{4}(1+\xi)(1+\eta)(\xi+\eta-1) \tag{4.108}$$

$$N_4(\xi,\eta) = \frac{1}{4}(1-\xi)(1+\eta)(-\xi+\eta-1) \tag{4.109}$$

4.7.3.1 Structure of Shape Function N_1

We illustrate the construction using N_1 as an example. We begin by writing:

$$N_1(\xi,\eta) = C(1-\xi)(1-\eta)P(\xi,\eta) \tag{4.110}$$

The factors $(1-\xi)(1-\eta)$ enforce vanishing along the edges $\xi = 1$ and $\eta = 1$, and hence at nodes (2), (3), (4), (6), and (7). However, N_1 must also vanish at the mid-edge nodes (5) and (8). This requirement determines the polynomial $P(\xi,\eta)$.

4.7.3.2 Determination of $P(\xi, \eta)$

To enforce zero values at nodes (5) and (8), we require a linear function that vanishes at $(0, -1)$ (node 5) and $(-1, 0)$ (node 8). Assuming $P(\xi, \eta) = a + b\xi + c\eta$ and substituting the nodal conditions, we find:

$$a = -1, \quad b = -1, \quad c = -1 \implies P(\xi, \eta) = -(1 + \xi + \eta) \tag{4.111}$$

4.7.3.3 Determination of Constant C

Now we have:

$$N_1(\xi, \eta) = C(1 - \xi)(1 - \eta)(-\xi - \eta - 1) \tag{4.112}$$

To satisfy $N_1(-1, -1) = 1$, substitution yields $C = 1/4$. Thus:

$$N_1 = \frac{1}{4}(1 - \xi)(1 - \eta)(-\xi - \eta - 1) \tag{4.113}$$

4.7.3.4 Edge Behavior

Along the bottom edge ($\eta = -1$), the shape function reduces to:

$$N_1(\xi, -1) = \frac{1}{4}(1 - \xi)(2)(1 - \xi) = \frac{1}{2}(1 - \xi)^2 \tag{4.114}$$

This matches the standard quadratic Lagrange interpolation.

The same procedure yields the shape functions for nodes (2), (3), and (4).

4.7.4 Mid-Edge Node Shape Functions

The mid-edge shape functions are constructed to satisfy the Kronecker delta property (value 1 at the corresponding node and 0 elsewhere) and quadratic variation along the edges:

$$\begin{aligned} N_5(\xi, \eta) &= \frac{1}{2}(1 - \xi^2)(1 - \eta) \\ N_6(\xi, \eta) &= \frac{1}{2}(1 + \xi)(1 - \eta^2) \\ N_7(\xi, \eta) &= \frac{1}{2}(1 - \xi^2)(1 + \eta) \\ N_8(\xi, \eta) &= \frac{1}{2}(1 - \xi)(1 - \eta^2) \end{aligned} \tag{4.115}$$

4.7.5 Final Q8 Shape Functions Summary

$$\boxed{\begin{aligned} N_1 &= \frac{1}{4}(1-\xi)(1-\eta)(-\xi-\eta-1), \\ N_2 &= \frac{1}{4}(1+\xi)(1-\eta)(\xi-\eta-1), \\ N_3 &= \frac{1}{4}(1+\xi)(1+\eta)(\xi+\eta-1), \\ N_4 &= \frac{1}{4}(1-\xi)(1+\eta)(-\xi+\eta-1), \\ N_5 &= \frac{1}{2}(1-\xi^2)(1-\eta), \\ N_6 &= \frac{1}{2}(1+\xi)(1-\eta^2), \\ N_7 &= \frac{1}{2}(1-\xi^2)(1+\eta), \\ N_8 &= \frac{1}{2}(1-\xi)(1-\eta^2) \end{aligned}} \tag{4.116}$$

More details with pictorial illustration can be found in Section 7.7.2 of Ref. [12].

4.7.6 Polynomial Basis

The Q8 element spans a reduced polynomial space with bases of:

$$1, \xi, \eta, \xi\eta, \xi^2, \eta^2, \xi^2\eta, \xi\eta^2 \tag{4.117}$$

The same isoparametric mapping used for Q4 elements applies to Q8, and the Jacobian matrix is computed in the same manner.

4.8 Shape Functions for T4 Elements

For a four-node tetrahedral element (T4), the shape functions can be constructed directly in the physical coordinate system (x, y, z), although a natural coordinate formulation is also possible. The physical-coordinate-based formulation is particularly useful because it naturally leads to the concept of **volume (barycentric) coordinates**—a direct extension of the area coordinates used for triangular elements—which can be systematically extended to higher-order elements.

This approach provides:

1. A direct geometric interpretation of the shape functions.
2. A natural pathway to higher-order tetrahedral elements (e.g., T10).

4.8.1 Element Description

Consider a tetrahedral element with four nodes in the physical coordinate system, as shown in Fig.4.8.

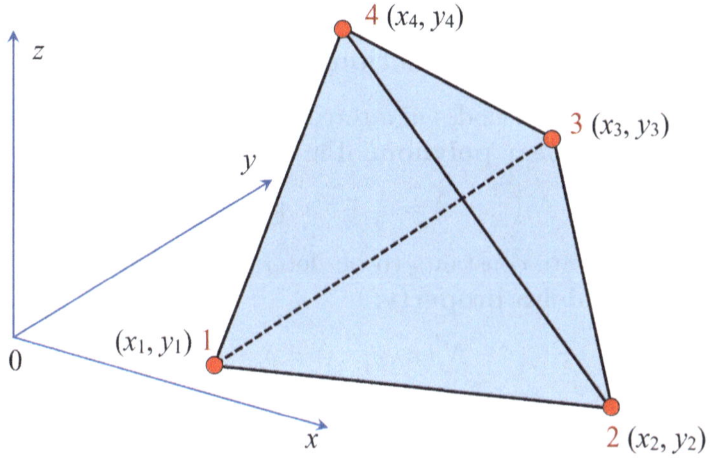

Figure 4.8: A tetrahedral element with six straight edges and four nodes in three-dimensional space. The shape functions, and hence the approximated field, vary linearly along the edges and throughout the interior of the element.

4.8.2 Local Node Numbering

Local node numbering at element level follows the following counterclockwise right-hand rule:

The first three nodes (1→2→3) are ordered counterclockwise on a chosen base face, while node 4 is positioned such that the node ordering follows the right-hand rule. This convention ensures a positive element volume (see Section 5.7.11) [6].

The nodal coordinates for these four nodes are given by:

$$\begin{aligned} &\text{Node (1): } (x_1, y_1, z_1), \quad \text{Node (2): } (x_2, y_2, z_2) \\ &\text{Node (3): } (x_3, y_3, z_3), \quad \text{Node (4): } (x_4, y_4, z_4) \end{aligned} \tag{4.118}$$

The volume of the tetrahedron can be computed as [10]:

$$V_e = \frac{1}{6} \begin{vmatrix} 1 & x_1 & y_1 & z_1 \\ 1 & x_2 & y_2 & z_2 \\ 1 & x_3 & y_3 & z_3 \\ 1 & x_4 & y_4 & z_4 \end{vmatrix} \tag{4.119}$$

We assume that the four nodes are distinct and not coplanar, so that the element volume is strictly positive ($V_e > 0$). For a field variable $u(x, y, z)$, the finite element approximation is:

$$u^h(x, y, z) = \sum_{i=1}^{4} N_i(x, y, z)u_i \tag{4.120}$$

Our goal is to construct the shape functions $N_i(x, y, z)$ directly in terms of (x, y, z).

4.8.3 Shape Function Construction

Since the T4 element has four nodes in a three-dimensional domain, each shape function is assumed to be a **linear polynomial** in (x, y, z):

$$N_i(x, y, z) = a_i + b_i x + c_i y + d_i z \tag{4.121}$$

where a_i, b_i, c_i, and d_i are constants to be determined. Each shape function must satisfy the **Kronecker delta property**:

$$N_i(x_j, y_j, z_j) = \delta_{ij} \tag{4.122}$$

This leads to a system of four linear equations for each N_i, which can be solved to determine the coefficients.

4.8.4 Matrix Formulation

The system arising from Eq.(4.122) can be written in matrix form as:

$$\underbrace{\begin{bmatrix} 1 & x_1 & y_1 & z_1 \\ 1 & x_2 & y_2 & z_2 \\ 1 & x_3 & y_3 & z_3 \\ 1 & x_4 & y_4 & z_4 \end{bmatrix}}_{\mathbf{P}} \begin{bmatrix} a_i \\ b_i \\ c_i \\ d_i \end{bmatrix} = \begin{bmatrix} \delta_{i1} \\ \delta_{i2} \\ \delta_{i3} \\ \delta_{i4} \end{bmatrix} \tag{4.123}$$

Since $V_e \neq 0$, the moment matrix $\mathbf{P}$ is invertible, and the coefficients can be uniquely determined for each of the four shape functions.

4.8.5 Volume Coordinates

Similar to area coordinates for triangles, we define **volume (barycentric) coordinates** for the tetrahedron. Consider an arbitrary point $P(x, y, z)$ inside the element, as shown in Fig.4.9.

Define the sub-volume V_1:

$$V_1 \equiv V_{P234} = \frac{1}{6}\begin{vmatrix} 1 & x & y & z \\ 1 & x_2 & y_2 & z_2 \\ 1 & x_3 & y_3 & z_3 \\ 1 & x_4 & y_4 & z_4 \end{vmatrix} \tag{4.124}$$

Similarly:

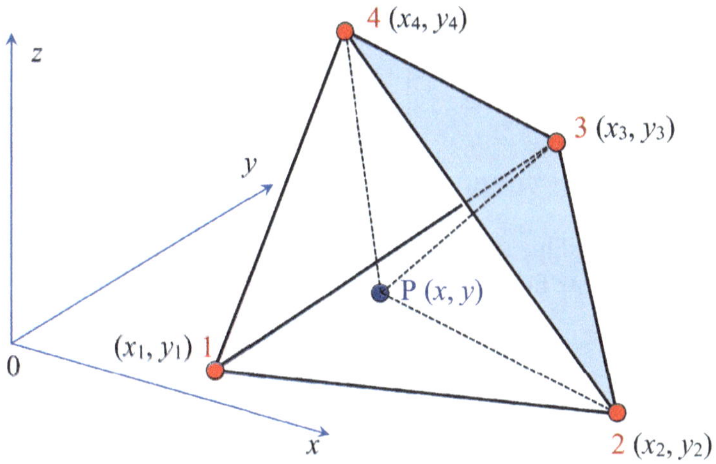

Figure 4.9: A sub-tetrahedron formed by replacing node (1) with an interior point P. Its volume is denoted by V_{P234}.

- $V_2 \equiv V_{P134}$ is obtained by replacing node (2) coordinates with (x, y, z) in Eq.(4.119).
- $V_3 \equiv V_{P124}$ is obtained by replacing node (3) coordinates with (x, y, z).
- $V_4 \equiv V_{P123}$ is obtained by replacing node (4) coordinates with (x, y, z).

The **volume coordinates** are defined as:

$$L_i(x, y, z) = \frac{V_i(x, y, z)}{V_e} \tag{4.125}$$

These coordinates satisfy:

- **Kronecker delta property:** $L_i(x_j, y_j, z_j) = \delta_{ij}$
- **Partition of unity:** $\sum_{i=1}^{4} L_i(x, y, z) = 1$

Thus, the volume coordinates are precisely the shape functions:

$$N_i(x, y, z) = L_i(x, y, z) = \frac{V_i(x, y, z)}{V_e} \tag{4.126}$$

4.8.6 Explicit Form of Shape Functions

The four shape functions can be written compactly using determinants:

$$N_1(x, y, z) = \frac{1}{6V_e} \begin{vmatrix} 1 & x & y & z \\ 1 & x_2 & y_2 & z_2 \\ 1 & x_3 & y_3 & z_3 \\ 1 & x_4 & y_4 & z_4 \end{vmatrix} N_2(x, y, z) = \frac{1}{6V_e} \begin{vmatrix} 1 & x_1 & y_1 & z_1 \\ 1 & x & y & z \\ 1 & x_3 & y_3 & z_3 \\ 1 & x_4 & y_4 & z_4 \end{vmatrix} \tag{4.127}$$

$$N_3(x,y,z) = \frac{1}{6V_e}\begin{vmatrix} 1 & x_1 & y_1 & z_1 \\ 1 & x_2 & y_2 & z_2 \\ 1 & x & y & z \\ 1 & x_4 & y_4 & z_4 \end{vmatrix} N_4(x,y,z) = \frac{1}{6V_e}\begin{vmatrix} 1 & x_1 & y_1 & z_1 \\ 1 & x_2 & y_2 & z_2 \\ 1 & x_3 & y_3 & z_3 \\ 1 & x & y & z \end{vmatrix} \tag{4.128}$$

Linear Completeness This set of shape functions is linearly complete and can exactly reproduce any linear field:

$$u(x,y,z) = a_0 + a_1 x + a_2 y + a_3 z \tag{4.129}$$

Clearly, T4 elements are capable of passing the patch test.

4.8.7 Derivatives of Shape Functions

Since N_i are linear functions of x, y, and z, their partial derivatives:

$$\frac{\partial N_i}{\partial x}, \quad \frac{\partial N_i}{\partial y}, \quad \frac{\partial N_i}{\partial z} \tag{4.130}$$

are constants throughout the element.

4.9 Shape Functions for T10 Elements

With the volume coordinates $L_i(x,y,z)$ established for the T4 element, higher-order tetrahedral elements can be constructed systematically. The simplest and most useful extension is the **ten-node tetrahedral element (T10)**, which provides **quadratic interpolation**.

4.9.1 Element Description

A T10 element consists of 4 corner nodes and 6 mid-edge nodes, as shown in Fig.4.10.

The nodes are categorized as:

$$\begin{aligned} &\text{Corner nodes:} && (1),(2),(3),(4) \\ &\text{Mid-edge nodes:} && (5),(6),(7),(8),(9),(10) \end{aligned} \tag{4.131}$$

Each mid-edge node lies halfway along an edge:

$$\begin{array}{lll} (5):\text{edge 1-2} & (6):\text{edge 2-3} & (7):\text{edge 3-1} \\ (8):\text{edge 1-4} & (9):\text{edge 2-4} & (10):\text{edge 3-4} \end{array} \tag{4.132}$$

The field approximation using a T10 element is:

$$u^h(x,y,z) = \sum_{i=1}^{10} N_i(x,y,z) u_i \tag{4.133}$$

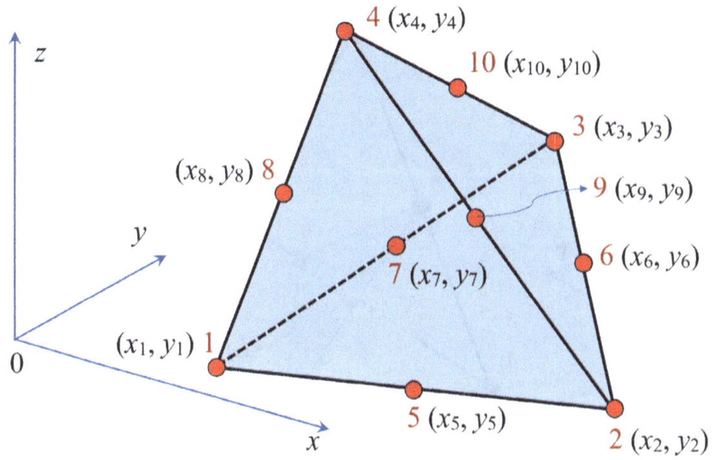

Figure 4.10: A quadratic tetrahedral element with six straight edges and ten nodes in three-dimensional space. The shape functions, and hence the approximated field, vary quadratically along the edges and throughout the interior of the element.

4.9.2 Shape Function Construction

The T10 shape functions are constructed as **quadratic polynomials** in L_i coordinates. The logic is identical to that used for T6 (triangular) elements. The extension from 2D to 3D is automatically handled by the fourth basis coordinate L_4. Geometrically, the area coordinate lines used in 2D are replaced by **volume coordinate planes** in 3D.

Figure 4.11 illustrates the volume coordinate planes ($L_1 = 1, L_1 = 1/2, L_1 = 0$) required to construct N_1. By enforcing the Kronecker delta property at all ten nodes, we obtain the following shape functions:

Corner Node Shape Functions:

$$\begin{aligned} N_1 &= L_1(2L_1 - 1) \quad N_2 = L_2(2L_2 - 1) \\ N_3 &= L_3(2L_3 - 1) \quad N_4 = L_4(2L_4 - 1) \end{aligned} \tag{4.134}$$

Mid-Edge Shape Functions:

$$\begin{aligned} N_5 \equiv N_{12} &= 4L_1L_2 \quad N_6 \equiv N_{23} = 4L_2L_3 \\ N_7 \equiv N_{31} &= 4L_3L_1 \quad N_8 \equiv N_{14} = 4L_1L_4 \\ N_9 \equiv N_{24} &= 4L_2L_4 \quad N_{10} \equiv N_{34} = 4L_3L_4 \end{aligned} \tag{4.135}$$

4.9.3 Verification of Key Properties

Kronecker Delta Property At a corner node (e.g., node 1):

- $L_1 = 1$ and $L_{2,3,4} = 0$. Thus, $N_1 = 1(2 - 1) = 1$, and all others vanish.

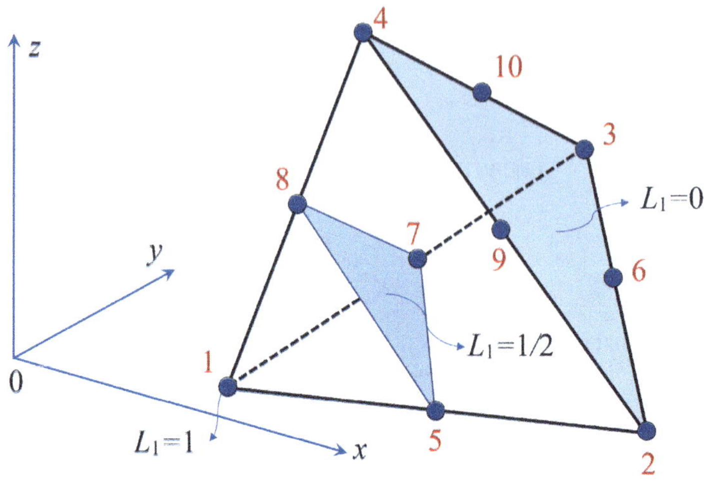

Figure 4.11: Planes associated with volume coordinates L_1 used for constructing the quadratic shape function N_1.

- At an edge midpoint (e.g., node 5): $L_1 = L_2 = 1/2$ and $L_{3,4} = 0$. Thus, $N_5 = 4(1/2)(1/2) = 1$, while $N_1 = (1/2)(1-1) = 0$.

Partition of Unity The sum of all shape functions equals unity:

$$\sum_{i=1}^{10} N_i = 1 \tag{4.136}$$

This is satisfied because the summation of the quadratic terms in L_i reduces to $(\sum L_i)^2 = (1)^2$ through the properties of barycentric coordinates.

4.9.4 Quadratic Completeness

The T10 element spans the complete quadratic polynomial space in 3D:

$$\{1,\ x,\ y,\ z,\ x^2,\ y^2,\ z^2,\ xy,\ yz,\ zx\} \tag{4.137}$$

This allows the element to exactly reproduce any quadratic field:

$$\begin{aligned} u(x,y,z) &= a_0 + a_1 x + a_2 y + a_3 z + a_4 x^2 \\ &\quad + a_5 y^2 + a_6 z^2 + a_7 xy + a_8 yz + a_9 zx \end{aligned} \tag{4.138}$$

4.9.5 Key Observation

The construction highlights a powerful principle of finite element analysis: > Higher-order tetrahedral elements are obtained systematically by taking higher-degree polynomials of the volume coordinates.

- **T4:** Linear in L_i

- **T10:** Quadratic in L_i
- **T20:** Cubic in L_i

This consistency makes volume coordinates the preferred tool for implementing high-order 3D elements.

4.9.6 Derivatives of Shape Functions for T10 Elements

Since the T10 shape functions are expressed in terms of the **volume coordinates** $L_i(x, y, z)$, their spatial derivatives are calculated using the chain rule:

$$\frac{\partial N}{\partial x} = \sum_{k=1}^{4} \frac{\partial N}{\partial L_k} \frac{\partial L_k}{\partial x} \tag{4.139}$$

To evaluate this, we require two components:

1. **Polynomial Derivatives:** $\frac{\partial N}{\partial L_k}$ (derivatives of the shape functions with respect to the barycentric coordinates).
2. **Geometric Gradients:** $\frac{\partial L_k}{\partial x}, \frac{\partial L_k}{\partial y}, \frac{\partial L_k}{\partial z}$ (constants determined by the element geometry).

4.9.6.1 1. Derivatives with Respect to Volume Coordinates

For Corner Nodes ($N_i = 2L_i^2 - L_i$):

$$\frac{\partial N_i}{\partial L_i} = 4L_i - 1, \quad \frac{\partial N_i}{\partial L_j} = 0 \quad (j \neq i) \tag{4.140}$$

For Mid-Edge Nodes ($N_{ij} = 4L_iL_j$):

$$\frac{\partial N_{ij}}{\partial L_i} = 4L_j, \quad \frac{\partial N_{ij}}{\partial L_j} = 4L_i, \quad \frac{\partial N_{ij}}{\partial L_k} = 0 \quad (k \neq i, j) \tag{4.141}$$

4.9.6.2 2. Derivatives of Volume Coordinates in Physical Space

From the linear T4 formulation, the volume coordinates are linear functions of the physical coordinates:

$$L_i(x, y, z) = a_i + b_ix + c_iy + d_iz \tag{4.142}$$

The gradients are therefore **constants** within the element:

$$\frac{\partial L_i}{\partial x} = b_i, \quad \frac{\partial L_i}{\partial y} = c_i, \quad \frac{\partial L_i}{\partial z} = d_i \tag{4.143}$$

4.9.6.3 Final Expressions in Physical Coordinates

For the **corner nodes**, the spatial derivatives are:

$$\frac{\partial N_i}{\partial x} = (4L_i - 1)\frac{\partial L_i}{\partial x}, \quad \frac{\partial N_i}{\partial y} = (4L_i - 1)\frac{\partial L_i}{\partial y}, \quad \frac{\partial N_i}{\partial z} = (4L_i - 1)\frac{\partial L_i}{\partial z} \tag{4.144}$$

For the **mid-edge nodes**, the spatial derivatives follow the product rule:

$$\frac{\partial N_{ij}}{\partial x} = 4\left(L_i\frac{\partial L_j}{\partial x} + L_j\frac{\partial L_i}{\partial x}\right) \tag{4.145}$$

$$\frac{\partial N_{ij}}{\partial y} = 4\left(L_i\frac{\partial L_j}{\partial y} + L_j\frac{\partial L_i}{\partial y}\right) \tag{4.146}$$

$$\frac{\partial N_{ij}}{\partial z} = 4\left(L_i\frac{\partial L_j}{\partial z} + L_j\frac{\partial L_i}{\partial z}\right) \tag{4.147}$$

4.9.6.4 Compact Vector Representation

Let the gradient of the volume coordinate be denoted as:

$$\nabla L_i = \begin{bmatrix} \frac{\partial L_i}{\partial x} & \frac{\partial L_i}{\partial y} & \frac{\partial L_i}{\partial z} \end{bmatrix} \tag{4.148}$$

The gradients of the shape functions can then be written compactly:

- **Corner Nodes:** $\nabla N_i = (4L_i - 1)\nabla L_i$
- **Mid-Edge Nodes:** $\nabla N_{ij} = 4\left(L_i \nabla L_j + L_j \nabla L_i\right)$

4.9.7 Implementation Insight

In numerical implementation, the process is streamlined as follows:

1. **Precompute** the constant vector ∇L_i once for the entire element.
2. **Evaluate** the volume coordinates L_i at the specific Gauss integration points.
3. **Assemble** the shape function gradients ∇N_i using the compact formulas above.

4.9.8 Remarks

- **Accuracy:** T10 elements provide significantly improved accuracy over T4 elements by allowing for higher-order interpolation.
- **Strain Field:** Unlike the T4 element, the strain field in a T10 element is no longer constant; it varies linearly within the element.
- **Generalization:** This volume-coordinate formulation generalizes naturally to even higher-order tetrahedral elements (e.g., T20).

4.10 Shape Functions for H8 Elements

When an eight-node hexahedral element (H8) is used, the shape functions are constructed in a natural coordinate system, similar to the Q4 element. This formulation provides:

1. **Continuity** of the field approximation across element faces.
2. A **simple reference geometry** (a cube), which enables the systematic construction of higher-order shape functions.

4.10.1 Local Node Numbering

The local node numbering of an H8 element follows a counterclockwise right-hand-rule rule:

The first four nodes ($1 \to 2 \to 3 \to 4$) are numbered counterclockwise on a chosen base face. Node 5 is then positioned directly above node 1 such that the edge ($1 \to 5$) follows the direction prescribed by the right-hand rule with respect to the base face. The remaining three nodes are numbered counterclockwise as ($5 \to 6 \to 7 \to 8$) on the opposite face, maintaining the same orientation as the base face.

4.10.2 Element Description

Consider a hexahedral element with eight nodes in the natural coordinate system (ξ, η, ζ), as shown in Fig.4.12.

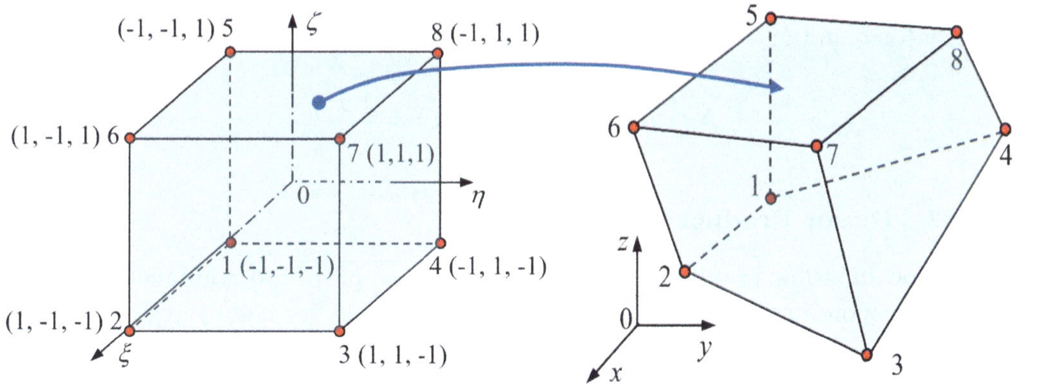

Figure 4.12: A hexahedral element with eight nodes and six flat faces in three-dimensional space. The shape functions, and hence the approximated field, vary linearly over each face and trilinearly throughout the interior of the element.

The element domain in the natural coordinates is defined as:

$$\Omega_e = [-1, 1] \times [-1, 1] \times [-1, 1] \tag{4.149}$$

The natural coordinates of the eight nodes are:

$$\begin{aligned} &(1) :(-1,-1,-1), && (2) :(1,-1,-1), \\ &(3) :(1,1,-1), && (4) :(-1,1,-1), \\ &(5) :(-1,-1,1), && (6) :(1,-1,1), \\ &(7) :(1,1,1), && (8) :(-1,1,1) \end{aligned} \tag{4.150}$$

For a field variable $u(x, y, z)$, the finite element approximation is:

$$u^h(\xi, \eta, \zeta) = \sum_{i=1}^{8} N_i(\xi, \eta, \zeta) u_i \tag{4.151}$$

Our goal is to construct the shape functions $N_i(\xi, \eta, \zeta)$.

4.10.3 Shape Function Construction

Because the H8 element is a cube in natural space, the shape functions can be constructed using **tensor products** of 1D linear Lagrange polynomials.

4.10.3.1 1D Linear Shape Functions

In each natural dimension, we define the standard linear basis:

$$N^-(\xi) = \frac{1-\xi}{2}, \quad N^+(\xi) = \frac{1+\xi}{2} \tag{4.152}$$

Similarly for η and ζ:

$$N^\pm(\eta) = \frac{1 \pm \eta}{2}, \quad N^\pm(\zeta) = \frac{1 \pm \zeta}{2} \tag{4.153}$$

4.10.3.2 Tensor Product Construction

Each shape function is obtained by selecting the appropriate combination of $(\pm)$ factors corresponding to the node's location. For example, for node 1 at $(-1, -1, -1)$:

$$N_1 = N^-(\xi) N^-(\eta) N^-(\zeta) = \frac{1}{8}(1-\xi)(1-\eta)(1-\zeta) \tag{4.154}$$

Applying this logic to all nodes:

$$\begin{aligned} N_1 &= \tfrac{1}{8}(1-\xi)(1-\eta)(1-\zeta) & N_2 &= \tfrac{1}{8}(1+\xi)(1-\eta)(1-\zeta) \\ N_3 &= \tfrac{1}{8}(1+\xi)(1+\eta)(1-\zeta) & N_4 &= \tfrac{1}{8}(1-\xi)(1+\eta)(1-\zeta) \\ N_5 &= \tfrac{1}{8}(1-\xi)(1-\eta)(1+\zeta) & N_6 &= \tfrac{1}{8}(1+\xi)(1-\eta)(1+\zeta) \\ N_7 &= \tfrac{1}{8}(1+\xi)(1+\eta)(1+\zeta) & N_8 &= \tfrac{1}{8}(1-\xi)(1+\eta)(1+\zeta) \end{aligned} \tag{4.155}$$

4.10.4 Final Shape Functions

This set of shape functions can be written compactly for a node located at (ξ_i, η_i, ζ_i) as:

$$\boxed{N_i(\xi, \eta, \zeta) = \frac{1}{8}(1 + \xi_i\xi)(1 + \eta_i\eta)(1 + \zeta_i\zeta)} \tag{4.156}$$

4.10.5 Derivatives of Shape Functions

The partial derivatives are obtained directly by differentiating the tensor product. For node i:

$$\frac{\partial N_i}{\partial \xi} = \frac{\xi_i}{8}(1 + \eta_i\eta)(1 + \zeta_i\zeta) \tag{4.157}$$

$$\frac{\partial N_i}{\partial \eta} = \frac{\eta_i}{8}(1 + \xi_i\xi)(1 + \zeta_i\zeta) \tag{4.158}$$

$$\frac{\partial N_i}{\partial \zeta} = \frac{\zeta_i}{8}(1 + \xi_i\xi)(1 + \eta_i\eta) \tag{4.159}$$

4.10.6 Key Properties

- **Kronecker Delta Property:** $N_i(\xi_j, \eta_j, \zeta_j) = \delta_{ij}$.
- **Partition of Unity:** $\sum_{i=1}^{8} N_i = 1$.
- **Trilinear Behavior:** Each shape function is linear in ξ, η, or ζ when the other two coordinates are fixed.

4.10.7 Polynomial Completeness

4.10.7.1 Trilinear Completeness in Natural Coordinates

The H8 element spans the trilinear polynomial space:

$$\{1,\ \xi,\ \eta,\ \zeta,\ \xi\eta,\ \eta\zeta,\ \zeta\xi,\ \xi\eta\zeta\} \tag{4.160}$$

This allows the element to exactly reproduce trilinear fields in natural coordinates.

4.10.7.2 Linear Completeness in Physical Coordinates

The H8 element linearly complete in the physical coordinates. The proof is exactly the same as that given for the Q4 elements.

Therefore, H8 elements are capable to pass the patch test.

4.10.8 Isoparametric Mapping for H8 Elements

In the isoparametric formulation, the same shape functions $N_i(\xi, \eta, \zeta)$ are used to interpolate both the geometry and the field variables. The mapping from the natural coordinates (ξ, η, ζ) to the physical coordinates (x, y, z) is defined as:

$$\begin{aligned} x(\xi, \eta, \zeta) &= \sum_{i=1}^{8} N_i(\xi, \eta, \zeta) x_i \\ y(\xi, \eta, \zeta) &= \sum_{i=1}^{8} N_i(\xi, \eta, \zeta) y_i \\ z(\xi, \eta, \zeta) &= \sum_{i=1}^{8} N_i(\xi, \eta, \zeta) z_i \end{aligned} \tag{4.161}$$

In compact vector form:

$$\mathbf{x}(\xi, \eta, \zeta) = \sum_{i=1}^{8} N_i(\xi, \eta, \zeta) \mathbf{x}_i \tag{4.162}$$

where $\mathbf{x}_i = [x_i, y_i, z_i]^T$ represents the nodal coordinates in physical space.

4.10.9 Key Properties of Isoparametric Mapping

(a) Exact Node Mapping At any node i, the natural coordinates $(\xi, \eta, \zeta)_i$ map exactly to the physical coordinates $(x, y, z)_i$. This is guaranteed by the Kronecker delta property of the shape functions.

(b) Partition of Unity Since $\sum_{i=1}^{8} N_i = 1$, the mapping correctly preserves rigid-body translations. A constant displacement of all nodes results in the same constant displacement throughout the element volume.

(c) Trilinear Geometry and Face Correspondence The mapping is trilinear; each physical coordinate varies linearly with one natural coordinate when the other two are held constant. Consequently:

- Lines parallel to the ξ, η, or ζ axes in natural space map to straight lines in physical space.
- The six faces of the reference cube map to planar quadrilateral faces in the physical element.
- There is a one-to-one face correspondence:

$$\text{face}_{i-j-k-l,\text{H8}} \iff \text{face}_{i-j-k-l,\text{reference}}. \tag{4.163}$$

(d) Face Continuity Because element faces remain planar and the interpolation along an edge or face depends only on the nodes belonging to that edge or face, adjacent H8 elements connect perfectly. The approximated field is C^0 continuous across element boundaries, satisfying the requirements for a conforming FEM mesh.

(e) Volume Correspondence and Scaling For a convex H8 element, the mapping maintains a consistent domain correspondence between the reference cube and the physical element. The physical volume is related to the natural volume through the determinant of the Jacobian matrix, $|\mathbf{J}|$. If the element becomes severely distorted (e.g., non-convex), this correspondence may fail.

(f) Bijectivity For the mapping to be bijective (a unique one-to-one correspondence between every point), the Jacobian determinant must remain strictly positive ($|\mathbf{J}| > 0$) everywhere within the element. This ensures that the element does not "fold" onto itself.

4.10.10 Jacobian Matrix

The Jacobian matrix $\mathbf{J}$ relates the natural coordinate gradients to the physical coordinate gradients and is defined as:

$$\mathbf{J} = \begin{bmatrix} \frac{\partial x}{\partial \xi} & \frac{\partial y}{\partial \xi} & \frac{\partial z}{\partial \xi} \\ \frac{\partial x}{\partial \eta} & \frac{\partial y}{\partial \eta} & \frac{\partial z}{\partial \eta} \\ \frac{\partial x}{\partial \zeta} & \frac{\partial y}{\partial \zeta} & \frac{\partial z}{\partial \zeta} \end{bmatrix} \tag{4.164}$$

The individual components are calculated as:

$$\frac{\partial x}{\partial \xi} = \sum_{i=1}^{8} \frac{\partial N_i}{\partial \xi} x_i, \quad \frac{\partial x}{\partial \eta} = \sum_{i=1}^{8} \frac{\partial N_i}{\partial \eta} x_i, \quad \frac{\partial x}{\partial \zeta} = \sum_{i=1}^{8} \frac{\partial N_i}{\partial \zeta} x_i \tag{4.165}$$

with analogous expressions for the derivatives of y and z.

4.10.11 Remark

- The Jacobian for H8 elements is generally **not constant** due to the nature of trilinear mapping.
- Geometric distortion (e.g., highly skewed or "brick" elements becoming "wedge" shaped) may degrade the Jacobian, affecting the accuracy of numerical integration and strain results.
- Monitoring $|\mathbf{J}|$ at Gauss points is a standard quality check in finite element solvers to ensure a valid formulation.

4.11 Shape Functions for H20 Elements

When a twenty-node hexahedral element (H20) is employed, the shape functions are defined in the natural coordinate system, similar to the Q8 element for 2D solids. This formulation offers:

1. **Quadratic variation** of the field along all edges.
2. **Improved accuracy** compared to H8 elements.
3. A **systematic extension** of serendipity-style interpolation over a cubic reference domain.

4.11.1 Element Description

Consider a hexahedral element with twenty nodes in the natural coordinate system (ξ, η, ζ), as shown in Fig.3.2. The reference domain is defined as:

$$\Omega_e = [-1, 1] \times [-1, 1] \times [-1, 1] \tag{4.166}$$

The nodes consist of:

- **8 corner nodes** (identical to H8).
- **12 mid-edge nodes** (one positioned at the midpoint of each edge).

Natural Coordinates

Corner nodes:

$$\begin{aligned} &(1): (-1, -1, -1), && (2): (1, -1, -1), \\ &(3): (1, 1, -1), && (4): (-1, 1, -1), \\ &(5): (-1, -1, 1), && (6): (1, -1, 1), \\ &(7): (1, 1, 1), && (8): (-1, 1, 1) \end{aligned} \tag{4.167}$$

Mid-edge nodes:

$$\begin{aligned} &(9): (0, -1, -1), && (10): (1, 0, -1), \\ &(11): (0, 1, -1), && (12): (-1, 0, -1), \\ &(13): (0, -1, 1), && (14): (1, 0, 1), \\ &(15): (0, 1, 1), && (16): (-1, 0, 1), \\ &(17): (-1, -1, 0), && (18): (1, -1, 0), \\ &(19): (1, 1, 0), && (20): (-1, 1, 0) \end{aligned} \tag{4.168}$$

For a field variable $u(x, y, z)$, the finite element approximation is:

$$u^h(\xi, \eta, \zeta) = \sum_{i=1}^{20} N_i(\xi, \eta, \zeta) u_i \tag{4.169}$$

4.11.2 Shape Function Construction

The H20 element is a **quadratic serendipity element** in 3D. Its construction ensures:

- The **Kronecker delta property** at all 20 nodes.

- **Quadratic variation** along all edges.
- **Inter-element compatibility** (C^0 continuity).

The construction principles are analogous to the 2D Q8 element; further details can be found in Ref. [12].

4.11.2.1 Corner Node Shape Functions

For the corner nodes, the shape functions take the form:

$$N_i = \frac{1}{8}(1+\xi_i\xi)(1+\eta_i\eta)(1+\zeta_i\zeta)(\xi_i\xi+\eta_i\eta+\zeta_i\zeta-2) \tag{4.170}$$

where (ξ_i, η_i, ζ_i) are the natural coordinates of node i. For example, for **Node 1** at $(-1,-1,-1)$:

$$N_1 = \frac{1}{8}(1-\xi)(1-\eta)(1-\zeta)(-\xi-\eta-\zeta-2) \tag{4.171}$$

4.11.2.2 Mid-Edge Shape Functions

Each mid-edge shape function is constructed using a quadratic term in the direction of the edge and linear terms in the transverse directions.

- **Edges parallel to ξ:** $N_i = \frac{1}{4}(1-\xi^2)(1+\eta_i\eta)(1+\zeta_i\zeta)$
- **Edges parallel to η:** $N_i = \frac{1}{4}(1-\eta^2)(1+\xi_i\xi)(1+\zeta_i\zeta)$
- **Edges parallel to ζ:** $N_i = \frac{1}{4}(1-\zeta^2)(1+\xi_i\xi)(1+\eta_i\eta)$

For example, for **Node 9** at $(0,-1,-1)$:

$$N_9 = \frac{1}{4}(1-\xi^2)(1-\eta)(1-\zeta) \tag{4.172}$$

4.11.3 Final H20 Shape Functions Summary

$$\boxed{\begin{aligned}
&\text{Corner nodes:}\\
&N_i = \frac{1}{8}(1+\xi_i\xi)(1+\eta_i\eta)(1+\zeta_i\zeta)(\xi_i\xi+\eta_i\eta+\zeta_i\zeta-2)\\
&\text{Mid-edge nodes:}\\
&N_i = \frac{1}{4}(1-\xi^2)(1+\eta_i\eta)(1+\zeta_i\zeta) \quad \text{(parallel to } \xi)\\
&N_i = \frac{1}{4}(1-\eta^2)(1+\xi_i\xi)(1+\zeta_i\zeta) \quad \text{(parallel to } \eta)\\
&N_i = \frac{1}{4}(1-\zeta^2)(1+\xi_i\xi)(1+\eta_i\eta) \quad \text{(parallel to } \zeta)
\end{aligned}} \tag{4.173}$$

4.11.4 Isoparametric Mapping and Jacobian

The physical geometry is interpolated using the same shape functions:

$$x = \sum_{i=1}^{20} N_i x_i, \quad y = \sum_{i=1}^{20} N_i y_i, \quad z = \sum_{i=1}^{20} N_i z_i \tag{4.174}$$

The **Jacobian matrix J** is calculated as:

$$\mathbf{J} = \begin{bmatrix} \frac{\partial x}{\partial \xi} & \frac{\partial y}{\partial \xi} & \frac{\partial z}{\partial \xi} \\ \frac{\partial x}{\partial \eta} & \frac{\partial y}{\partial \eta} & \frac{\partial z}{\partial \eta} \\ \frac{\partial x}{\partial \zeta} & \frac{\partial y}{\partial \zeta} & \frac{\partial z}{\partial \zeta} \end{bmatrix} \tag{4.175}$$

The Jacobian is nonlinear and varies spatially within the element, reflecting the higher-order geometric flexibility of the quadratic formulation.

4.11.5 Polynomial Basis

The H20 element spans a reduced quartic polynomial space in natural coordinates:

$$\begin{aligned} &\{1,\ \xi,\ \eta,\ \zeta,\ \xi\eta,\ \eta\zeta,\ \zeta\xi,\ \xi^2,\ \eta^2,\ \zeta^2, \\ &\xi\eta\zeta,\ \xi^2\eta,\ \xi^2\zeta,\ \eta^2\zeta,\ \eta^2\xi,\ \zeta^2\eta,\ \zeta^2\xi, \xi^2\eta\zeta,\ \eta^2\zeta\xi,\ \zeta^2\eta\xi\}. \end{aligned} \tag{4.176}$$

4.11.6 Remarks

- The H20 element is the direct **3D extension of the 2D Q8 element**.
- It provides significantly higher accuracy than the linear H8 element without requiring an interior node.
- It is particularly well-suited for problems characterized by smooth solution fields and complex curved geometries.

4.12 Grant Summary

We have developed the shape functions for the major finite elements used in one-, two-, and three-dimensional function approximations, together with their derivatives. These shape functions are constructed to guarantee the following two critical features:

1. The shape functions satisfy the fundamental properties required for finite element approximations, including the Kronecker delta property, partition of unity, and at least linear completeness. These properties ensure the **capability to approximate** continuous field functions accurately as the mesh is refined.
2. By enforcing compatibility at element interfaces, any function approximated over the entire domain using these elemental shape functions possesses the so-called C^0 **continuity**. This means that the field function remains continuous

throughout the domain, although its derivatives may be discontinuous across a finite number of interfaces. Consequently, the first derivatives of these functions are square-integrable.

These two essential properties ensure that the approximated functions are **admissible**, and are in a so-called Sobolev space $H^1(\Omega)$ that is a subspace in the Hilbert space, thereby enabling the successful construction of finite element models. In the following chapter, we shall introduce FEM theories and formulations that guarantee the realization of the **approximation capability** of these shape functions.

4.13 Exercises

4.13.1 1D and 2D Elements

1: For a 1D quadratic element with three nodes at $\xi = -1, 0, 1$, derive the shape function N_2 (the mid-node) using the Lagrange interpolation formula. Verify that $N_2(0) = 1$.

2: Consider a 2D linear triangular element (T3). Prove that the shape functions N_i are identical to the area coordinates L_i. Show that $L_1 + L_2 + L_3 = 1$ must hold for any point within the triangle.

3: For a Q4 (bilinear quadrilateral) element, calculate the value of the shape functions at the element centroid ($\xi = 0, \eta = 0$). What does this result imply about the weight of each node at the center?

4: A T6 (quadratic triangle) element has a mid-edge node at $L_1 = 1/2, L_2 = 1/2, L_3 = 0$. Using the formula $N_k = 4L_iL_j$, show that this shape function vanishes at all three corner nodes.

5: Explain the "Serendipity" concept in the context of a Q8 element. Why do we prefer the Q8 element over a 9-node Lagrangian quadrilateral in many engineering applications?

4.13.2 3D Tetrahedral Elements (T4 and T10)

6: Calculate the volume V_e of a T4 element with nodes at: Node 1: $(0, 0, 0)$, Node 2: $(2, 0, 0)$, Node 3: $(0, 2, 0)$, Node 4: $(0, 0, 2)$.

7: For the T4 element in Exercise 6, determine the gradient ∇N_1. Verify that the gradient is constant throughout the element.

8: For a T10 element, write the expression for the corner shape function N_4 in terms of the volume coordinate L_4. Evaluate N_4 at the mid-edge node 10 (the midpoint of edge 3-4).

9: Prove the Partition of Unity for a T4 element: $\sum_{i=1}^{4} N_i = 1$.

10: Describe why the strain field in a T10 element varies linearly, while in a T4 element it is constant. How does this affect "locking" behavior in nearly incompressible materials?

4.13.3 3D Hexahedral Elements (H8 and H20)

11: Derive the derivative $\partial N_1/\partial\xi$ for an H8 element. At which locations within the element does this derivative reach its maximum and minimum values?

12: An H8 element is mapped from natural space to physical space. If the physical nodes form a perfect cube of side length a, show that the Jacobian determinant $|J|$ is constant and equal to $a^3/8$.

13: For the H20 serendipity element, evaluate the corner shape function N_1 at the center of the element $(\xi = 0, \eta = 0, \zeta = 0)$.

14: Consider the mid-edge node 17 of an H20 element located at $(-1, -1, 0)$. Write its shape function N_{17} and verify the Kronecker delta property at its own coordinates.

15: Critical Thinking: In 3D isoparametric mapping, why is it essential for the Jacobian determinant $|J|$ to remain positive? Describe a physical scenario where an H8 element might fail this check (e.g., node "fold-over").

Chapter 5

FEM Theory and Formulation

Building upon the discretization techniques and Python implementations introduced in Chapter 3, and the element-wise shape function formulations established in Chapter 4, this chapter develops the fundamental theory required to construct the global FEM system equations. We transition here from local approximation to the integral statements that govern the entire physical domain of a solid.

5.1 Governing Equation System

5.1.1 Equilibrium Equation

Consider the strong form of the dynamic equilibrium equation for linear elastodynamics defined over a domain Ω in $\mathbb{R}^d$ (where $d = 1, 2,$ or 3):

$$\mathbf{L}_d^\top \mathbf{C} \mathbf{L}_d \, \mathbf{u} + \mathbf{b} = \rho \ddot{\mathbf{u}} \quad \text{in } \Omega \tag{5.1}$$

Where the physical quantities are defined as:

- $\mathbf{u}$: Displacement vector field consisting of component displacement functions.
- $\mathbf{L}_d$: Differential operator matrix mapping displacement to strain ($\varepsilon = \mathbf{L}_d \mathbf{u}$).
- $\mathbf{C}$: Material constitutive (stiffness) matrix containing linear elastic material constants.
- $\mathbf{b}$: Body force vector containing component force functions per unit volume.
- ρ: Mass density, a scalar representing the material's physical property.
- $\ddot{\mathbf{u}}$: Acceleration vector ($\partial^2 \mathbf{u}/\partial t^2$).

Notice that for $\mathbf{u}$ to satisfy Eq. (5.1), it must be spatially differentiable at least to the second order. Hence, Eq. (5.1) is called the **strong form** of the equilibrium equation.

Note that $\mathbf{L}_d^\top \mathbf{C} \mathbf{L}_d$ is a **self-adjoint** (or premium symmetric) differential operator.

5.1.2 Boundary Conditions

The boundary Γ of the domain Ω is partitioned into two disjoint subsets, Γ_u and Γ_t, such that $\Gamma = \Gamma_u \cup \Gamma_t$ and $\Gamma_u \cap \Gamma_t = \emptyset$.

I. Displacement (Essential/Dirichlet) Boundary Condition Prescribed displacements on the boundary:

$$\mathbf{u} = \bar{\mathbf{u}} \quad \text{on } \Gamma_u \tag{5.2}$$

II. Traction (Natural/Neumann) Boundary Condition Prescribed surface forces, where $\mathbf{L}_n$ is the matrix of outward unit normals:

$$\mathbf{L}_n^\top \sigma = \mathbf{L}_n^\top \mathbf{C}\mathbf{L}_d \mathbf{u} = \bar{\mathbf{t}} \quad \text{on } \Gamma_t \tag{5.3}$$

5.2 Weak Form of the Equilibrium Equation

This section derives the weak form for solid mechanics problems. We begin with a general formulation known as the **Weighted Residual Form**.

5.2.1 Weighted Residual Form

Directly solving the strong form in Eq. (5.1) is analytically intractable for complex geometries or complicated boundary conditions. We seek an approximate solution by relaxing the requirement that the differential equation holds exactly at every point, requiring it instead to hold in an average (integral) sense.

Let $\delta\mathbf{w}$ denote an arbitrary variation of the admissible displacement field that satisfies the displacement boundary conditions and remains compatible throughout the domain. The quantity $\delta\mathbf{w}$ is commonly referred to as the **test function**.

The admissibility conditions require that:

- $\delta\mathbf{w} = \mathbf{0}$ on Γ_u, since the displacement $\mathbf{u}$ is prescribed there.
- $\delta\mathbf{w}$ is sufficiently smooth and belongs to the Sobolev space $H^1(\Omega)$, meaning that both the function and its first derivatives are square-integrable. This ensures displacement compatibility throughout the domain.

The weighted residual statement requires:

$$\int_\Omega \delta\mathbf{w}^\top \underbrace{(\mathbf{L}_d^\top \mathbf{C}\mathbf{L}_d \mathbf{u} + \mathbf{b} - \rho\ddot{\mathbf{u}})}_{\mathbf{r}} \, d\Omega = 0 \tag{5.4}$$

where $\mathbf{r}$ is the vector of residuals. The weighted residual statement dictates that the inner product of $\delta\mathbf{w}$ and $\mathbf{r}$ vanishes, meaning that $\delta\mathbf{w}$ and $\mathbf{r}$ must be **orthogonal** to each other [9]. Because this inner product defines a projection, the weighted residual statement acts as an orthogonal projection.

Equation (5.4) provides the following physical and mathematical insights:

1. If Eq. (5.1) is exactly satisfied by $\mathbf{u}$, the residual $\mathbf{r}$ is zero, meaning $\mathbf{u}$ is the exact solution. Therefore, the weighted residual statement Eq. (5.4) is entirely consistent with the strong form PDE.

2. If $\mathbf{u}$ does not exactly satisfy Eq. (5.1), $\mathbf{r}$ is nonzero. Equation (5.4) forces the residual to be orthogonal to a chosen set of test functions $\delta\mathbf{w}$. By properly constructing this set of test functions, we establish a mechanism to minimize error and control the quality of the approximated solution $\mathbf{u}$. Thus, the weighted residual statement offers a practical framework for finding a meaningful approximate solution when an exact solution is unavailable.

The problem then reduces to:

1. How to construct an appropriate set of test functions $\delta\mathbf{w}$.

2. How to approximate $\mathbf{u}$ (referred to as the **trial function**) such that the resulting solution is physically meaningful.

Importantly, these functions must be defined over arbitrary, potentially irregular domains while strictly satisfying the boundary conditions, considering that $\mathbf{L}_d^\top \mathbf{C} \mathbf{L}_d$ is a **self-adjoint** differential operator. Additionally, the approximated $\mathbf{u}$ in the strong form must exhibit sufficient continuity to be twice differentiable.

Note that since the weight residual is measured in the global integral form, we are essentially looking for approximated solutions that are in overall equilibrium.

5.2.2 Principle of Virtual Work

To reduce the continuity requirements on the displacement field $\mathbf{u}$ from C^1 continuity to C^0 continuity, we apply the following integral identity from continuum mechanics, Eq.(2.48), proven in Chapter 2:

$$-\int_\Omega \delta\mathbf{w}^\top \mathbf{L}_d^\top \mathbf{q}\, d\Omega = \int_\Omega (\mathbf{L}_d \delta\mathbf{w})^\top \mathbf{q}\, d\Omega - \int_\Gamma \delta\mathbf{w}^\top \mathbf{L}_n^\top \mathbf{q}\, d\Gamma \tag{5.5}$$

Substituting the constitutive relation

$$\mathbf{q} = \sigma = \mathbf{C}\mathbf{L}_d\mathbf{u}$$

into Eq.(5.5) and applying it to Eq.(5.4) yields

$$\int_\Omega (\mathbf{L}_d \delta\mathbf{w})^\top \mathbf{C}\mathbf{L}_d\mathbf{u}\, d\Omega = \int_\Omega \delta\mathbf{w}^\top \mathbf{b}\, d\Omega + \int_\Gamma \delta\mathbf{w}^\top \mathbf{t}\, d\Gamma - \int_\Omega \rho\, \delta\mathbf{w}^\top \ddot{\mathbf{u}}\, d\Omega. \tag{5.6}$$

The boundary integral can be decomposed over the essential and natural boundaries:

$$\int_\Gamma \delta\mathbf{w}^\top \mathbf{t}\, d\Gamma = \int_{\Gamma_u} \delta\mathbf{w}^\top \mathbf{t}\, d\Gamma + \int_{\Gamma_t} \delta\mathbf{w}^\top \mathbf{t}\, d\Gamma. \tag{5.7}$$

The admissible virtual displacement field must satisfy the homogeneous essential boundary conditions. Therefore,

1. On Γ_u, the displacement field is prescribed, and thus the virtual displacement vanishes:

   $$\delta\mathbf{w} = \mathbf{0}.$$

   Consequently,

   $$\int_{\Gamma_u} \delta\mathbf{w}^\top \mathbf{t}\, d\Gamma = 0.$$

2. On Γ_t, the traction is prescribed as

   $$\mathbf{t} = \bar{\mathbf{t}}.$$

   Therefore,

   $$\int_{\Gamma_t} \delta\mathbf{w}^\top \mathbf{t}\, d\Gamma = \int_{\Gamma_t} \delta\mathbf{w}^\top \bar{\mathbf{t}}\, d\Gamma.$$

Substituting these boundary conditions into Eq.(5.6) gives

$$\boxed{\underbrace{\int_\Omega (\mathbf{L}_d \delta\mathbf{w})^\top \mathbf{C}\mathbf{L}_d\mathbf{u}\, d\Omega}_{W_{\text{int}}} = \underbrace{\int_\Omega \delta\mathbf{w}^\top \mathbf{b}\, d\Omega + \int_{\Gamma_t} \delta\mathbf{w}^\top \bar{\mathbf{t}}\, d\Gamma}_{W_{\text{ext}}} + \underbrace{\int_\Omega \delta\mathbf{w}^\top (-\rho\ddot{\mathbf{u}})\, d\Omega}_{W_{\text{iner}}}.} \tag{5.8}$$

Equation (5.8) is the **Principle of Virtual Work (PVW)**. It states that, for every admissible virtual displacement field $\delta\mathbf{w}$, the internal virtual work equals the virtual work of the external forces and inertial forces.

In the context of continuum mechanics, $\delta\mathbf{w}$ represents a virtual displacement field. Under this interpretation:

1. The first two terms on the right-hand side represent the virtual work of the external forces: the body force $\mathbf{b}$ acting throughout the domain and the prescribed traction $\bar{\mathbf{t}}$ acting on the boundary Γ_t. Together, these contributions constitute the external virtual work W_{ext}.
2. The final term represents the virtual work of the inertial force $-\rho\ddot{\mathbf{u}}$ acting throughout the domain. This contribution is denoted by W_{iner}.
3. The left-hand side represents the internal virtual work W_{int}, arising from the stress field

$$\sigma = \mathbf{C}\mathbf{L}_d\mathbf{u}$$

acting through the virtual strain field

$$\delta\varepsilon = \mathbf{L}_d\delta\mathbf{w}.$$

In the context of continuum mechanics, $\delta\mathbf{w}$ represents a virtual displacement field. Under this interpretation:

1. The first two terms on the right-hand side represent the virtual work of the external forces: the body force $\mathbf{b}$ acting over the domain and the prescribed traction $\bar{\mathbf{t}}$ acting on the boundary Γ_t (collectively denoted by W_{ext}).
2. The final term represents the virtual work of the inertial force $-\rho\ddot{\mathbf{u}}$ acting over the domain (denoted by W_{iner}).
3. The left-hand side represents the internal virtual work associated with the stress field $\sigma = \mathbf{C}\mathbf{L}_d\mathbf{u}$ acting through the virtual strain field $\delta\varepsilon = \mathbf{L}_d\delta\mathbf{w}$ (denoted by W_{int}).

Equation (5.8) is the mathematical expression of the **Principle of Virtual Work** for a constrained solid body in dynamic equilibrium. It can be written concisely as:

$$\boxed{W_{\text{int}} = W_{\text{ext}} + W_{\text{iner}}} \tag{5.9}$$

While this relationship is often asserted directly from physical principles, we have derived it rigorously here from the weighted residual method using the integral identity.

5.2.3 Weak Form

Equation (5.8) is called the **weak form** because the maximum spatial derivative order required of the trial function $\mathbf{u}$ is reduced from second-order to first-order. This shift is achieved by transferring one order of differentiation to the test function $\delta\mathbf{w}$ via the integral identity. This **weakens** the differentiability requirements on $\mathbf{u}$, expanding the solution domain to a much broader class of approximate functions.

When the trial function $\mathbf{u}$ and the test function $\delta\mathbf{w}$ are approximated using the same set of shape functions, the resulting formulation is called the **Galerkin weak form**.

In this case, the resulting system matrix is symmetric because of the self-adjoint property of $\mathbf{L}_d^\top \mathbf{C} \mathbf{L}_d$. This approach represents an energy-orthogonal projection and is commonly referred to as a **Galerkin projection**.

5.2.4 Conditions for the Use of Weak Form

For the weak form integrals to be mathematically well-defined, all energy terms must remain bounded. The conditions for the validity of the weak form are outlined below:

1. **Internal Strain Energy Integral:** Because the test function $\delta\mathbf{w}$ belongs to the Sobolev space $H^1(\Omega)$, the trial function $\mathbf{u}$ must also belong to $H^1(\Omega)$ to ensure the integral is bounded. Standard finite element shape functions (discussed in Chapter 4) feature C^0 continuity and possess square-integrable first derivatives, satisfying $H^1(\Omega)$ requirements. Consequently, it is mathematically valid to use these shape functions to approximate $\mathbf{u}$.

2. **Inertial/Kinetic Energy Integral:** If both $\delta\mathbf{w}$ and $\mathbf{u}$ (along with its temporal derivative $\ddot{\mathbf{u}}$) belong to $H^1(\Omega)$, this domain integral is unconditionally bounded.

3. **External Body Force Work Integral:** Since $\delta\mathbf{w} \in H^1(\Omega)$, the distributed body force vector $\mathbf{b}$ must be square-integrable over the domain ($\mathbf{b} \in L^2(\Omega)$) to guarantee a bounded energy contribution.

4. **External Traction Work Integral:** Since $\delta\mathbf{w} \in H^1(\Omega)$, the prescribed traction vector $\bar{\mathbf{t}}$ must be square-integrable along the Neumann boundary profile ($\bar{\mathbf{t}} \in L^2(\Gamma_t)$).

   Key Takeaway: Shape functions constructed via the procedures in Chapter 4 can be rigorously used to establish FEM models for 1D, 2D, and 3D solids with optimal convergence, provided that the external load functions are square-integrable.

Detailed discussions and concrete examples regarding these function spaces can be found in Ref. [11].

5.2.5 Interpretation

- **Essential vs. Natural:** The displacement boundary condition is "essential" because it must be explicitly enforced within the trial function's approximation space. The traction boundary condition is "natural" because it is automatically (implicitly) satisfied through the integral formulation itself.

- **Symmetry:** The internal work term is symmetric with respect to $\mathbf{u}$ and $\delta\mathbf{w}$, provided that the constitutive matrix $\mathbf{C}$ is symmetric, which is standard for linear elastic materials.

- **Scalar Fields:** For scalar problems (e.g., heat transfer, potential flow), the operators simplify significantly; for instance, $\mathbf{L}_d$ reduces to the gradient operator

∇. Comprehensive derivations for scalar fields are available in Chapter 2 of Ref. [7].

5.2.6 Energy principles

Using the variational approach, the same weak form can also be derived from an energy principle. In this approach, an energy functional is constructed, and the equilibrium state of the solid is obtained by requiring the functional to be stationary. The resulting stationary condition leads directly to the same weak form.

For dynamic problems, **Hamilton's principle** is used, which yields a formulation equivalent to Eq.(5.8). For static problems, the **principle of total potential energy** is employed, leading to a formulation without the dynamic term.

When an energy principle is used in solid mechanics, the strong form, Eq.(5.1), is not invoked explicitly. In this sense, the energy principle serves as an alternative statement of equilibrium [4]. The detailed FEM formulation process based on energy principles can be found in Ref. [12].

5.2.7 Weak Statement

In the context of functional analysis, the governing equations are frequently cast as a **weak statement** using functional spaces. Let V be the continuous solution space and $V_h \subset V$ be the discrete finite element subspace. The continuous weak statement of the solid mechanics problem requires finding a function $\mathbf{u} \in V$ such that:

$$a(\mathbf{u}, \mathbf{v}) = L(\mathbf{v}) \qquad \forall \mathbf{v} \in V$$

where: $a(\mathbf{u}, \mathbf{v})$ is the **bilinear form**, representing the inner product of $\mathbf{u}$ and $\mathbf{v}$ scaled by the material stiffness (strain energy). It is explicitly defined as:

$$a(\mathbf{u}, \mathbf{v}) = \int_\Omega (\mathbf{L}_d\mathbf{v})^\top \mathbf{C}(\mathbf{L}_d\mathbf{u})\, d\Omega \tag{5.10}$$

is the **linear form**, representing the inner product of $\mathbf{v}$ with the given external and inertial forces:

$$L(\mathbf{v}) = \int_\Omega \mathbf{v}^\top \mathbf{b},\, d\Omega + \int_{\Gamma_t} \mathbf{v}^\top \bar{\mathbf{t}},\, d\Gamma + \int_\Omega \mathbf{v}^\top(-\rho\ddot{\mathbf{u}}),\, d\Omega$$

The discrete Galerkin formulation restricts the test space to the finite-dimensional subspace V_h. It seeks a discrete solution $\mathbf{u}_h \in V_h$ such that:

$$a(\mathbf{u}_h, \mathbf{v}_h) = L(\mathbf{v}_h) \qquad \forall \mathbf{v}_h \in V_h \tag{5.11}$$

Subtracting the discrete weak form from the continuous weak form (choosing $\mathbf{v} = \mathbf{v}_h \in V_h \subset V$) yields the **Galerkin Orthogonality** condition:

$$a(\mathbf{u} - \mathbf{u}_h, \mathbf{v}_h) = 0 \quad \forall \mathbf{v}_h \in V_h$$

This indicates that the approximation error $(\mathbf{u}-\mathbf{u}_h)$ is orthogonal to the entire discrete space V_h, meaning the solution $\mathbf{u}_h$ minimizes the error in the energy norm.

A close examination reveals the explicit mapping between these parallel frameworks:

1. The test function $\mathbf{v}$ is equivalent to the virtual displacement $\delta\mathbf{w}$, and $\mathbf{u}_h$ is equivalent to the approximated trial displacement $\mathbf{u}$.
2. The discrete space V_h contains all admissible variations $\delta\mathbf{w}$.
3. The functional weak statement is completely equivalent to the Principle of Virtual Work.

While terms like *weak form*, *weak statement*, *energy principle*, and *principle of virtual work* are often used in the literature, this analysis demonstrates that **they are mathematically identical formulations**. You may employ whichever terminology best serves your immediate conceptual or communicative needs.

5.3 Finite Element Approximation

To use the weak form, we need to approximate both the displacement trial and test functions.

5.3.1 Shape Function Matrix

Assume the solid is discretized with N_e elements and N_n nodes following the procedure discussed in Chapter 3. We approximate the displacement field $\mathbf{u}$ using a discrete set of nodal values multiplied by global shape functions:

$$\boxed{\mathbf{u}^h = \mathbf{N}\mathbf{D}} = \begin{bmatrix} \mathbf{N}_1 & \mathbf{N}_2 & \cdots & \mathbf{N}_{N_n} \end{bmatrix} \begin{bmatrix} \mathbf{d}_1 \\ \mathbf{d}_2 \\ \vdots \\ \mathbf{d}_{N_n} \end{bmatrix} \tag{5.12}$$

where:

- $\mathbf{D}$ is the unknown global nodal displacement vector consisting displacement vectors for all the nodes $\mathbf{d}_i$ for $i = 1, 2, \ldots, N_n$. For 3D problems for example, $\mathbf{d}_i = [u_i \ v_i \ w_i]^\top$.
- $\mathbf{N}$ is a matrix containing the nodal shape functions defined over the entire problem domain for the FEM function space, which is a subspace in $H^1(\Omega)$. It

consists of **nodal sub-matrix** $\mathbf{N}_i$ $i = 1, 2, ..., N_n$. For 3D problems for example, $\mathbf{N}_i = \begin{bmatrix} N_i & 0 & 0 \\ 0 & N_i & 0 \\ 0 & 0 & N_i \end{bmatrix}$, in which N_i is the union of nodal shape functions of all elements connected to node i that are determined at the element level in Chapter 4. This nodal sub-matrix formulation is useful in deriving FEM element matrices in the following chapters.

A key property of these basis functions is their **compact support**. A global nodal shape function is strictly equal to unity at its corresponding node, drops linearly or quadratically to zero at all neighboring nodes, and remains zero throughout the rest of the mesh domain. Consequently, it is only non-zero over the specific elements connected to that node. The global matrix $\mathbf{N}$ is constructed as a union of the element-level nodal shape functions obtained in Chapter 4 for various element geometries.

Figure 5.1 shows an example for a 1D domain, illustrating the global nodal shape functions for Node 1 and Node 5. The interior nodal shape function N_5 is non-zero only over the two elements immediately adjacent to it (one on its left and one on its right). The boundary nodal shape function N_1 is on the boundary, and hence it is non-zero only over the single element on its right.

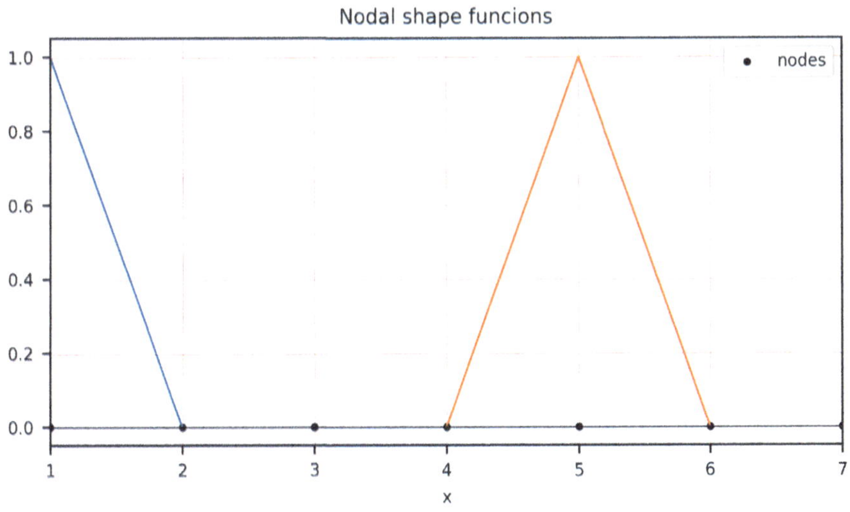

Figure 5.1: Two nodal shape functions locally supported in a 1D domain.

Figure 5.2 shows an example within a 2D mesh for the shape function of Node I. It is non-zero only over the six elements directly sharing and connected to Node I.

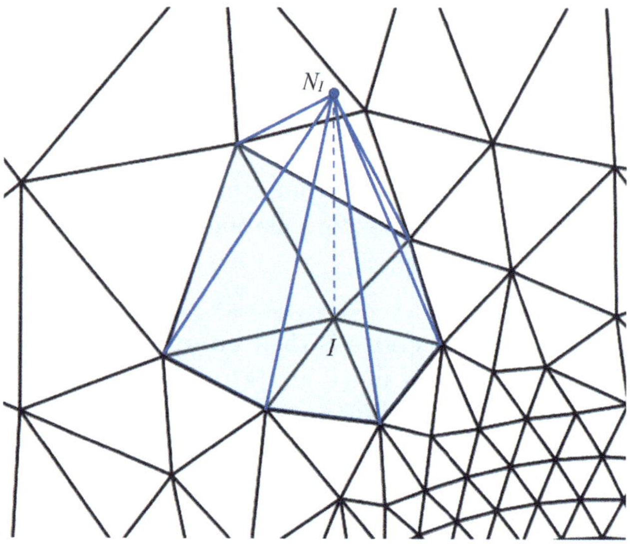

Figure 5.2: Shape function for node I in a 2D domain.

Following the **Galerkin method**, the test function (or virtual displacement variation) is chosen from the exact same functional space as the trial functions, utilizing $\mathbf{N}$ as the interpolating basis functions:

$$\delta \mathbf{w}^h = \mathbf{N}\delta\mathbf{D} \tag{5.13}$$

5.3.2 Strain–Displacement Matrix

The strain vector field ε is given by applying the differential operator to the approximated displacement field:

$$\varepsilon = \mathbf{L}_d\mathbf{u}^h = \mathbf{L}_d\mathbf{N}\mathbf{D} = \mathbf{B}\mathbf{D} \tag{5.14}$$

where the **strain-displacement matrix B** is defined as:

$$\mathbf{B} = \mathbf{L}_d\mathbf{N} \tag{5.15}$$

Since the global shape function matrix $\mathbf{N}$ is known for a given mesh of elements, the matrix $\mathbf{B}$ is also completely known.

Equation (5.15) can be written explicitly as:

$$\begin{aligned} \mathbf{B} = \mathbf{L}_d\mathbf{N} &= \begin{bmatrix} \mathbf{L}_d\mathbf{N}_1 & \mathbf{L}_d\mathbf{N}_2 & \cdots & \mathbf{L}_d\mathbf{N}_{N_n} \end{bmatrix} \\ &= \begin{bmatrix} \mathbf{B}_1 & \mathbf{B}_2 & \cdots & \mathbf{B}_{N_n} \end{bmatrix} \end{aligned} \tag{5.16}$$

with

$$\mathbf{B}_i = \mathbf{L}_d \mathbf{N}_i \tag{5.17}$$

which is **nodal sub-matrix** for node i. This nodal submatrix $\mathbf{B}_i$ will be used in deriving analytical forms of FEM element matrices in the following chapters.

5.4 Application of the Weak Form

To transition from the continuous weak form to the discretized system of equations, we substitute the approximations from Eq. (5.12) and Eq. (5.13) into the standard weak form expression established in Eq. (5.8):

$$\int_{\Omega} (\mathbf{B}\delta\mathbf{D})^{\top}\mathbf{C}(\mathbf{B}\mathbf{D})\, d\Omega = \int_{\Omega} (\mathbf{N}\delta\mathbf{D})^{\top}\mathbf{b}\, d\Omega + \int_{\Gamma_t} (\mathbf{N}\delta\mathbf{D})^{\top}\bar{\mathbf{t}}\, d\Gamma - \int_{\Omega} \rho(\mathbf{N}\delta\mathbf{D})^{\top}\mathbf{N}\ddot{\mathbf{D}}\, d\Omega \tag{5.18}$$

Using the matrix transposition property $(\mathbf{XY})^{\top} = \mathbf{Y}^{\top}\mathbf{X}^{\top}$, we rewrite the terms as:

$$\int_{\Omega} \delta\mathbf{D}^{\top}\mathbf{B}^{\top}\mathbf{C}\mathbf{B}\mathbf{D}\, d\Omega = \int_{\Omega} \delta\mathbf{D}^{\top}\mathbf{N}^{\top}\mathbf{b}\, d\Omega + \int_{\Gamma_t} \delta\mathbf{D}^{\top}\mathbf{N}^{\top}\bar{\mathbf{t}}\, d\Gamma - \int_{\Omega} \rho\delta\mathbf{D}^{\top}\mathbf{N}^{\top}\mathbf{N}\ddot{\mathbf{D}}\, d\Omega \tag{5.19}$$

Recognizing that the nodal vectors $\mathbf{D}$, $\ddot{\mathbf{D}}$, and $\delta\mathbf{D}$ are discrete values independent of the spatial integration coordinates, they can be pulled outside of the domain integrals. Moving all terms to one side allows us to factor out the arbitrary nodal variation $\delta\mathbf{D}^{\top}$:

$$\delta\mathbf{D}^{\top}\left[\int_{\Omega} \mathbf{B}^{\top}\mathbf{C}\mathbf{B}\, d\Omega\, \mathbf{D} - \int_{\Omega} \mathbf{N}^{\top}\mathbf{b}\, d\Omega - \int_{\Gamma_t} \mathbf{N}^{\top}\bar{\mathbf{t}}\, d\Gamma + \int_{\Omega} \rho\mathbf{N}^{\top}\mathbf{N}\, d\Omega\, \ddot{\mathbf{D}}\right] = 0 \tag{5.20}$$

Since $\delta\mathbf{D}$ is completely arbitrary and admissible, for Eq. (5.20) to hold true always for any virtual variation, the term inside the square brackets must vanish identically. This gives:

$$\boxed{\underbrace{\int_{\Omega} \mathbf{B}^{\top}\mathbf{C}\mathbf{B}\, d\Omega}_{\mathbf{K}}\, \mathbf{D} - \underbrace{\left(\int_{\Omega} \mathbf{N}^{\top}\mathbf{b}\, d\Omega + \int_{\Gamma_t} \mathbf{N}^{\top}\bar{\mathbf{t}}\, d\Gamma\right)}_{\mathbf{F}} + \underbrace{\int_{\Omega} \rho\mathbf{N}^{\top}\mathbf{N}\, d\Omega}_{\mathbf{M}}\, \ddot{\mathbf{D}} = 0} \tag{5.21}$$

Which directly yields the foundational spatially discrete algebraic system of finite element equations:

$$\boxed{\mathbf{M}\ddot{\mathbf{D}} + \mathbf{K}\mathbf{D} = \mathbf{F}} \tag{5.22}$$

where:

- **M**: Global **Mass Matrix**
- **K**: Global **Stiffness Matrix**
- **F**: Global **External Force Vector**

5.4.1 Stiffness Matrix

The global stiffness matrix **K** represents the internal restoration forces of the elastic solid per unit displacement:

$$\mathbf{K} = \int_{\Omega} \mathbf{B}^{\top}\mathbf{C}\mathbf{B}\, d\Omega \tag{5.23}$$

- **Symmetry:** Because the material constitutive matrix **C** is symmetric ($\mathbf{C} = \mathbf{C}^{\top}$), the product $\mathbf{B}^{\top}\mathbf{C}\mathbf{B}$ ensures that **K** is symmetric by construction ($\mathbf{K} = \mathbf{K}^{\top}$) under the Galerkin framework together with the **self-adjoint** property of the differential operator $\mathbf{L}_d^T\mathbf{C}\mathbf{L}_d$ in the strong form Eq.(5.1).
- **Integrability:** Since the global shape functions **N** are continuous across element boundaries (belonging to the Sobolev space $H^1(\Omega)$), their spatial derivatives in **B** are bounded. Thus, the integrand $\mathbf{B}^{\top}\mathbf{C}\mathbf{B}$ is fully integrable over the domain Ω. In practice, this integration is performed numerically at the element level using Gauss quadrature schemes.

The integrand in Eq.(5.23) is an outer-product, and can be explicitly written as

$$\begin{aligned}\mathbf{B}^{\top}\mathbf{C}\mathbf{B} &= \begin{bmatrix}\mathbf{B}_1^{\top}\\ \mathbf{B}_2^{\top}\\ \vdots\\ \mathbf{B}_{N_n}^{\top}\end{bmatrix}\mathbf{C}\begin{bmatrix}\mathbf{B}_1 & \mathbf{B}_2 & \cdots & \mathbf{B}_{N_n}\end{bmatrix}\\ &= \begin{bmatrix}\mathbf{B}_1^{\top}\mathbf{C}\mathbf{B}_1 & \mathbf{B}_1^{\top}\mathbf{C}\mathbf{B}_2 & \cdots & \mathbf{B}_1^{\top}\mathbf{C}\mathbf{B}_{N_n}\\ \mathbf{B}_2^{\top}\mathbf{C}\mathbf{B}_1 & \mathbf{B}_2^{\top}\mathbf{C}\mathbf{B}_2 & \cdots & \mathbf{B}_2^{\top}\mathbf{C}\mathbf{B}_{N_n}^{\top}\\ \vdots & \vdots & \ddots & \vdots\\ \mathbf{B}_{N_n}^{\top}\mathbf{C}\mathbf{B}_1 & \mathbf{B}_{N_n}^{\top}\mathbf{C}\mathbf{B}_2 & \cdots & \mathbf{B}_{N_n}^{\top}\mathbf{C}\mathbf{B}_{N_n}\end{bmatrix}\end{aligned} \tag{5.24}$$

The global stiffness matrix **K** can now be written in blocks:

$$\mathbf{K} = \begin{bmatrix}\mathbf{K}_{11} & \mathbf{K}_{12} & \cdots & \mathbf{K}_{N_nN_n}\\ \mathbf{K}_{21} & \mathbf{K}_{22} & \cdots & \mathbf{K}_{2N_n}\\ \vdots & \cdots & \ddots & \vdots\\ \mathbf{K}_{N_n1} & \mathbf{K}_{N_n2} & \cdots & \mathbf{K}_{N_nN_n}\end{bmatrix} \tag{5.25}$$

with

$$\mathbf{K}_{ij} = \int_{\Omega} \mathbf{B}_i^{\top}\mathbf{C}\mathbf{B}_j d\Omega \tag{5.26}$$

which is a **nodal sub-matrix** corresponding to node i and node j.

5.4.2 Mass Matrix

The global mass matrix $\mathbf{M}$ accounts for the inertial properties of the continuum model:

$$\mathbf{M} = \int_\Omega \rho \mathbf{N}^\top \mathbf{N} \, d\Omega \tag{5.27}$$

- **Symmetry:** Like the stiffness matrix, the term $\mathbf{N}^\top\mathbf{N}$ inherently guarantees that $\mathbf{M}$ is symmetric by construction.
- **Evaluation:** Since $\mathbf{N}$ is explicitly known once the element mesh topology is defined, and is continuous over the domain, $\mathbf{N}^\top\mathbf{N}$ is unconditionally integrable. While numerical integration is standard, analytical (exact) closed-form integrations can be derived for simple element geometries, as demonstrated in subsequent chapters. This form is strictly known as the **consistent mass matrix**.

The global mass matrix $\mathbf{M}$ can now be written in blocks:

$$\mathbf{M} = \begin{bmatrix} \mathbf{M}_{11} & \mathbf{M}_{12} & \cdots & \mathbf{M}_{N_n N_n} \\ \mathbf{M}_{21} & \mathbf{M}_{22} & \cdots & \mathbf{M}_{2N_n} \\ \vdots & \cdots & \ddots & \vdots \\ \mathbf{M}_{N_n 1} & \mathbf{M}_{N_n 2} & \cdots & \mathbf{M}_{N_n N_n} \end{bmatrix} \tag{5.28}$$

with

$$\mathbf{M}_{ij} = \int_\Omega \rho \mathbf{N}_i^\top \mathbf{N}_j d\Omega \tag{5.29}$$

which is a nodal sub-matrix corresponding to node i and node j.

5.4.3 Force Vector

The global external force vector $\mathbf{F}$ aggregates all external structural actions transformed into equivalent nodal forces:

$$\mathbf{F} = \int_\Omega \mathbf{N}^\top \mathbf{b} \, d\Omega + \int_{\Gamma_t} \mathbf{N}^\top \bar{\mathbf{t}} \, d\Gamma \tag{5.30}$$

The first term transforms the volumetric body forces $\mathbf{b}$ into kinematically equivalent nodal loads, while the second term converts the surface tractions $\bar{\mathbf{t}}$ applied along the Neumann boundary Γ_t. Because the shape functions are fully defined by the mesh layout, these integrals are readily computed numerically or analytically.

The global force vector $\mathbf{F}$ can also be written explicitly as

$$\mathbf{F} = \int_\Omega \begin{bmatrix} \mathbf{N}_1 \\ \mathbf{N}_2 \\ \vdots \\ \mathbf{N}_{N_n} \end{bmatrix} \mathbf{b} \, d\Omega + \int_{\Gamma_t} \begin{bmatrix} \mathbf{N}_1 \\ \mathbf{N}_2 \\ \vdots \\ \mathbf{N}_{N_n} \end{bmatrix} \bar{\mathbf{t}} \, d\Gamma = \begin{bmatrix} \mathbf{F}_1 \\ \mathbf{F}_2 \\ \vdots \\ \mathbf{F}_{N_n} \end{bmatrix} \tag{5.31}$$

with

$$\mathbf{F}_i = \int_{\Omega} \mathbf{N}_i^{\top} \mathbf{b}\, d\Gamma + \int_{\Gamma_t} \mathbf{N}_i^{\top} \bar{\mathbf{t}}\, d\Gamma \tag{5.32}$$

which is a nodal sub-vector corresponding to node i.

5.4.4 Points to Note

1. All these matrices and vectors can be **readily computed** when the shape functions are found.
2. **Variational Consistency:** This formulation, derived here essentially via the Galerkin week form, is mathematically equivalent to the system obtained using **Hamilton's Principle** (a variational approach detailed in Ref. [12]).
3. **Physical Energy Principle:** For solid mechanics applications, the statement of the weak form directly reflects the fundamental **Principle of Virtual Work**, matching internal virtual work with external virtual work.
4. **Galerkin Advantages:** Choosing the test function space to match the trial function space preserves both the mathematical symmetry of the operator and guarantees optimal convergence properties.
5. **Boundary Condition Handling:** Essential (Dirichlet) boundary conditions are not yet embedded in Eq. (5.22); they must be explicitly enforced by modifying the algebraic system before solving. Conversely, natural (Neumann) boundary conditions are implicitly satisfied through the boundary integral term of the force vector.

   Note on Time Integration: Equation (5.22) is a coupled system of second-order ordinary differential equations (ODEs) with respect to the temporal coordinate t. To compute a solution in the time domain, finite difference time-stepping algorithms (such as the Newmark-β or Central Difference types of methods) are utilized to transform the continuous ODE system into an iterative sequence of linear algebraic equations. This book will not discuss about this techniques. Interested reader are referred to Ref. [12].

5.5 Element-Level Formulation

For computational efficiency and geometric flexibility, the global stiffness matrix, mass matrix, and force vector are not evaluated directly over the entire domain Ω. Instead, they are computed independently at the **element level** over local subdomains Ω_e and then systematically assembled into global quantities.

This domain decomposition approach is mathematically valid provided that the finite element mesh satisfies two fundamental criteria:

- **Geometric Compatibility:** The elements must fit together perfectly, leaving no gaps and creating no overlaps between adjacent element boundaries.
- **Interpolation Compatibility:** The chosen approximation fields must maintain displacement continuity (C^0 continuity) across inter-element interfaces.

These essential requirements are rigorously ensured through proper mesh generation techniques (as discussed in Chapter 3) and the mathematically sound construction of localized shape functions (as established in Chapter 4).

5.5.1 Element Matrices and Vectors

The global domain integrals are broken down into a sum of integrals over individual element subdomains Ω_e. For each individual element, we define its localized properties using lowercase boldface notation to distinguish them from global matrices:

5.5.2 Element Stiffness Matrix

The element stiffness matrix $\mathbf{k}_e$ captures the localized strain energy capacity of an individual element:

$$\mathbf{k}_e = \int_{\Omega_e} \mathbf{B}_e^\top \mathbf{C} \mathbf{B}_e \, d\Omega \tag{5.33}$$

5.5.3 Element Mass Matrix

The element mass matrix $\mathbf{m}_e$ represents the distributed inertial mass attributed to the element's local volume:

$$\mathbf{m}_e = \int_{\Omega_e} \rho \mathbf{N}_e^\top \mathbf{N}_e \, d\Omega \tag{5.34}$$

5.5.4 Element Force Vector

The element nodal force vector $\mathbf{f}_e$ collects the local body forces and any surface tractions acting directly on the element's boundaries:

$$\mathbf{f}_e = \int_{\Omega_e} \mathbf{N}_e^\top \mathbf{b} \, d\Omega + \int_{\Gamma_t \cap \partial\Omega_e} \mathbf{N}_e^\top \bar{\mathbf{t}} \, d\Gamma \tag{5.35}$$

Note that the boundary traction integral is computed only for elements whose local boundaries $\partial\Omega_e$ intersect with the global traction boundary Γ_t.

5.5.5 Energy Equivalence

The nodal force vector $\mathbf{f}_e$ obtained using Eq.(5.35) is **energetically equivalent**, meaning that the work done by $\mathbf{f}_e$ under any admissible virtual nodal displacement $\delta\mathbf{d}_e$ of the element is same as the work done the displacement field resulted from the same arbitrary nodal displacement $\delta\mathbf{d}_e$. This can be proven using the body force as an example.

The work done by the nodal force vector is $(\delta\mathbf{d}_e)^\top\mathbf{f}_e$, which can be evaluated using Eq.(5.35):

$$\begin{aligned}(\delta\mathbf{d}_e)^\top\mathbf{f}_e &= (\delta\mathbf{d}_e)^\top\int_{\Omega_e}\mathbf{N}_e^\top\mathbf{b}\,d\Omega = \int_{\Omega_e}(\delta\mathbf{d}_e)^\top\mathbf{N}_e^\top\mathbf{b}\,d\Omega \\ &= \int_{\Omega_e}[\mathbf{N}_e(\delta\mathbf{d}_e)]^\top\mathbf{b}\,d\Omega = \int_{\Omega_e}(\delta\mathbf{u}_h)_e^\top\mathbf{b}\,d\Omega\end{aligned} \tag{5.36}$$

The term $\int_{\Omega_e}(\delta\mathbf{u}_h)_e^\top\mathbf{b}\,d\Omega$ is exactly the work done by the body force over the element under the displacement field $(\delta\mathbf{u}_h)_e$ that is resulted from the arbitrary nodal displacement $\delta\mathbf{d}_e$. This completes the proof.

The energetically equivalence can be understood intuitively, because the nodal force vector are derived following strictly the energy principle.

These element matrices $\mathbf{k}_e$ and $\mathbf{m}_e$, and the force vector $\mathbf{f}_e$ can also be written explicitly corresponding to the nodes of the element.

5.5.6 Numerical Integration and Isoparametric Mapping

Evaluating the element integrals directly in physical coordinates over arbitrary, distorted geometries is highly challenging. Therefore, numerical quadrature known as **Gaussian Quadrature** also known as **Gauss-Legendre integration** [12] is utilized. This process relies on three interrelated pillars:

1. **Reference Geometry:** Elements are mapped from their distorted physical shapes in the global coordinate system (x, y, z) to a highly regular reference domain (e.g., a unit square or a bi-unit cube) defined by natural coordinates (ξ, η, ζ).

2. **Natural Coordinate System:** Shape functions $\mathbf{N}_e$ are formulated explicitly within these natural coordinates, making their mathematical expressions invariant to the actual physical shape of the element.

3. **The Jacobian Matrix:** To transform the integration limits and differential volume elements from the physical domain to the reference domain, a spatial transformation tensor known as the **Jacobian matrix J** is established. The differential volume transforms via its determinant:

$$d\Omega = \det(\mathbf{J})\,d\xi\,d\eta\,d\zeta \tag{5.37}$$

This allows a general 2D element stiffness integral to be computed systematically using discrete sampling points (Gauss points) and corresponding weights W_i:

$$\begin{aligned}\mathbf{k}_e &= \int_{-1}^{1}\int_{-1}^{1}\mathbf{B}_e^\top(\xi,\eta)\mathbf{C}\mathbf{B}_e(\xi,\eta)\det(\mathbf{J})\,d\xi\,d\eta \\ &\approx \sum_i\sum_j W_iW_j\mathbf{B}_e^\top(\xi_i,\eta_j)\mathbf{C}\mathbf{B}_e(\xi_i,\eta_j)\det(\mathbf{J}(\xi_i,\eta_j))\end{aligned} \tag{5.38}$$

The Gaussian Quadrature is particularly chosen in FEM, due to its efficiency in the natural coordinate settings [12].

5.6 The Assembly Procedure

5.6.1 Consistent Summation

The global structural system is reconstructed by compiling the localized contributions from all elements through a **degree-of-freedom (DOF) consistent summation**:

$$\mathbf{K} = \sum_e \mathbf{k}_e, \quad \mathbf{M} = \sum_e \mathbf{m}_e, \quad \mathbf{F} = \sum_e \mathbf{f}_e \tag{5.39}$$

Here, the summation operator $\sum_e$ does not denote simple matrix addition. Instead, it represents a formal Boolean mapping operation. A local-to-global connectivity matrix (or topology array) expands the small element-level matrices (e.g., 4×4 or 6×6) into the dimensions of the global system ($N_{\text{DOF}} \times N_{\text{DOF}}$), placing each local term into its exact destination coordinate.

5.6.2 DOF Mapping and Sparsity

During the structural assembly phase:

- Each element matrix is initially calculated in an isolated, **local DOF ordering** based entirely on its local node numbers (e.g., nodes 1, 2, 3, 4).
- A global connectivity array translates these local identifiers into their unique **global DOF locations**.

This systematic mapping ensures that contributions from multiple elements sharing a common node are correctly accumulated (summed) into the same global matrix address. Because a single node is physically connected to only a tiny fraction of the thousands of elements within a large mesh, the resulting global matrices (**K** and **M**) possess distinct mathematical traits:

- They are **highly sparse** (filled mostly with zeros), i.e., $\mathbf{K}_{ij} = 0$, if nodes i and j are far apart.
- They form a highly structured sparsity pattern, typically resulting in a **banded** or **block-sparse** distribution centered along the main diagonal.

This sparse structure is computationally vital. It allows specialized sparse direct solvers (such as Skyline, Frontal, or MUMPS) and iterative algorithms (such as Conjugate Gradient) to bypass the zero entries entirely, saving immense amounts of memory and reducing solution times from $O(n^3)$ to nearly $O(n)$. Many of these solvers are used in Numpy linear algebra modules, some of which will be used in our FEM solvers in the following chapters.

5.6.3 Points to note

1. **Implicit Continuity:** The assembly procedure enforces inter-element displacement continuity implicitly. By forcing shared nodes to share the exact same global DOF value, the elements are mathematically "glued" together. This is made possible because of the assumed displacements are continuous.

2. **Symmetry Preservation:** Because the transformation arrays map rows and columns identically, the structural symmetry of $\mathbf{k}_e$ is perfectly preserved during global assembly, ensuring $\mathbf{K} = \mathbf{K}^\top$.

3. **Mass Matrix Options:** The structural mass distribution depends heavily on your formulation choice. The expression in Eq. (5.34) yields a **consistent mass matrix**, which mirrors the exact mathematical distribution of the shape functions. Alternatively, engineers often use a **lumped mass matrix**, which diagonalizes $\mathbf{M}$ by allocating the element mass directly to the nodes, vastly accelerating explicit dynamic time-integration solvers.

5.7 The Static Limit

The full governing equation system developed in Eq. (5.22) accounts for the transient elastodynamics of a solid body where inertial forces ($\mathbf{M}\ddot{\mathbf{D}}$) play an active role.

For steady-state or **static problems**, loads are applied so slowly that acceleration effects become negligible ($\ddot{\mathbf{D}} \approx \mathbf{0}$). Dropping the inertial term reduces the second-order differential equation system to a timeless, linear algebraic system:

$$\boxed{\mathbf{KD} = \mathbf{F}} \tag{5.40}$$

In this static limit, the problem simplifies from a complex time-marching integration scheme into a direct matrix inversion task. Resolving this linear system yields the final static nodal displacements $\mathbf{D}$ of the elastic solids under the given workloads.

In the following chapters, these FEM formulations will be implemented with Python codes for solids in 1D, 2D, and 3D, respectively.

5.8 Exercises

1. **Strong Form to Weak Form Derivation**

   Consider a 1D bar of length L, cross-sectional area A, and Young's modulus E, subjected to a distributed axial body force $b(x)$. The governing strong-form equilibrium equation is:

$$EA\frac{d^2u}{dx^2} + b(x) = 0 \quad \text{for } x \in (0, L)$$

The boundary conditions are $u(0) = 0$ (fixed end) and $EA\frac{du}{dx}\Big|_{x=L} = \bar{t}$ (applied pulling traction). 1. Define the appropriate space for the test function $\delta w(x)$. 2. Derive the complete weak form of this differential equation using the weighted residual method and integration by parts.

2. **Properties of the Differential Operator ($\mathbf{L}_d$)**

   In a 2D plane strain elasticity problem, the displacement vector is $\mathbf{u} = [u_x, u_y]^\top$ and the engineering strain vector is defined as $\varepsilon = [\varepsilon_{xx}, \varepsilon_{yy}, \gamma_{xy}]^\top$. 1. Explicitly write out the matrix entries of the differential strain-displacement operator $\mathbf{L}_d$ such that $\varepsilon = \mathbf{L}_d\mathbf{u}$. 2. If an element uses linear shape functions where $\mathbf{u}^h = \mathbf{ND}$, explain why the resulting strain field inside the element remains constant.

3. **Mathematical Spaces and Continuity**

   Explain why shape functions that are continuous across element boundaries (C^0 continuity) are legally admissible in the weak form equation derived in this chapter, but would fail if substituted directly into the strong form equilibrium equation (Eq. (5.1)). Frame your answer around the derivative demands of both forms.

4. **Compact Support and Sparsity**

   Suppose you have a 1D domain divided sequentially into 5 linear elements (numbered 1 to 5) connected by 6 nodes (numbered 1 to 6). 1. Sketch or mathematically define the global nodal shape function $N_3(x)$ associated with Node 3. 2. Identify the exact element domains over which $N_3(x)$ is non-zero. 3. If we evaluate the global stiffness matrix component $K_{16} = \int_\Omega \mathbf{B}_1^\top \mathbf{C}\mathbf{B}_6 \, d\Omega$, prove why its value must equal zero without performing any integration.

5. **Symmetry of the Stiffness Matrix (K)**

   Prove analytically that the element stiffness matrix $\mathbf{k}_e = \int_{\Omega_e} \mathbf{B}_e^\top \mathbf{C}\mathbf{B}_e \, d\Omega$ is symmetric ($\mathbf{k}_e = \mathbf{k}_e^\top$). State the exact underlying requirement that the material constitutive matrix $\mathbf{C}$ must satisfy for this proof to hold true.

6. **Consistent vs. Lumped Mass Matrices**

   Consider a single 1D bar element of length l, uniform cross-sectional area A, and constant mass density ρ. The local linear shape functions are $N_1(\xi) = \frac{1-\xi}{2}$ and $N_2(\xi) = \frac{1+\xi}{2}$ for $\xi \in [-1, 1]$. 1. Calculate the 2×2 **consistent mass matrix** $\mathbf{m}_e = \int_{\Omega_e} \rho \mathbf{N}_e^\top \mathbf{N}_e \, d\Omega$ using the coordinate mapping $dx = \frac{l}{2} d\xi$. 2. The **lumped mass matrix** strategy simplifies this matrix into a diagonal form by distributing the total element mass equally to the two nodes. Write down this lumped mass matrix and briefly discuss its computational advantage in explicit time-integration algorithms.

7. **Isoparametric Mapping and the Jacobian**

A 1D linear element is mapped from physical coordinates $x \in [x_1, x_2]$ to the natural coordinate domain $\xi \in [-1, 1]$. 1. Write down the coordinate interpolation expression $x(\xi) = N_1(\xi)x_1 + N_2(\xi)x_2$. 2. Derive the 1D Jacobian operator $J = \frac{dx}{d\xi}$ and verify that its determinant is equal to $l/2$ (where $l = x_2 - x_1$ is the element length). 3. Transform the integral $\int_{x_1}^{x_2} f(x)\, dx$ completely into natural coordinates.

8. **Essential vs. Natural Boundary Conditions**

   When assembling the global structural system equation $\mathbf{KD}+\mathbf{M\ddot{D}} = \mathbf{f}$: 1. Why are the traction conditions $\bar{\mathbf{t}}$ already accounted for within the structure of the force vector $\mathbf{f}$? 2. Why is it impossible to solve a static problem ($\mathbf{KD} = \mathbf{f}$) if no displacement boundary conditions ($\mathbf{u} = \bar{\mathbf{u}}$) are applied? Explain this in terms of the mathematical invertibility (singularity) of the unconstrained stiffness matrix $\mathbf{K}$.

9. **Global Assembly Logic**

   Imagine an element e is connected to global nodes i and j. Its local element stiffness matrix is calculated as:

$$\mathbf{k}_e = \begin{bmatrix} k_{11} & k_{12} \\ k_{21} & k_{22} \end{bmatrix}$$

   Assuming each node possesses only 1 degree of freedom, explicitly demonstrate how the components of $\mathbf{k}_e$ are mapped and added into the larger global stiffness matrix $\mathbf{K}$. Write out a symbolic snapshot of rows and columns i and j of the global matrix.

10. **The Semi-Discrete System and Time Integration (optional)**

   The dynamic system equation derived in this chapter is:

$$\mathbf{M\ddot{D}}(t) + \mathbf{KD}(t) = \mathbf{f}(t)$$

   1. Why is this equation referred to as a **semi-discrete** system rather than a fully discrete algebraic system? 2. If we approximate the acceleration vector at time step n using the central difference scheme:

$$\ddot{\mathbf{D}}_n \approx \frac{\mathbf{D}_{n+1} - 2\mathbf{D}_n + \mathbf{D}_{n-1}}{\Delta t^2}$$

   Substitute this approximation into the semi-discrete equation at time step n and rearrange it to isolate the unknown future displacement vector $\mathbf{D}_{n+1}$ on the left-hand side.

Chapter 6

FEM Solver for 1D Solids

```python
import importlib
from sympy import init_printing
import numpy as np
import sympy as sp
import matplotlib.pyplot as plt                              # For plots
from IPython.display import display, Math
from sympy import symbols, lambdify, latex, Matrix, Function

init_printing(use_unicode=True)     # For latex-like quality printing
np.set_printoptions(precision=4,suppress=True) # Digits in print-out
```

Chapter 3 introduced domain discretization for solid mechanics problems, including Python-based mesh generation. Chapter 4 developed the construction of shape functions for various element types, enabling the approximation of displacement fields. Chapter 5 presented FEM theory and the formation of discrete finite element equations.

This chapter presents a systematic derivation of the finite element formulation for a 1D two-node linear bar element, including:

- Displacement interpolation and the properties of linear shape functions
- Strain–displacement relations and the constant strain state
- Constitutive behavior based on Hooke's law
- Derivation of the element stiffness and mass matrices via energy principles
- Formulation of external nodal force vectors from body forces and surface tractions

These formulations are subsequently applied to develop a **1D FEM solver** to analyze a structural **1D bar** fixed at one end and subjected to axial loading, establishing a foundation for multi-dimensional finite element analysis.

This chapter covers the **core foundational principles** of the finite element method using the simplest one-dimensional axial bar configuration. Although the structural model itself is geometrically basic, this comprehensive analysis has explicitly revealed the major fundamental properties:

1. Nodal solution exactness (unique to 1D)
2. Reproducibility of FEM models (applicable to all dimensions)
3. Optimal convergence rates of Galerkin-based models (applicable to all dimensions).
4. Regularity analysis and convergence rates loss (applicable to all dimensions)

A thorough mastery of these one-dimensional mechanics will greatly accelerate your comprehension of the multi-dimensional formulations and models presented in the subsequent chapters for two-dimensional (2D) and three-dimensional (3D) continua.

6.1 Problem Setup

Consider a **1D bar element** of length l_e aligned along the local x-axis, as illustrated in Fig. 4.1. The geometric and physical domain of the element is bounded by its two endpoints, designated as nodes.

- **Node 1:** Positioned at the local coordinate $x = 0$, with an axial displacement denoted by u_1.
- **Node 2:** Positioned at the local coordinate $x = l_e$, with an axial displacement denoted by u_2.

The material and geometric properties of the element are defined by:

- **Young's Modulus (E):** Represents the material stiffness.
- **Cross-Sectional Area (A):** Assumed to be constant over the element length.
- **Mass Density (ρ):** Represents the material mass per unit volume.

6.2 Element Matrix Formulation

The state of deformation within the element is uniquely determined by its nodal degrees of freedom. The element nodal displacement vector $\mathbf{d}_e$ is expressed in vector notation as:

$$\mathbf{d}_e = \begin{bmatrix} u_1 \\ u_2 \end{bmatrix} \tag{6.1}$$

6.2.1 Displacement Approximation

To evaluate the continuous displacement field $u(x)$ anywhere within the element domain $x \in [0, l_e]$, a spatial interpolation scheme is constructed using the discrete nodal values. The displacement field is approximated via **linear shape functions**:

$$u(x) = N_1(x)u_1 + N_2(x)u_2 \tag{6.2}$$

where the **nodal shape function** N_1 is for node 1, and N_2 is for node 2:

$$N_1(x) = 1 - \frac{x}{l_e}, \qquad N_2(x) = \frac{x}{l_e} \tag{6.3}$$

These shape functions satisfy the essential **Kronecker delta property**, ensuring that the interpolated displacement matches the exact nodal values at the boundaries:

$$N_i(x_j) = \delta_{ij} = \begin{cases} 1 & \text{if } i = j \\ 0 & \text{if } i \neq j \end{cases} \tag{6.4}$$

They satisfy also the partition of unity property ($\sum_{i=1}^{2} N_i(x) = 1$), which is a strict prerequisite for capturing rigid-body translations without inducing spurious strain states.

Furthermore, these shape functions have the linearity property, meaning that it is capable of produce an arbitrary linear displacement field.

In compact matrix form, Eq.(6.2) is rewritten as:

$$u(x) = \underset{1\times 2}{\mathbf{N}_e}\ \underset{2\times 1}{\mathbf{d}_e} \tag{6.5}$$

where the element shape function matrix $\mathbf{N}_e$ is formed by the two nodal shape functions:

$$\mathbf{N}_e = \begin{bmatrix} N_1(x) & N_2(x) \end{bmatrix} \tag{6.6}$$

6.2.2 Strain–Displacement Relation

Under small deformation assumptions in 1D continuum mechanics, the axial engineering strain ε is defined by the spatial derivative of the axial displacement field:

$$\varepsilon = \frac{du}{dx} \tag{6.7}$$

Substituting the discretized displacement approximation into this kinematic relation yields:

$$\varepsilon = \frac{d}{dx}\left(\mathbf{N}_e\mathbf{d}_e\right) = \left(\frac{d\mathbf{N}_e}{dx}\right)\mathbf{d}_e \tag{6.8}$$

This maps the continuous strain field directly to the discrete nodal displacements via the strain–displacement matrix $\mathbf{B}_e$:

$$\varepsilon = \underset{1\times 2}{\mathbf{B}_e}\ \underset{2\times 1}{\mathbf{d}_e} \tag{6.9}$$

where the operator matrix $\mathbf{B}_e$ is defined by applying the differential operator vector $\mathbf{L}_d = \frac{d}{dx}$ to the shape function vector:

$$\mathbf{B}_e = \mathbf{L}_d\mathbf{N}_e = \frac{d\mathbf{N}_e}{dx} = \begin{bmatrix} \frac{dN_1}{dx} & \frac{dN_2}{dx} \end{bmatrix} \tag{6.10}$$

Evaluating the derivatives of Eq. (6.6) with respect to x, we arrive at the explicit strain–displacement matrix for the 1D bar element:

$$\underset{1\times 2}{\mathbf{B}_e} = \begin{bmatrix} \underbrace{-\frac{1}{l_e}}_{B_1} & \underbrace{\frac{1}{l_e}}_{B_2} \end{bmatrix} \tag{6.11}$$

where B_1 is for node 1, and B_2 is for node 2. Both are constant depending only on the geometry (the length) of the element.

> **Key Observation:** Because the spatial derivatives of the linear shape functions are independent of the coordinate x, the strain–displacement matrix $\mathbf{B}_e$ is constant throughout the element domain. Consequently, the internal strain field ε exhibits a **uniform strain distribution**. This specific class of element is therefore fundamentally categorized as a **Constant Strain Element (CSE)**.

6.2.3 Constitutive Relation

For a 1D bar under axial loading, the constitutive behavior of the material is governed by the one-dimensional Hooke's law for linear isotropic elasticity. This relation maps the kinematic axial strain to the corresponding internal axial stress:

$$\sigma = E\varepsilon \tag{6.12}$$

where E represents the Young's modulus of the material. Within the generalized matrix framework of the finite element method, the material constitutive matrix $\mathbf{C}$ simplifies to a scalar quantity:

$$\mathbf{C} = E \tag{6.13}$$

6.2.4 Element Stiffness Matrix

The element stiffness matrix characterizes the internal resistance of the element to deformation. Derived systematically from the weak form of the governing differential equation (or equivalently, via the Principle of Minimum Total Potential Energy discussed in the previous chapter), the general formulation is expressed as:

$$\mathbf{k}_e = \int_{\Omega_e} \mathbf{B}_e^T \mathbf{C} \mathbf{B}_e \, d\Omega \tag{6.14}$$

For a 1D bar element with a uniform cross-sectional area A, the differential volume element is $d\Omega = A\,dx$. Substituting the scalar constitutive relation yields:

$$\underset{2\times 2}{\mathbf{k}_e} = \int_0^{l_e} EA \underset{2\times 1}{\mathbf{B}_e^T} \underset{1\times 2}{\mathbf{B}_e} \, dx \tag{6.15}$$

As demonstrated in the previous section, the strain–displacement matrix $\mathbf{B}_e$ is independent of the spatial coordinate x for a linear element. Assuming the material and geometric properties (E and A) are also uniform across the element length, the integrand becomes constant and can be factored outside the integral:

$$\mathbf{k}_e = EA\mathbf{B}_e^T\mathbf{B}_e \int_0^{l_e} dx = EAl_e\mathbf{B}_e^T\mathbf{B}_e \tag{6.16}$$

Recalling the explicit definition of $\mathbf{B}_e$ from Eq. (6.11), the matrix product $\mathbf{B}_e^T\mathbf{B}_e$ is expanded as:

$$\mathbf{B}_e^T\mathbf{B}_e = \begin{bmatrix} -\frac{1}{l_e} \\ \frac{1}{l_e} \end{bmatrix} \begin{bmatrix} -\frac{1}{l_e} & \frac{1}{l_e} \end{bmatrix} = \frac{1}{l_e^2} \begin{bmatrix} 1 & -1 \\ -1 & 1 \end{bmatrix} \tag{6.17}$$

Evaluating the product yields the final, explicit form of the **element stiffness matrix** for a 2-node linear bar:

$$\boxed{\mathbf{k}_e = \frac{EA}{l_e} \begin{bmatrix} 1 & -1 \\ -1 & 1 \end{bmatrix}} \tag{6.18}$$

6.2.5 Properties of $\mathbf{k}_e$

- **Symmetry:** The matrix is symmetric ($\mathbf{k}_e = \mathbf{k}_e^T$). This is guaranteed by the self-adjoint property of the underlying differential operator and the Galerkin projection used in the variational formulation.
- **Singularity:** The determinant of the matrix is identically zero ($\det(\mathbf{k}_e) = 0$). This singularity indicates that the element is unconstrained in space, meaning that displacement boundary conditions have not yet been enforced to eliminate rigid-body motion.

- **Zero Row and Column Sums:** The sum of elements in any row or column equals zero. Physically, this guarantees that a rigid-body translation (where $u_1 = u_2$) results in zero internal nodal forces ($\mathbf{k}_e \mathbf{d}_e = \mathbf{0}$), confirming that no spurious strain energy is generated during rigid displacement.

6.2.6 Element Mass Matrix

For dynamic analysis, vibration, or wave propagation problems, the inertial properties of the element must be represented. The formula for the consistent element mass matrix is (see chapter 5):

$$\underset{2\times 2}{\mathbf{m}_e} = \int_{\Omega_e} \rho \mathbf{N}_e^T \mathbf{N}_e \, d\Omega \tag{6.19}$$

Substituting $d\Omega = A\,dx$ and expanding the product of the shape function row vector $\mathbf{N}_e$ yields:

$$\begin{aligned} \mathbf{m}_e &= \rho A \int_0^{l_e} \begin{bmatrix} N_1(x) \\ N_2(x) \end{bmatrix} \begin{bmatrix} N_1(x) & N_2(x) \end{bmatrix} dx \\ &= \rho A \int_0^{l_e} \begin{bmatrix} N_1^2 & N_1 N_2 \\ N_1 N_2 & N_2^2 \end{bmatrix} dx \end{aligned} \tag{6.20}$$

Moving the integral inside the matrix, and evaluating these definite integrals using the explicit linear shape functions results in:

$$\int_0^{l_e} \left(1 - \frac{x}{l_e}\right)^2 dx = \frac{l_e}{3} \int_0^{l_e} \left(\frac{x}{l_e}\right)^2 dx = \frac{l_e}{3} \int_0^{l_e} \left(1 - \frac{x}{l_e}\right)\left(\frac{x}{l_e}\right) dx = \frac{l_e}{6} \tag{6.21}$$

Assembling these components provides the explicit consistent element mass matrix:

$$\boxed{\mathbf{m}_e = \frac{\rho A l_e}{6} \begin{bmatrix} 2 & 1 \\ 1 & 2 \end{bmatrix}} \tag{6.22}$$

6.2.7 Properties of $\mathbf{m}_e$

- **Symmetry and Positive Definiteness:** The matrix is symmetric and strictly positive definite, which satisfies the physical requirement that any non-zero velocity field must produce positive kinetic energy.

- **Conservation of Mass:** Summing all terms in the matrix yields the exact total physical mass of the bar element, verifying that the continuous mass is fully preserved in the discrete formulation:

$$\sum_{i=1}^{2} \sum_{j=1}^{2} m_{ij} = \frac{\rho A l_e}{6}(2 + 1 + 1 + 2) = \rho A l_e \tag{6.23}$$

6.2.8 Lumped Mass Matrix

In explicit dynamic integration schemes, calculating the inverse of a large, coupled global mass matrix is computationally restrictive. To bypass this, a diagonal approximation known as the lumped mass matrix is frequently employed. It is constructed by splitting the total element mass to the two nodes of the element:

$$\mathbf{m}_e^{\text{lumped}} = \frac{\rho A l_e}{2} \begin{bmatrix} 1 & 0 \\ 0 & 1 \end{bmatrix} \tag{6.24}$$

Physical Interpretation: The continuous mass distribution is simplified into two discrete point masses, where each node carries exactly half of the total physical mass of the element. This eliminates the cross-coupling terms and simplifies inversion, but introduces a slight algorithmic damping effect that alters high-frequency wave propagation behaviors.

6.2.9 Nodal Force Vector

In an FEM model, external **distributed loads** over the element must be transformed into energetically equivalent **discrete forces** acting at the nodal degrees of freedom.

6.2.9.1 Body Force $b(x)$

The equivalent nodal force vector originating from a distributed body force $b(x)$ is determined via the virtual work done across the element domain (see Chapter 5):

$$\mathbf{f}_{b,e} = \int_0^{l_e} \mathbf{N}_e^T A b(x)\, dx \tag{6.25}$$

For a **uniformly distributed** constant body force intensity ($b(x) = b$), the integral simplifies directly:

$$\mathbf{f}_{b,e} = Ab \int_0^{l_e} \begin{bmatrix} 1 - \frac{x}{l_e} \\ \frac{x}{l_e} \end{bmatrix} dx \tag{6.26}$$

Moving the integral inside the matrix, and evaluating the components yields the classical statically equivalent distribution:

$$\boxed{\mathbf{f}_{b,e} = \frac{Abl_e}{2} \begin{bmatrix} 1 \\ 1 \end{bmatrix}} \tag{6.27}$$

Physical Insight: The total body force Abl_e is partitioned evenly between the two nodes of this linear element. This is true, as long as $\mathbf{N}_e$ is linear and the distributed force is constant.

6.2.9.2 Concentrated End Traction

When a point load or an external boundary traction vector T is applied directly at a node, it enters the nodal force vector directly. For instance, if an axial tensile force T acts at node 2 ($x = l_e$), we simply have:

$$\mathbf{f}_{t,e} = \begin{bmatrix} N_1(l_e) \\ N_2(l_e) \end{bmatrix} T = \begin{bmatrix} 0 \\ 1 \end{bmatrix} T = \begin{bmatrix} 0 \\ T \end{bmatrix} \tag{6.28}$$

Here we used the Kronecker delta property of the shape function.

6.2.10 Points to Note

1. The two-node linear element produces a **piecewise constant strain and stress field** within each element.
2. It reproduces **exactly** any linear displacement field and, therefore, passes the patch test. This guarantees convergence toward the exact solution for bar problems without body force, i.e., $b(x) = 0$.
3. It ensures **global equilibrium**. Although the approximate stress field may be discontinuous across element interfaces, equilibrium is satisfied in an average sense over the entire bar structure.

The following examples will confirm these properties.

6.3 Example: A Loaded Bar with Uniform Body Force

We now apply the finite element formulations developed in the preceding sections to solve a classical one-dimensional boundary value problem (BVP) in elastostatics. This example highlights how the discrete system approximates a continuous field and illustrates the underlying properties of the Galerkin approach.

6.3.1 Problem Statement

Consider a 1D structural bar of length L, fixed rigidly at its left boundary ($x = 0$) and subjected to the following concurrent loading conditions:

1. A **uniformly distributed constant body force** b (defined as force per unit volume) acting along the entire longitudinal domain of the bar.
2. A **concentrated tensile point force** P applied directly at the free right boundary ($x = L$).

The material behavior and geometric characteristics are defined by a constant Young's modulus E and a uniform cross-sectional area A.

For this 1D bar problem, the strong-form governing equation is a second-order ordinary differential equation (ODE) with a constant source term (see Chapter 2). Integrating the equation twice yields the exact analytical solution for the displacement field, which is a second-order (quadratic) polynomial [4]. The complete set of exact analytical expressions is summarized in the table below.

6.3.2 Exact Solution Summary [4]

Physical Quantity	Analytical Expression	Physical Description
Governing Equation	$\frac{d^2u}{dx^2} = -\frac{b}{E}$	Strong form axial equilibrium with constant body force intensity
Displacement $u(x)$	$u(x) = -\frac{b}{2E}x^2 + \left(\frac{P}{EA} + \frac{bL}{E}\right)x$	Quadratic spatial variation along the longitudinal axis
Strain $\varepsilon(x)$	$\varepsilon(x) = \frac{du}{dx} = \frac{P}{EA} + \frac{b}{E}(L - x)$	Linear distribution satisfying kinematic compatibility
Stress $\sigma(x)$	$\sigma(x) = E\varepsilon(x) = \frac{P}{A} + b(L - x)$	Linearly decreasing profile from the fixed root to the tip
Stress at Root ($x = 0$)	$\sigma(0) = \frac{P}{A} + bL$	Maximum normal stress, occurring at the fully constrained boundary
Stress at Tip ($x = L$)	$\sigma(L) = \frac{P}{A}$	Boundary stress matching the applied external traction vector
Tip Displacement $u(L)$	$u(L) = \frac{PL}{EA} + \frac{bL^2}{2E}$	Total global elongation of the bar under combined loading
Reaction Force R	$R = -(P + AbL)$	Concentrated structural reaction balancing total external loads

These exact solutions will be used for comparison studies with FEM solutions.

6.3.3 Observations from the Exact Solutions

1. The exact displacement field is **quadratic**.
2. Both the exact strain and stress fields vary **linearly** across the domain. Their maximum magnitudes occur at the **fixed root** ($x = 0$), where the structure

must support both the concentrated tip load P and the total integrated effect of the distributed body force.

3. **Global equilibrium:** The analytical reaction force R exactly balances the total external load acting on the structure, consisting of the concentrated load P and the total distributed body force AbL.

6.3.4 FEM Solution with One Element

We first model the entire bar domain using a single linear finite element of length $l_e = L$. This choice allows us to inspect the performance of the formulation under the coarsest possible discretization.

6.3.4.1 Single-element Model

The single-element mesh contains two discrete nodes aligned with the global coordinate boundaries:

- **Node 1 ($x = 0$):** Fully constrained due to the fixed root condition, meaning $u_1 = 0$.
- **Node 2 ($x = L$):** Positioned at the free tip, with an unknown axial displacement u_2.

The corresponding element nodal displacement vector is defined as:

$$\mathbf{d}_e = \begin{bmatrix} u_1 \\ u_2 \end{bmatrix} \tag{6.29}$$

6.3.4.2 Element Stiffness Matrix

Applying the formulation for a uniform cross-section and material domain, the element stiffness matrix matches the global matrix before boundary conditions are enforced:

$$\mathbf{k}_e = \frac{EA}{L} \begin{bmatrix} 1 & -1 \\ -1 & 1 \end{bmatrix} \tag{6.30}$$

6.3.4.3 Element Force Vector

The total nodal force vector is obtained by superimposing the contributions from the distributed internal domain forces and the external discrete boundary loads.

(a) Body Force Contribution
Evaluating the energetically equivalent nodal force vector for a constant body force intensity b over the element domain $l_e = L$ yields:

$$\mathbf{f}^{(b)} = \frac{AbL}{2} \begin{bmatrix} 1 \\ 1 \end{bmatrix} \tag{6.31}$$

Clearly, the total body force Abl_e is equally split to the two nodes of the bar (element).

(b) Concentrated Point Force
The external tensile point load P is applied directly to the degree of freedom at the tip (node 2):

$$\mathbf{f}^{(P)} = \begin{bmatrix} 0 \\ P \end{bmatrix} \tag{6.32}$$

(c) Reaction Force at the Root Since the root of the bar is constrained, a reaction force must develop there to enforce the prescribed displacement boundary condition. The magnitude of this reaction force is generally unknown prior to solving the problem. We therefore denote the reaction force at Node 1 by R, giving the corresponding nodal force vector

$$\mathbf{f}^{(R)} = \begin{bmatrix} R \\ 0 \end{bmatrix} \tag{6.33}$$

(d) Total Element Force Vector
Summing the distinct load contributions gives the net force vector acting on the element:

$$\mathbf{f}_e = \mathbf{f}^{(b)} + \mathbf{f}^{(P)} + \mathbf{f}^{(R)} = \begin{bmatrix} \frac{AbL}{2} + R \\ \frac{AbL}{2} + P \end{bmatrix} \tag{6.34}$$

6.3.4.4 Global System of Equations

Since the model consists of only one element, the element matrix equations represent the unconstrained global system directly:

$$\mathbf{k}_e \mathbf{d}_e = \mathbf{f}_e \tag{6.35}$$

or

$$\frac{EA}{L} \begin{bmatrix} 1 & -1 \\ -1 & 1 \end{bmatrix} \begin{bmatrix} u_1 \\ u_2 \end{bmatrix} = \begin{bmatrix} \frac{AbL}{2} + R \\ \frac{AbL}{2} + P \end{bmatrix} \tag{6.36}$$

6.3.4.5 Application of Essential Boundary Conditions

The structural constraint at the left support requires enforcing the homogeneous essential boundary condition:

$$u_1 = 0 \tag{6.37}$$

Applying this constraint simplifies the second row equation of the system matrix in Eq. (6.36) to:

$$\frac{EA}{L}(u_2 - 0) = \frac{AbL}{2} + P \tag{6.38}$$

This is the reduced equation to find the unknown u_2. It can be seen that, although the reaction force R is unknown, its value is not required to solve the system. This is because the displacement u_1 is prescribed the first equilibrium equation is therefore excluded from the set of unknown equations.

This illustrates a fundamental **complementary characteristic** of solid mechanics problems: when the displacement is prescribed at a boundary, the corresponding reaction force is generally unknown and must be determined as part of the solution. Conversely, when a force (traction) is prescribed, the corresponding displacement is unknown [4].

6.3.4.6 Solution for Displacement Field

Solving for the single unknown yields:

$$u_2 = \frac{L}{EA}\left(\frac{AbL}{2} + P\right) = \boxed{\frac{PL}{EA} + \frac{bL^2}{2E}} \tag{6.39}$$

Substituting this nodal value and $u_1 = 0$ back into Eq.(6.2) gives

$$\boxed{u(x) = \left(\frac{P}{EA} + \frac{bL}{2E}\right) x} \tag{6.40}$$

This single-element FEM model produces a linear displacement field. Since the exact analytical solution is quadratic, the FEM solution is clearly an approximation.

6.3.4.7 Exact Nodal Value Property

However, comparing the discrete tip displacement u_2 in Eq.(6.39) with the exact analytical solution evaluated at $x = L$ shows that they are identical. This reveals an important property of standard finite element formulations in one dimension:

> **The Nodal Exactness Property:** For one-dimensional structural and solid mechanics problems governed by self-adjoint ordinary differential equations, the standard Galerkin finite element method produces mathematically exact solutions at the discrete nodal locations.

We will show later that this property holds:

1. regardless of the mesh density,
2. regardless of the order of the 1D elements (such as quadratic elements), and
3. regardless of the body force type, provided that its exact energy-equivalent nodal force vector can be obtained.

This property does **not** generally hold for two- and three-dimensional problems.

6.3.4.8 Reaction Force at the Constrained Root

With the active nodal displacement solved, the structural reaction force R_1 developed at the fixed support (Node 1) is recovered from the first row equation of the global system in Eq. (6.36):

$$\frac{AbL}{2} + R = \frac{EA}{L}(u_1 - u_2) \tag{6.41}$$

Enforcing the kinematic constraint $u_1 = 0$ and substituting the derived value for u_2 yields:

$$R = -\frac{EA}{L}\left[\frac{L}{EA}\left(\frac{AbL}{2} + P\right)\right] - \frac{AbL}{2} \tag{6.42}$$

$$\boxed{R = -P - AbL} \tag{6.43}$$

which is the exact solution for the reaction force.

6.3.4.9 Strain and Stress Fields

The engineering axial strain remains constant across the single linear element:

$$\varepsilon = \frac{du}{dx} = \frac{u_2 - u_1}{L} = \frac{u_2}{L} \tag{6.44}$$

$$\varepsilon = \frac{P}{EA} + \frac{bL}{2E} \tag{6.45}$$

Applying Hooke's law, the corresponding constant axial normal stress is:

$$\boxed{\sigma = E\varepsilon = \frac{P}{A} + \frac{bL}{2}} \tag{6.46}$$

6.3.4.10 Summary of Accuracy and Limitations

1. **Displacement Field:** The tip displacement is **exact**. However, the internal field is linear, underestimating the true quadratic curvature induced by non-zero body forces.

2. **Stress Field:** The calculated stress is **approximate**, producing a single constant value over the entire domain. Physically, this constant value matches the exact analytical linear stress distribution at the element's midpoint ($x = L/2$).

3. **Reproducibility:** If the distributed body force is zero ($b = 0$), the exact analytical solution simplifies to a linear profile. In that specific case, the single-element finite element solution matches the exact analytical continuum solution everywhere across the entire domain.

To capture strong gradients or linear stress profiles when $b \neq 0$, a multi-element discretization (h-refinement) or higher-order shape functions (p-refinement) are required.

6.3.5 FEM Solution with Two Elements

We now solve the identical boundary value problem using a two-element discretization. This case study shall confirm the nodal exactness property and also highlight how localized element matrices are integrated into a unified global system while preserving degree-of-freedom (DOF) consistency.

6.3.5.1 Two-element Model

The continuous domain of length L is partitioned into two identical, linear elements of equal length $l_e = L/2$, as illustrated schematically in Fig. 6.1.

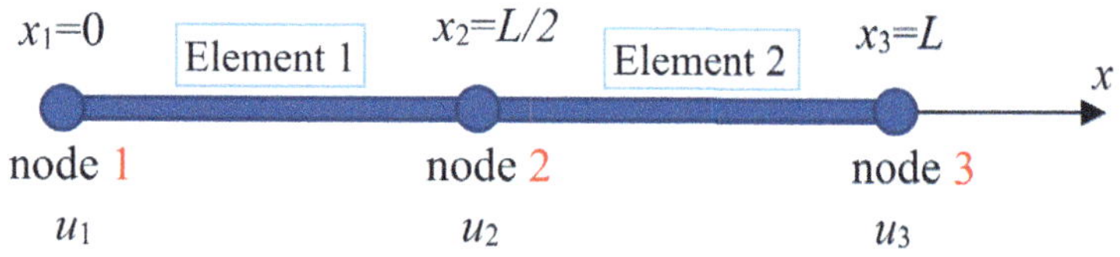

Figure 6.1: A uniform bar discretized with two equal linear elements.

:

- **Element 1 Domain:** $x \in [0, L/2]$, defined by its local boundary nodes (1, 2).
- **Element 2 Domain:** $x \in [L/2, L]$, defined by its local boundary nodes (2, 3).

This layout establishes three unique global nodes across the span, associated with the following kinematic boundary conditions and variables:

- **Node 1 ($x = 0$):** Fully constrained root support, yielding $u_1 = 0$.
- **Node 2 ($x = L/2$):** Internal structural node, possessing an unknown displacement u_2.
- **Node 3 ($x = L$):** Free tip boundary node, possessing an unknown displacement u_3.

The global displacement vector containing all structural degrees of freedom is expressed as:

$$\mathbf{d} = \begin{bmatrix} u_1 \\ u_2 \\ u_3 \end{bmatrix} \tag{6.47}$$

6.3.5.2 Element Stiffness Matrix

Because the two elements share identical physical properties (E, A) and geometric lengths $(l_e = L/2)$, their localized element stiffness matrices are mathematically identical:

$$\mathbf{k}^{(1)} = \mathbf{k}^{(2)} = \frac{EA}{l_e}\begin{bmatrix} 1 & -1 \\ -1 & 1 \end{bmatrix} = \frac{2EA}{L}\begin{bmatrix} 1 & -1 \\ -1 & 1 \end{bmatrix} \tag{6.48}$$

6.3.5.3 Element Body Force Vector

Similarly, calculating the consistent nodal load vectors for a uniform body force intensity b across each localized domain of length $l_e = L/2$ yields:

$$\mathbf{f}^{(1)} = \mathbf{f}^{(2)} = \frac{Abl_e}{2}\begin{bmatrix} 1 \\ 1 \end{bmatrix} = \frac{AbL}{4}\begin{bmatrix} 1 \\ 1 \end{bmatrix} \tag{6.49}$$

Consistent with the principles of linear interpolation, the total distributed mass/force within each discrete sub-domain $(AbL/2)$ is split equally between its respective bounding nodes.

6.3.5.4 Assembly of the Global Stiffness Matrix

The transition from localized element matrices to the interconnected global matrix requires mapped transformation based on element connectivity topology:

- **Element 1** maps to global DOFs **(1, 2)**:

$$k_{11}^{(1)} \to K_{11}, \quad k_{12}^{(1)} \to K_{12}, \quad k_{21}^{(1)} \to K_{21}, \quad k_{22}^{(1)} \to K_{22} \tag{6.50}$$

- **Element 2** maps to global DOFs **(2, 3)**:

$$k_{11}^{(2)} \to K_{22}, \quad k_{12}^{(2)} \to K_{23}, \quad k_{21}^{(2)} \to K_{32}, \quad k_{22}^{(2)} \to K_{33} \tag{6.51}$$

Because global node 2 serves as the shared physical interface connecting both elements, its corresponding main diagonal entry receives additive contributions from both localized domains $(K_{22} = k_{22}^{(1)} + k_{11}^{(2)})$.

A mathematically formal and intuitive way to visualize this assembly architecture is to expand each localized matrix $\mathbf{k}^{(e)}$ to the full dimensions of the global structural topology (3×3) by mapping unreached degrees of freedom with zero entries:

For Element 1:

$$\mathbf{k}_{e1} = \frac{2EA}{L}\begin{bmatrix} 1 & -1 & 0 \\ -1 & 1 & 0 \\ 0 & 0 & 0 \end{bmatrix} \tag{6.52}$$

For Element 2:

$$\mathbf{k}_{e2} = \frac{2EA}{L}\begin{bmatrix} 0 & 0 & 0 \\ 0 & 1 & -1 \\ 0 & -1 & 1 \end{bmatrix} \tag{6.53}$$

.

Summing these expanded representations yields the complete, unconstrained global stiffness matrix:

$$\mathbf{K} = \mathbf{k}_{e1} + \mathbf{k}_{e2} = \frac{2EA}{L}\begin{bmatrix} 1 & -1 & 0 \\ -1 & 2 & -1 \\ 0 & -1 & 1 \end{bmatrix} \tag{6.54}$$

> **Algorithmic Note:** This intuitive expansion and addition process serves as an educational visualization technique but is computationally inefficient for large systems due to excessive zero-allocation storage operations [12]. In actual finite element codes, this structural mapping is executed via **direct assembly**—looping through elements and adding entries directly into the global array using a local-to-global pointer array (topology matrix). This optimized direct assembly technique is implemented in the Python scripts provided at the end of this chapter.

6.3.5.5 Assembly of the Global Force Vector

Assembled Body Force Contribution

Mapping the localized consistent element body forces to their matching global coordinates yields:

- Element 1 contributing to global DOFs (1, 2)
- Element 2 contributing to global DOFs (2, 3)

Summing these overlapping vectors leads to the unified global body force vector:

$$\mathbf{f}^{(b)} = \begin{bmatrix} \frac{AbL}{4} \\ \frac{AbL}{4} + \frac{AbL}{4} \\ \frac{AbL}{4} \end{bmatrix} = \begin{bmatrix} \frac{AbL}{4} \\ \frac{AbL}{2} \\ \frac{AbL}{4} \end{bmatrix} \tag{6.55}$$

Following the assembly of these two linear elements, the internal node 2 accumulates twice the force intensity compared to the boundary nodes. Because the finite element method maintains **energy equivalence**, the work performed by these discrete point forces equals the exact continuous work done by the true distributed load under any admissible virtual displacement field, as proven in the previous chapter.

Concentrated Tip Force

The discrete external point load P acts exclusively at the tip of the structure, mapping directly to global Node 3:

$$\mathbf{f}^{(P)} = \begin{bmatrix} 0 \\ 0 \\ P \end{bmatrix} \tag{6.56}$$

> **Tip:** In engineering practice, when a concentrated point load acts on an internal or boundary location within a solid body, a node is deliberately placed at that precise spatial coordinate during mesh generation to permit direct allocation into the force vector.

Combined Global Force Vector

Superimposing the body force vector and the concentrated point load yields the final global force vector:

$$\mathbf{f} = \mathbf{f}^{(b)} + \mathbf{f}^{(P)} + \mathbf{f}^{(R)} = \begin{bmatrix} \frac{AbL}{4} + R \\ \frac{AbL}{2} \\ \frac{AbL}{4} + P \end{bmatrix} \tag{6.57}$$

The global system equation becomes

$$\frac{2EA}{L} \begin{bmatrix} 1 & -1 & 0 \\ -1 & 2 & -1 \\ 0 & -1 & 1 \end{bmatrix} = \begin{bmatrix} u_1 \\ u_2 \\ u_3 \end{bmatrix} = \begin{bmatrix} \frac{AbL}{4} + R \\ \frac{AbL}{2} \\ \frac{AbL}{4} + P \end{bmatrix} . \tag{6.58}$$

6.3.5.6 Application of Essential Boundary Conditions

Enforcing the constraint at the root support ($u_1 = 0$) permits the elimination of the first row and column of the global matrix equation. This reduction condenses the system into a stable, non-singular 2×2 matrix system with two active nodes 2 and 3:

$$\frac{2EA}{L} \begin{bmatrix} 2 & -1 \\ -1 & 1 \end{bmatrix} \begin{bmatrix} u_2 \\ u_3 \end{bmatrix} = \begin{bmatrix} \frac{AbL}{2} \\ \frac{AbL}{4} + P \end{bmatrix} \tag{6.59}$$

As shown again, the unknown R is not in the reduced system equation containing only the unknowns.

6.3.5.7 Algebraic Solution of the Discretized System

Multiplying both sides by the scalar factor $\frac{L}{2EA}$ isolates the algebraic matrix system:

$$\begin{bmatrix} 2 & -1 \\ -1 & 1 \end{bmatrix} \begin{bmatrix} u_2 \\ u_3 \end{bmatrix} = \begin{bmatrix} \frac{bL^2}{4E} \\ \frac{bL^2}{8E} + \frac{PL}{2EA} \end{bmatrix} \tag{6.60}$$

Using standard matrix inversion or Gaussian elimination (or other advanced linear algebraic equation solver for large systems), the foregoing equation can be solved:

$$\boxed{u_2 = \frac{PL}{2EA} + \frac{3bL^2}{8E}} \tag{6.61}$$

$$\boxed{u_3 = \frac{PL}{EA} + \frac{bL^2}{2E}} \tag{6.62}$$

These are the final nodal displacement solutions found for this example problem.

6.3.5.8 Analytical Comparison and Convergence Analysis

Displacement Field Evaluation

Evaluating the exact analytical quadratic displacement field derived earlier at the two active nodes reveals:

- At the internal node ($x = L/2$):

$$\begin{aligned} u_{\text{exact}}\left(\frac{L}{2}\right) &= -\frac{b}{2E}\left(\frac{L}{2}\right)^2 + \left(\frac{P}{EA} + \frac{bL}{E}\right)\left(\frac{L}{2}\right) \\ &= \frac{PL}{2EA} + \frac{3bL^2}{8E} \end{aligned} \tag{6.63}$$

- At the free tip node ($x = L$):

$$\begin{aligned} u_{\text{exact}}(L) &= -\frac{b}{2E}(L)^2 + \left(\frac{P}{EA} + \frac{bL}{E}\right)(L) \\ &= \frac{PL}{EA} + \frac{bL^2}{2E} \end{aligned} \tag{6.64}$$

The finite element nodal solutions u_2 and u_3 match the exact solutions identically. This reinforces the **Nodal Exactness Property** of 1D FEM model, demonstrating that the solution at the nodes remains exact regardless of whether a one-element or two-element model is deployed.

Furthermore, if the body force is removed ($b = 0$), the finite element solution matches the exact analytical field across the entire domain, confirming that the use of extra element can increase of the computational effort, the results will still be exact: **use of extra elements will not hurt the solution accuracy in theory**.

Stress and Strain Field Evaluation

While the displacement profile is continuous and piecewise linear, the derived axial stress field $\sigma = E\frac{du}{dx}$ is constant within each individual element, producing a discontinuous step-profile at the internal node:

- **Element 1 Domain ($x \in [0, L/2]$):**

$$\sigma_1 = E\left(\frac{u_2 - u_1}{l_e}\right) = E\left(\frac{u_2 - 0}{L/2}\right) = \boxed{\frac{P}{A} + \frac{3}{4}bL} \tag{6.65}$$

- **Element 2 Domain ($x \in [L/2, L]$):**

$$\sigma_2 = E\left(\frac{u_3 - u_2}{l_e}\right) = E\left(\frac{u_3 - u_2}{L/2}\right) = \boxed{\frac{P}{A} + \frac{1}{4}bL} \tag{6.66}$$

Comparing these discrete values to the exact linear stress distribution ($\sigma_{\text{exact}}(x) = \frac{P}{A} + b(L - x)$) reveals that the finite element model yields the mathematically exact stress value at the **midpoint of each element**:

- Element 1 Midpoint ($x = L/4$): $\sigma_{\text{exact}}(L/4) = \frac{P}{A} + \frac{3}{4}bL = \sigma_1$
- Element 2 Midpoint ($x = 3L/4$): $\sigma_{\text{exact}}(3L/4) = \frac{P}{A} + \frac{1}{4}bL = \sigma_2$

Readers are encouraged to check whether the reaction force R obtained from this two-element FEM model is exact or not.

> **Summary:** Moving from a single-element to a two-element model preserves nodal exactness while improving the internal displacement approximation and capturing the declining stress trend via a multi-step constant profile. This highlights the standard convergence of h-refinement using low-order elements.
>
> For structural problems smooth fields, an alternative strategy to enhance gradient accuracy is increasing the interpolation order of the elements (p-refinement).

6.3.6 FEM Solution with One Quadratic Element

To demonstrate the application of higher-order interpolation bases, we solve the identical boundary value problem using a single **three-node quadratic bar element** over the global domain $[0, l_e]$ (where $l_e = L$).

The structural layout consists of three distinct nodal points:

- **Node 1 ($x = 0$):** Fully constrained root support, $u_1 = 0$.
- **Node 2 ($x = l_e/2$):** Internal midpoint node, possessing an unknown displacement u_2.
- **Node 3 ($x = l_e$):** Free tip boundary node, possessing an unknown displacement u_3.

The element displacement vector enclosing the complete structural degrees of freedom is expressed as:

$$\mathbf{d}_e = \begin{bmatrix} u_1 \\ u_2 \\ u_3 \end{bmatrix} \tag{6.67}$$

6.3.6.1 Quadratic 1D Shape Functions

By mapping the physical coordinates to a standardized natural coordinate system defined by the non-dimensional parameter $\xi \in [-1, 1]$, the lagrangian quadratic shape functions are constructed as (see Chapter 4):

$$\begin{aligned} N_1(\xi) &= \frac{1}{2}\xi(\xi - 1) \\ N_2(\xi) &= 1 - \xi^2 \\ N_3(\xi) &= \frac{1}{2}\xi(\xi + 1) \end{aligned} \tag{6.68}$$

> **Methodological Note:** Although a direct physical-coordinate derivation is mathematically possible for this simple 1D problem, implementing the natural coordinate system is essential for developing general finite element formulations, treating non-uniform geometry, and executing automated numerical integration schemes such as Gauss–Legendre quadrature.

6.3.6.2 Isoparametric Mapping

The coordinate transformation between the physical space x and the natural space ξ is established by enforcing mapping pairs at the key structural nodes:

- **At Node 1:** Physical $x = 0$ maps to natural $\xi = -1$
- **At Node 2:** Physical $x = l_e/2$ maps to natural $\xi = 0$
- **At Node 3:** Physical $x = l_e$ maps to natural $\xi = 1$

This transformation topology is illustrated schematically in Fig. 6.2.

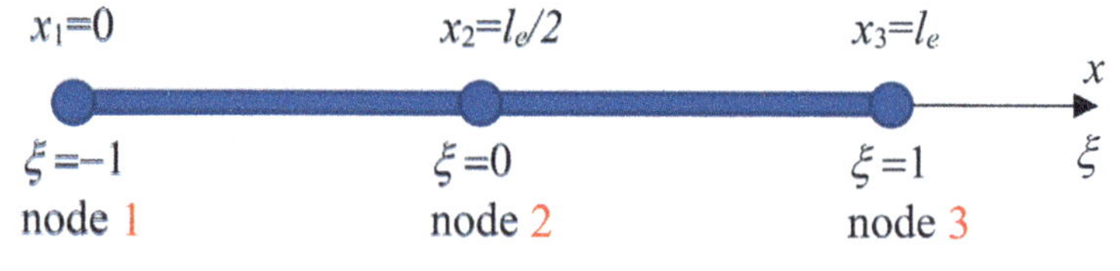

Figure 6.2: The mapping scheme between physical (x) and natural (ξ) coordinate spaces for a three-node quadratic 1D bar element.

Applying standard isoparametric interpolation, the coordinate mapping relation is given explicitly by:

$$x(\xi) = \sum_{i=1}^{3} N_i(\xi)x_i = \frac{l_e}{2}(1+\xi) \tag{6.69}$$

Differentiating this mapping yields the 1D Jacobian operator (J) and its inverse scaling relationship:

$$J = \frac{dx}{d\xi} = \frac{l_e}{2}, \qquad \frac{d\xi}{dx} = \frac{2}{l_e} \tag{6.70}$$

.

6.3.6.3 Strain–Displacement Matrix

Under small-strain assumptions, the axial engineering strain is defined by the spatial derivative:

$$\varepsilon = \frac{du}{dx} \tag{6.71}$$

Transforming the differential operator from physical to natural coordinates via the chain rule yields:

$$\frac{d}{dx} = \frac{d\xi}{dx}\frac{d}{d\xi} = \frac{2}{l_e}\frac{d}{d\xi} \tag{6.72}$$

The continuous internal strain field is linked directly to the discrete nodal displacement vector by:

$$\varepsilon = \underset{1\times 3}{\mathbf{B}_e(\xi)}\ \underset{3\times 1}{\mathbf{d}_e} \tag{6.73}$$

where the strain–displacement matrix $\mathbf{B}_e(\xi)$ is derived by differentiating the shape functions with respect to the natural coordinate:

$$\begin{aligned}\mathbf{B}_e(\xi) &= \frac{2}{l_e}\begin{bmatrix}\frac{dN_1}{d\xi} & \frac{dN_2}{d\xi} & \frac{dN_3}{d\xi}\end{bmatrix} \\ &= \begin{bmatrix}\frac{2\xi-1}{l_e} & -\frac{4\xi}{l_e} & \frac{2\xi+1}{l_e}\end{bmatrix} \\ &= \begin{bmatrix}\mathbf{B}_{e1} & \mathbf{B}_{e1} & \mathbf{B}_{e1}\end{bmatrix}\end{aligned} \tag{6.74}$$

The components $\mathbf{B}_{ei}$ correspond to the nodal submatrix of strain-displacement for each individual node i. For this 1D configuration, these nodal submatrix reduce to scalar algebraic functions. In higher-dimensional continuum mechanics or multi-DOF structural models, these components will be matrices of coupled spatial functions; boldface notation is preserved here to maintain system consistency.

> **Key Kinematic Observation:** Unlike the linear element formulation, the strain–displacement matrix $\mathbf{B}_e(\xi)$ is a linear function of the position variable ξ. This explicit dependency allows the single quadratic element to capture a **non-uniform, linearly varying strain field** directly within its internal domain.

6.3.6.4 Element Stiffness Matrix

The mathematical structure of the element stiffness matrix is defined by the variational volume integral:

$$\underset{3\times 3}{\mathbf{k}_e} = \int_{\Omega_e} \mathbf{B}_e^T E A \mathbf{B}_e \, d\Omega \tag{6.75}$$

Transforming this integral into natural coordinates by substituting the differential volume relation $d\Omega = A\,dx = A\frac{l_e}{2}d\xi$ yields:

$$\mathbf{k}_e = \int_{-1}^{1} \mathbf{B}_e^T E A \mathbf{B}_e \left(\frac{l_e}{2}\right) d\xi = \frac{EAl_e}{2}\int_{-1}^{1} \mathbf{B}_e^T \mathbf{B}_e \, d\xi \tag{6.76}$$

Expanding the vector inner product gives:

$$\underset{3\times 3}{\mathbf{k}_e} = \frac{EAl_e}{2}\begin{bmatrix} \int_{-1}^{1} \mathbf{B}_{e1}^{\top}\mathbf{B}_{e1}\,d\xi & \int_{-1}^{1} \mathbf{B}_{e1}^{\top}\mathbf{B}_{e2}\,d\xi & \int_{-1}^{1} \mathbf{B}_{e1}^{\top}\mathbf{B}_{e3}\,d\xi \\ \int_{-1}^{1} \mathbf{B}_{e2}^{\top}\mathbf{B}_{e1}\,d\xi & \int_{-1}^{1} \mathbf{B}_{e2}^{\top}\mathbf{B}_{e2}\,d\xi & \int_{-1}^{1} \mathbf{B}_{e2}^{\top}\mathbf{B}_{e3}\,d\xi \\ \int_{-1}^{1} \mathbf{B}_{e3}^{\top}\mathbf{B}_{e1}\,d\xi & \int_{-1}^{1} \mathbf{B}_{e3}^{\top}\mathbf{B}_{e2}\,d\xi & \int_{-1}^{1} \mathbf{B}_{e3}^{\top}\mathbf{B}_{e3}\,d\xi \end{bmatrix} \tag{6.77}$$

Because the element is quadratic, the components of $\mathbf{B}_e$ are algebraic functions of ξ, requiring integration across the domain. The following Sympy code gets this done:

```
# Integration to obtain ke:

xi = sp.symbols('xi', real=True)
E, A, le = sp.symbols('E A l_e', positive=True, real=True)    # L = le

B = (2/le)*sp.Matrix([[xi-sp.Rational(1,2),    # Define B (1x3 vector)
                     -2*xi,
                     xi + sp.Rational(1,2)]])
BTB = B.T * B                                         # B^T*B (3x3 matrix)
I = sp.integrate(BTB, (xi, -1, 1))         # Integrate over xi   [-1, 1]

ke = (E*A*le/2) * I                           # Element stiffness matrix
ke = sp.simplify(ke)                                          # Simplify

display(Math(f" \\text{{$k_e$=}} {latex(sp.simplify(ke))}"))
```

$$k_e = \begin{bmatrix} \frac{7AE}{3l_e} & -\frac{8AE}{3l_e} & \frac{AE}{3l_e} \\ -\frac{8AE}{3l_e} & \frac{16AE}{3l_e} & -\frac{8AE}{3l_e} \\ \frac{AE}{3l_e} & -\frac{8AE}{3l_e} & \frac{7AE}{3l_e} \end{bmatrix}$$

The explicit higher-order element stiffness matrix can be written as:

$$\boxed{\mathbf{k}_e = \frac{EA}{3l_e}\begin{bmatrix} 7 & -8 & 1 \\ -8 & 16 & -8 \\ 1 & -8 & 7 \end{bmatrix}} \tag{6.78}$$

6.3.6.5 Element Force Vector

Consistent Body Force Allocation

The equivalent nodal load vector generated by a uniformly distributed continuous body force b is found using:

$$\mathbf{f}^{(b)} = \int_{\Omega_e} \mathbf{N}^T \rho A b \, d\Omega = \int_0^{l_e} \mathbf{N}^T A b \, dx \tag{6.79}$$

Transforming the integration limits to natural space via $dx = \frac{l_e}{2} d\xi$ provides:

$$\underset{3\times1}{\mathbf{f}^{(b)}} = \frac{Abl_e}{2}\begin{bmatrix} \int_{-1}^{1} \frac{1}{2}\xi(\xi-1)\, d\xi \\ \int_{-1}^{1} (1-\xi^2)\, d\xi \\ \int_{-1}^{1} \frac{1}{2}\xi(\xi+1)\, d\xi \end{bmatrix} \tag{6.80}$$

All these integrals inside the vector can be evaluated using the following Sympy snippet.

```
# Evaluate integrals:
xi, le, A, b = sp.symbols('xi l_e A b', real=True)          # Symbols

# Shape function vector (quadratic 1D element)
NT = sp.Matrix([ sp.Rational(1,2)*xi*(xi - 1),
                1 - xi**2,
                sp.Rational(1,2)*xi*(xi + 1) ])

J = le/2                                   # Jacobian for 1D case: dx/dxi
integrand = NT * A * b * J                                # xi   [-1, 1]

# Perform integration component-wise
fe = sp.integrate(integrand, (xi, -1, 1))
fe = sp.simplify(fe)     # Simplify result

print(fe)
#display(Math(f" \\text{{$f_e$=}} {latex(fe)}"))
```

```
Matrix([[A*b*l_e/6], [2*A*b*l_e/3], [A*b*l_e/6]])
```

The exact energy-equivalent load vector is found as:

$$\boxed{\mathbf{f}^{(b)} = \frac{Abl_e}{6}\begin{bmatrix}1\\4\\1\end{bmatrix}} \tag{6.81}$$

Mechanical Interpretation: The quadratic interpolation scheme allocates a larger share of the distributed body force ($\frac{4}{6}$ or $\frac{2}{3}$ of the total load) directly to the central internal node (Node 2). This distribution differs fundamentally from the two-element linear assembly model ($[1/4, 1/2, 1/4]^\top$), highlighting how the choice of shape function space alters the localized allocation of distributed fields while maintaining global work equivalence.

Concentrated End Point Force

The discrete external point load P acts directly at the tip position (Node 3), thus:

$$\mathbf{f}^{(P)} = \begin{bmatrix}0\\0\\P\end{bmatrix} \tag{6.82}$$

Total Combined Element Force Vector

Superimposing both external force components yields the unified nodal load vector:

$$\mathbf{f}_e = \mathbf{f}^{(b)} + \mathbf{f}^{(P)} + \mathbf{f}^{(R)} = \begin{bmatrix}\frac{Abl_e}{6} + R\\ \frac{2Abl_e}{3}\\ \frac{Abl_e}{6} + P\end{bmatrix} \tag{6.83}$$

6.3.6.6 Application of Essential Boundary Conditions

Enforcing the essential boundary condition at the fixed support requires setting $u_1 = 0$. Eliminating the row and column channels associated with the constrained degree of freedom reduces the system to a non-singular 2×2 algebraic matrix system governing active nodes 2 and 3:

$$\frac{EA}{3l_e}\begin{bmatrix}16 & -8\\ -8 & 7\end{bmatrix}\begin{bmatrix}u_2\\ u_3\end{bmatrix} = \begin{bmatrix}\frac{2Abl_e}{3}\\ \frac{Abl_e}{6} + P\end{bmatrix} \tag{6.84}$$

The resulting reduced stiffness matrix is symmetric, positive definite, and well-conditioned for direct inversion. The solution can be found using the following snippet.

6.3.6.7 SymPy Code For Solving the System Equation

```
## SymPy Code For Solving the System Equation
E, A, le, b, P = sp.symbols('E A l_e b P', real=True) #Define symbols
u2, u3 = sp.symbols('u2 u3', real=True)

K = (E*A)/(3*le) * sp.Matrix([[16, -8],                    # Define the K matrix
                              [-8,  7] ])
F = sp.Matrix([2*A*b*le/3, A*b*le/6 + P ])                      # Force Vector
D = sp.Matrix([u2, u3])                                        # Unknown vector

# Solve for nodal displacement D:
solution_D = sp.solve(K*D - F, (u2, u3))
u2_sol = sp.simplify(solution_D[u2])                         # Simplify results
u3_sol = sp.simplify(solution_D[u3])

display(Math(f" \\text{{$u_2$=}} {latex(sp.simplify(u2_sol))}"))
display(Math(f" \\text{{$u_3$=}} {latex(sp.simplify(u3_sol))}"))
```

$$u_2 = \frac{l_e\left(3Abl_e + 4P\right)}{8AE}$$

$$u_3 = \frac{l_e\left(Abl_e + 2P\right)}{2AE}$$

Setting $l_e = L$, and rearrange the terms, the explicit expressions for the active nodal displacements are:

$$\boxed{u_2 = \frac{PL}{2EA} + \frac{3bL^2}{8E}} \tag{6.85}$$

$$\boxed{u_3 = \frac{PL}{EA} + \frac{bL^2}{2E}} \tag{6.86}$$

6.3.6.8 Reproducibility of Quadratic Elements

1. **Displacement Field Match:** The continuous displacement interpolation within the element domain follows a quadratic path. Because the exact analytical solution for a bar under a uniform body force is also a quadratic polynomial, the single quadratic element **reproduces the exact displacement field everywhere** across the entire domain.

2. **Strain and Stress Match:** Differentiating this displacement field yields an exact linear strain distribution ($\varepsilon = \mathbf{B}_e\mathbf{d}_e$) and an **exact linear stress distribution** ($\sigma = E\varepsilon$). The model captures the maximum stress at the root and the minimum stress at the tip without requiring mesh refinement.

We can confirm (readers may give it a try) that the reaction force R obtained from this FEM model with one quadratic element is exact.

This exact agreement underscores a fundamental mathematical principle of the finite element method:

> **Reproducibility:** If the exact analytical solution to a governing differential equation is contained within the polynomial approximation space spanned by the chosen element shape functions, the Galerkin finite element projection will reproduce that exact analytical solution across the entire domain.

6.3.7 Comparative Performance Analysis

The table below provides a systematic comparison of the linear and quadratic finite element formulations against the exact analytical continuum solution for a 1D bar under uniform body force and tip traction loading.

Design and Field Features	1 Linear Element ($M = 1$)	2 Linear Elements ($M = 2$)	1 Quadratic Element ($M = 1$)	Exact Analytical Solution
Total Nodes	2	3	3	∞ (Continuum)
Active Degrees of Freedom	1	2	2	Continuous Function
Displacement Field Profile	Linear	Piecewise Linear	Quadratic	Quadratic
Nodal Displacements	Exact at tip node only	Mathematically Exact	Mathematically Exact	Exact Reference
Global Field Accuracy	Coarse Approximation	Good Approximation	**Exact Everywhere**	Exact Reference
Stress Field Profile	Constant	Piecewise Constant	Linear	Linear
Stress Values	Midpoint Exact Only	Midpoint Exact Only	**Exact Everywhere**	Exact Reference

6.4 Example: A 1D Bar with Sinusoidal Body Force

6.4.1 Problem Definition

To rigorously evaluate the convergence characteristics of linear (P_1) and quadratic (P_2) finite elements under non-polynomial loading conditions, we consider a 1D bar subjected to a spatially varying, sinusoidal load.

- **Domain:** Longitudinal axis $x \in [0, L]$.

- **Boundary Conditions:** A rigid constraint at the left boundary, $u(0) = 0$, and a concentrated tensile force at the right boundary, $\sigma(L)A = P$.
- **Material and Geometry:** Constant Young's modulus E and uniform cross-sectional area A.
- **Distributed Body Force:**

$$b(x) = b_0 \sin\left(\frac{\pi x}{2L}\right) \tag{6.87}$$

6.4.2 Exact Analytical Solution

Using the sinusoidal body force, the strong form equation becomes:

$$\frac{d^2u}{dx^2} = -\frac{b(x)}{E} = -\frac{b_0}{E}\sin\left(\frac{\pi x}{2L}\right) \tag{6.88}$$

Integrating this twice and enforcing $u(0) = 0$ along with the force boundary condition at $x = L$ provides the exact displacement solution [4]:

$$u(x) = \frac{4L^2 b_0}{\pi^2 E}\sin\left(\frac{\pi x}{2L}\right) + \frac{P}{EA}x \tag{6.89}$$

The derivative of the displacement function or the strain field is

$$\frac{du}{dx} = \frac{2Lb_0}{\pi E}\cos\left(\frac{\pi x}{2L}\right) + \frac{P}{EA} \tag{6.90}$$

Applying Hooke's law ($\sigma(x) = E\frac{du}{dx}$), the exact internal normal stress field across the bar becomes:

$$\sigma(x) = \frac{P}{A} + \frac{2Lb_0}{\pi}\cos\left(\frac{\pi x}{2L}\right) \tag{6.91}$$

Because the exact solution is not polynomial, FEM solution will not be exact in general. To obtained accurate solutions, more elements are needed. Thus, the primary objective of this case study is to quantify the numerical accuracy using

1. the global L^2 **error norm** for the displacement field, and
2. the global **energy error norm** for the stress field.

These metrics allow us to determine the empirical convergence rates as a function of nodal density (h-refinement) and compare them against theoretical limits.

6.4.3 FEM Solution Using One Quadratic Element

We begin by modeling the continuous domain using a single three-node quadratic element ($l_e = L$). Because the structural geometry, interpolation basis, and material properties match the previous example, the element stiffness matrix remains identical to the expression derived in Eq. (6.84). The introduction of the sinusoidal loading terms requires re-evaluating the consistent nodal body force vector.

To compute the force vector using the standardized natural coordinate system ($\xi \in [-1, 1]$), we execute a two-step transformation procedure: 1. Apply the geometric coordinate mapping relation: $x(\xi) = \frac{L}{2}(1 + \xi)$. 2. Substitute this mapping into the continuous load definition to express the body force in terms of the natural parameter ξ:

$$b(x) = b_0 \sin\left(\frac{\pi x}{2L}\right) \quad \Longrightarrow \quad b(\xi) = b_0 \sin\left(\frac{\pi}{4}(1+\xi)\right) \tag{6.92}$$

Inserting this natural coordinate representation along with the differential relationship $dx = \frac{L}{2} d\xi$ into the consistent force vector integral (Eq. 6.79) yields:

$$\mathbf{f}^{(b)} = \frac{Ab_0 L}{2} \begin{bmatrix} \int_{-1}^{1} N_1(\xi) \sin\left(\frac{\pi}{4}(1+\xi)\right) d\xi \\ \int_{-1}^{1} N_2(\xi) \sin\left(\frac{\pi}{4}(1+\xi)\right) d\xi \\ \int_{-1}^{1} N_3(\xi) \sin\left(\frac{\pi}{4}(1+\xi)\right) d\xi \end{bmatrix} \tag{6.93}$$

Because the integrand contains transcendental terms multiplied by polynomial shape functions, we leverage the symbolic computing library SymPy in Python to

1. evaluate these integrals exactly,
2. establish the discrete system equations, and
3. solve it for displacements.

```
xi, L, A, b0 = sp.symbols('xi L A b_0', real=True)              # le = L
x = L/2 * (1 + xi)                                          # Mapping: x(xi)

# Body force in physical coordinates → expressed in xi
b = b0 * sp.sin(sp.pi * x / (2*L))

# Shape functions (quadratic 1D element)
N = sp.Matrix([sp.Rational(1,2)*xi*(xi - 1),
               1 - xi**2,
               sp.Rational(1,2)*xi*(xi + 1) ])
J = L/2                                                       # Jacobian
integrand = N * A * b * J                                     # Integrand

# Integrate over xi   [-1, 1]
fe = sp.integrate(integrand, (xi, -1, 1))
```

```
fe = sp.simplify(fe)
display(Math(f" \\text{{$f_e$=}} {latex(fe)}"))

## SymPy Code For Solving the System Equation.

E, A, L, b0, P = sp.symbols('E A L b_0 P',real=True) # Define symbols
X = sp.symbols('X',real=True)          # symbol for x for exact solution
u2, u3 = sp.symbols('u2 u3', real=True)

K = (E*A)/(3*L) * sp.Matrix([[16, -8],             # Define the K matrix
                             [-8,  7] ])
F = sp.Matrix([fe[1], fe[2] + P ])                  # Force Vector
D = sp.Matrix([u2, u3])                               # Unknown vector

# Solve for nodal displacement D:
solution_D = sp.solve(K*D - F, (u2, u3))
u2_sol = sp.simplify(solution_D[u2])                 # Simplify results
u3_sol = sp.simplify(solution_D[u3])

display(Math(f" \\text{{FEM (one L3), $u_2$=}} {latex(u2_sol)}"))
display(Math(f" \\text{{FEM (one L3), $u_3$=}} {latex(u3_sol)}"))

# Compute the exact solution u(x):
u = (4*L**2*b0/(sp.pi**2 * E)) * sp.sin(sp.pi*X/(2*L)) + (P/(E*A))*X

# Evaluate at node 2 (X = L/2) and node 3 (X = L):
u2_ex = sp.simplify(u.subs(X, L/2))
u3_ex = sp.simplify(u.subs(X, L))
display(Math(f"\\text{{Exact $u_2$=}} {latex(u2_ex)}"))
display(Math(f"\\text{{Exact $u_3$=}} {latex(u3_ex)}"))
```

$$f_e = \begin{bmatrix} \frac{2ALb_0(-16+2\pi+\pi^2)}{\pi^3} \\ \frac{16ALb_0(4-\pi)}{\pi^3} \\ \frac{4ALb_0(-8+3\pi)}{\pi^3} \end{bmatrix}$$

FEM (one L3), $u_2 = \dfrac{L\left(-2\pi ALb_0 + 24ALb_0 + \pi^3 P\right)}{2\pi^3 AE}$

FEM (one L3), $u_3 = \dfrac{4L^2 b_0}{\pi^2 E} + \dfrac{LP}{AE}$

Exact $u_2 = \dfrac{2\sqrt{2}L^2 b_0}{\pi^2 E} + \dfrac{LP}{2AE}$

Exact $u_3 = \dfrac{4L^2 b_0}{\pi^2 E} + \dfrac{LP}{AE}$

6.4.4 Boundary Exactness Property

Analyzing the finite element solution using single quadratic element for a 1D bar under a sinusoidal body force reveals two fundamental behaviors:

- **Interior Field Approximation:** The finite element solution using a single quadratic element provides only an *approximation* to the continuous analytical solution at internal points, including the central node 2. This occurs because a second-order polynomial basis cannot exactly span the infinite-dimensional transcendental space of a sine function.
- **Boundary Exactness:** Despite the internal interpolation approximation, the calculated displacement at the free end of the bar (u_3 at $x = L$) matches the exact analytical continuum value identically. This support our earlier remarks that the nodal exactness holds regardless of the body force type, provided that its exact energy-equivalent nodal force vector can be obtained.

This boundary exactness is not accidental. It is a direct consequence of the Galerkin finite element formulation and the weak form of equilibrium. Although the finite element approximation may not reproduce the exact internal strain field, it satisfies global equilibrium exactly in the variational sense. As a result, certain global quantities—such as the displacement at the loaded boundary in a 1D bar—are recovered exactly.

The following analysis reveals what is really going on. First, let us find the shape function in the physical coordinate x using the following snippet.

```
x, L = sp.symbols('x, L ', real=True)                          # le = L
#x = L/2 * (1 + xi)                                      # mapping: x(xi)
xi = 2*x/L - 1                                      # the reverse mapping
# Shape functions in xi (given earlier)
N = sp.Matrix([[sp.Rational(1,2)*xi*(xi - 1),
                1 - xi**2,
                sp.Rational(1,2)*xi*(xi + 1) ]])
Nx = N.applyfunc(sp.expand)
display(Math(f" \\text{{Shape function matrix in }}x: {latex(Nx)}"))
```

Shape function matrix in x : $\left[1 - \frac{3x}{L} + \frac{2x^2}{L^2} \quad \frac{4x}{L} - \frac{4x^2}{L^2} \quad -\frac{x}{L} + \frac{2x^2}{L^2}\right]$

We obtain

$$\begin{aligned} N_1(x) &= 1 - \frac{3x}{L} + \frac{2x^2}{L^2} \\ N_2(x) &= \frac{4x}{L} - \frac{4x^2}{L^2} \\ N_3(x) &= -\frac{x}{L} + \frac{2x^2}{L^2} \end{aligned} \tag{6.94}$$

Equation (6.84) can then be written in the more general form

$$\frac{EA}{3l_e}\begin{bmatrix}16 & -8\\ -8 & 7\end{bmatrix}\begin{bmatrix}u_2\\ u_3\end{bmatrix} = \begin{bmatrix}F_2\\ F_3\end{bmatrix} \tag{6.95}$$

where F_2 and F_3 are the equivalent nodal forces generated by an arbitrary body force distribution $b(x)$. They are computed using the standard finite element load vector formula:

$$\begin{aligned} F_2 &= \int_0^L N_2(x)\, b(x)\, dx \\ F_3 &= \int_0^L N_3(x)\, b(x)\, dx + P \end{aligned} \tag{6.96}$$

Note that the matrix in Eq.(6.95) is the reduced stiffness matrix by setting the DBC $u_1 = 0$, and hence is invertible.

Next, we use the following SymPy code snippet to solve Eq. (6.95) via inversion to find u_2 and u_3:

```
F2, F3 = sp.symbols('F2, F3', real=True)                          # le = L
u2, u3 = sp.symbols('u2, u3', real=True)

K = (E*A)/(3*L) * sp.Matrix([[16, -8],                    # Define the K matrix
                             [-8,  7] ])
F = sp.Matrix([F2, F3])                                   # Force Vector

# Solve for nodal displacement D23 by inverting K:
D23 = K.inv()*F
display(Math(f" \\text{{$u_2$=}} {latex(D23[0])}"))
display(Math(f" \\text{{$u_3$=}} {latex(D23[1])}"))
```

$$u_2 = \frac{7F_2L}{16AE} + \frac{F_3L}{2AE}$$

$$u_3 = \frac{F_2L}{2AE} + \frac{F_3L}{AE}$$

Using the this result, Eq.(6.94), and (6.96), we have

$$
\begin{aligned}
u_3 &= \frac{L}{AE}\left(\frac{1}{2}F_2 + F_3\right) \\
&= \frac{L}{AE}\left(\int_0^L \frac{1}{2}N_2(x)b(x)dx + \int_0^L N_3(x)b(x)dx + P\right) \\
&= \frac{L}{AE}\left(\int_0^L \frac{2x}{L} - \frac{2x^2}{L^2} - \frac{x}{L} + \frac{2x^2}{L^2}\right) b(x)dx + P \\
&= \frac{1}{AE}\left(\int_0^L xb(x)dx + PL\right)
\end{aligned}
\tag{6.97}
$$

This is the exact displacement solution at the right end of the bar. It is the first-moment balance of all the externally applied force with respect to root of the bar. Remarkably, this result is obtained independently of the specific form of the body force distribution $b(x)$. Substituting the sinusoidal body force defined in Eq. (6.87) yields

$$
u_3 = \frac{4L^2 b_0}{\pi^2 E} + \frac{LP}{AE} \tag{6.98}
$$

which can be verified using the following snippet.

```
# Initialize symbolic variables
x, L, E, A, b0, P = sp.symbols('x L E A b_0 P', real=True)

# Define the continuous sinusoidal body force function b(x)
b = b0 * sp.sin(sp.pi * x / (2 * L))

Qb = sp.integrate(b * x, (x, 0, L))
Qb_simplified = sp.simplify(Qb)

# Evaluate the total tip displacement u(L):
uL = sp.simplify((Qb_simplified + P * L) / (E * A))

# Display the verified mathematical components
display(Math(f"\\int_0^L b(x) x \\, dx = {sp.latex(Qb_simplified)}"))
display(Math(f"u(L) = {sp.latex(uL.expand())}"))
```

$$
\int_0^L b(x)x\,dx = \frac{4L^2 b_0}{\pi^2}
$$

$$
u(L) = \frac{4L^2 b_0}{\pi^2 AE} + \frac{LP}{AE}
$$

It is important to note that, in the third step of Eq. (6.97), the quadratic terms in the integrand cancel exactly, leaving only the linear contribution. This cancellation reflects a fundamental property of the Galerkin finite element formulation known as

nodal exactness. The following derivation provides a general proof of this property using the **Principle of Virtual Work**.

6.4.4.1 Proof via the Principle of Virtual Work

For a body in equilibrium, the virtual work of the external forces must equal the virtual work of the internal stresses for every kinematically admissible virtual displacement field [4].

Consider the 1D elastic bar governed by

$$-\frac{d}{dx}\left(EA\frac{du}{dx}\right) = b(x) \tag{6.99}$$

subject to

$$u(0) = 0, \qquad EA\frac{du}{dx}\bigg|_{x=L} = P \tag{6.100}$$

where $b(x)$ is the distributed axial load and P is the end force.

To extract the tip displacement, choose the virtual displacement field

$$v(x) = x \tag{6.101}$$

which satisfies the homogeneous essential boundary condition

$$v(0) = 0 \tag{6.102}$$

and is therefore kinematically admissible. Since $v(x)$ is continuous and possesses a square-integrable first derivative, it belongs to the Sobolev space $H^1(0, L)$.

The weak form of equilibrium is

$$\int_0^L EA\frac{du}{dx}\frac{dv}{dx}\,dx = \int_0^L b(x)v(x)\,dx + Pv(L) \tag{6.103}$$

For Eq. (6.103) to be well defined, the standard weak-form regularity assumptions apply:

1. $u(x) \in H^1(0, L)$, so that $\dfrac{du}{dx}$ is square-integrable.
2. $b(x) \in L^2(0, L)$, so that the external virtual work integral exists.

For the chosen virtual displacement field,

$$\frac{dv}{dx} = 1, \qquad v(L) = L \tag{6.104}$$

Substituting into Eq. (6.103) gives

$$\int_0^L EA\frac{du}{dx}\,dx = \int_0^L b(x)x\,dx + PL \tag{6.105}$$

Since EA is constant,

$$EA\int_0^L \frac{du}{dx}\,dx = \int_0^L b(x)x\,dx + PL \tag{6.106}$$

Applying the fundamental theorem of calculus,

$$EA\left[u(L) - u(0)\right] = \int_0^L b(x)x\,dx + PL \tag{6.107}$$

Using the essential boundary condition $u(0) = 0$ yields

$$EA\,u(L) = \int_0^L b(x)x\,dx + PL \tag{6.108}$$

and therefore

$$u(L) = \frac{1}{EA}\left(\int_0^L b(x)x\,dx + PL\right) \tag{6.109}$$

This is the expression for arbitrary body force $b(x)$. For the special case of a uniform body force $b(x) = b$, Eq. (6.109) reduces to

$$u(L) = \frac{PL}{EA} + \frac{bL^2}{2E} \tag{6.110}$$

which matches the exact analytical solution given previously for constant body forces. For sinusoidal body force defined in Eq. (6.87), we have Eq. (6.98).

This result is significant because the tip displacement depends only on the global equilibrium moments of the applied loads, not on the detailed internal displacement field. The Galerkin finite element formulation enforces this weak equilibrium relation exactly. Consequently, even when the internal displacement and strain fields are only approximate, the nodal displacement at the loaded boundary is recovered exactly, provided that the admissible finite element space contains the linear field $v(x) = x$.

6.4.4.2 Nodal Exactness, Unique to 1D

Note that the nodal exactness property is largely unique to one-dimensional solids. In general, such exact nodal recovery does not hold for 2D and 3D problems. This distinction can be understood directly from the proof above.

In the 1D case, the weak-form integral reduces to a single spatial integral, which can be converted exactly into boundary evaluations through the fundamental theorem of calculus. Consequently, the equilibrium relation collapses to expressions involving only endpoint quantities.

For 2D and 3D solids, however, the weak form involves area and volume integrals. These integrals cannot, in general, be reduced to evaluations at a single node. At most, they can be transformed into equivalent boundary integrals through Green's Divergence Theorem or its higher-dimensional counterparts [9]. As a result, the exact recovery of nodal displacements observed in 1D finite element models does not generally extend to multidimensional solids.

6.4.4.3 Connection to Finite Element Approximations

In the standard Galerkin finite element method, the admissible trial and test spaces are constructed from the element shape functions. If the interpolation basis contains a complete first-order polynomial space, then the linear virtual displacement field

$$v(x) = x \tag{6.111}$$

is represented exactly within the finite element subspace. This condition is satisfied by all standard 1D elements of order $p \geq 1$, including linear, quadratic, and higher-order polynomial elements.

Because the finite element stiffness matrices and consistent load vectors are derived from the same virtual work functional, the discrete system satisfies the identical weak equilibrium relation:

$$W_{\text{int},h} = W_{\text{ext},h} \tag{6.112}$$

Substituting the admissible test function $v_h(x) = x$ into the discrete weak form yields the same first-moment equilibrium identity obtained in the continuous problem. Consequently,

$$u_h(L) = u(L) \tag{6.113}$$

Thus, the exact recovery of the boundary displacement is a direct consequence of the Galerkin projection preserving the linear virtual work mode associated with global equilibrium.

> **Summary:** Choosing the virtual displacement field $v(x) = x$ transforms the weak form into a global first-moment equilibrium statement. Since standard 1D finite element spaces reproduce this linear field exactly, the discrete system preserves the same equilibrium identity, leading to exact recovery of the boundary displacement.

6.4.4.4 Generalization to Structural Truss Systems

This variational equilibrium property extends naturally to structural truss systems composed of interconnected 1D bar members. Since each truss member behaves as a one-dimensional axial element, the same weak-form argument applies locally to every member.

Consequently, standard Galerkin truss elements satisfy exact member-level equilibrium relations whenever the interpolation basis contains a complete linear polynomial field.

> **Nodal Exactness in Trusses:** For pin-jointed truss structures subjected to discrete nodal loading, the standard Galerkin finite element method reproduces the exact analytical nodal displacements, provided each member uses at least a complete linear interpolation basis and the system is solved exactly.

This property arises because:

- the element interpolation reproduces constant strain states exactly,
- the discrete virtual work equations enforce exact member equilibrium,
- and the assembled algebraic system represents the exact global nodal force balance of the truss network.

As a result, for classical truss problems governed purely by axial member behavior, the finite element equations are not merely approximations of equilibrium—they are the exact equilibrium equations of the structural system itself.

6.4.4.5 Proof of Nodal Exactness in Structural Truss Systems (Optional)

To establish the mathematical basis of nodal exactness in truss systems, we construct a localized virtual displacement field that isolates the equilibrium equation of a single nodal degree of freedom. This serves as the discrete multi-degree-of-freedom analogue of the $v(x) = x$ test function used for the 1D bar.

Consider a pin-jointed truss structure discretized using standard linear bar elements. Assume:

- each member is linear elastic with constant axial rigidity EA_e,
- distributed loads are converted into consistent nodal forces,
- and essential boundary conditions are enforced exactly.

Let $\mathbf{u}$ denote the exact nodal displacement vector and $\mathbf{u}_h$ the finite element solution.

6.4.4.6 Step 1: Localized Virtual Field

Choose an unconstrained node i and Cartesian direction k. Define the virtual displacement vector

$$\mathbf{v}_j^{(i,k)} = \begin{cases} \mathbf{e}_k, & j = i \\ \mathbf{0}, & j \neq i \end{cases} \tag{6.114}$$

where $\mathbf{e}_k$ is the unit vector in direction k.

Interpolating this nodal field with the same finite element shape functions produces a piecewise-linear virtual displacement field that belongs exactly to the finite element test space V_h. Since the field vanishes at constrained nodes, it is kinematically admissible.

6.4.4.7 Step 2: Continuous Virtual Work

Applying the Principle of Virtual Work to the exact solution gives

$$W_{\text{int}}(\mathbf{u}, \mathbf{v}^{(i,k)}) = W_{\text{ext}}(\mathbf{v}^{(i,k)}) \tag{6.115}$$

Because the virtual field is zero everywhere except at node i, the external virtual work reduces to

$$W_{\text{ext}}(\mathbf{v}^{(i,k)}) = f_{i,k} \tag{6.116}$$

where $f_{i,k}$ is the external nodal force component.

The internal virtual work becomes

$$W_{\text{int}}(\mathbf{u}, \mathbf{v}^{(i,k)}) = \sum_e \frac{EA_e}{L_e} \Delta u_e \Delta v_e^{(i,k)} \tag{6.117}$$

which is precisely the (i,k) component of the global equilibrium operator:

$$W_{\text{int}}(\mathbf{u}, \mathbf{v}^{(i,k)}) = (\mathbf{K}\mathbf{u})_{i,k} \tag{6.118}$$

Hence,

$$(\mathbf{K}\mathbf{u})_{i,k} = f_{i,k} \tag{6.119}$$

6.4.4.8 Step 3: Finite Element Equilibrium

The Galerkin finite element solution satisfies the weak form for every admissible test function in V_h. Since $\mathbf{v}^{(i,k)} \in V_h$, we obtain

$$W_{\text{int}}(\mathbf{u}_h, \mathbf{v}^{(i,k)}) = W_{\text{ext}}(\mathbf{v}^{(i,k)}) \tag{6.120}$$

which yields

$$(\mathbf{K}\mathbf{u}_h)_{i,k} = f_{i,k} \tag{6.121}$$

Comparing Eqs. (6.119) and (6.121) gives

$$\mathbf{K}\mathbf{u} = \mathbf{K}\mathbf{u}_h \tag{6.122}$$

After enforcing the essential boundary conditions, the reduced stiffness matrix is non-singular. Therefore,

$$\mathbf{u}_h = \mathbf{u} \tag{6.123}$$

showing that the finite element solution reproduces the exact nodal displacements.

6.4.4.9 Interior Field Approximation

Although the nodal displacements are exact, the displacement field inside each element is generally approximate. The finite element solution enforces a polynomial interpolation within the element,

$$\mathbf{u}_h(x) = \mathbf{N}(x)\mathbf{d}_e \tag{6.124}$$

whereas the exact continuum solution may possess more complex spatial variation under distributed loading.

Thus, the exactness property is confined to the nodal degrees of freedom; the interior field remains an energy-optimal approximation.

> **Key Insight:** The finite element method can construct a virtual displacement field that isolates the equilibrium of any individual structural node. Because this field is represented exactly within the approximation space, the resulting nodal equilibrium equations are exact, leading to exact nodal displacements for truss systems.

6.5 Case Study: 1D Bar under Sinusoidal Loading with Two Linear Elements

To further illustrate the nodal exactness property under non-polynomial loading, consider the sinusoidally loaded bar discretized using two standard linear elements. An internal node is placed at the midpoint of the bar.

6.5.1 Problem Setup

- Each element uses a local natural coordinate $\xi \in [-1, 1]$
- Each element has length $l_e = L/2$
- The assembled system contains three global nodes
- The essential boundary condition $u_1 = 0$ is enforced
- The unknowns are u_2 and u_3

The finite element procedure is identical to the previous constant body force example. The primary difference lies in the evaluation of the equivalent nodal load vectors.

Because the body force varies sinusoidally along the global coordinate axis, the loading function must first be mapped into the natural coordinate system of each element before performing the consistent load integration.

6.5.2 Element 1 Load Vector ($x \in [0, L/2]$)

The coordinate mapping for Element 1 is

$$x(\xi) = \frac{l_e}{2}(1+\xi)\frac{L}{4}(1+\xi) \tag{6.125}$$

Substituting this into the loading function gives

$$b^{(1)}(\xi) = b_0 \sin\left(\frac{\pi}{8}(1+\xi)\right) \tag{6.126}$$

The consistent nodal load vector is therefore

$$\mathbf{f}^{(b,1)} = \int_{-1}^{1} \left[N_1(\xi)\ N_2(\xi)\right] b_0 \sin\left(\frac{\pi}{8}(1+\xi)\right)\left(\frac{L}{4}\right) d\xi \tag{6.127}$$

Evaluating the integral yields

$$\mathbf{f}^{(b,1)} = \frac{2b_0 L}{\pi^2}\begin{bmatrix} \pi - 2\sqrt{2} \\ 4 - \pi\sqrt{2} + 2\sqrt{2} \end{bmatrix} \tag{6.128}$$

6.5.3 Element 2 Load Vector ($x \in [L/2, L]$)

The coordinate mapping for Element 2 is

$$x(\xi) = \frac{L}{2} + \frac{l_e}{2}(1+\xi) = \frac{L}{4}(3+\xi) \tag{6.129}$$

Substituting into the loading function gives

$$b^{(2)}(\xi) = b_0 \sin\left(\frac{\pi}{8}(3+\xi)\right) \tag{6.130}$$

Here, we omit the reaction force at node 1, since it does not affect solving the unknowns. The corresponding consistent nodal load vector becomes

$$\mathbf{f}^{(b,2)} = \int_{-1}^{1} \begin{bmatrix} N_1(\xi) \\ N_2(\xi) \end{bmatrix} b_0 \sin\left(\frac{\pi}{8}(3+\xi)\right)\left(\frac{L}{4}\right) d\xi \tag{6.131}$$

Evaluating the integral gives

$$\mathbf{f}^{(b,2)} = \frac{2b_0L}{\pi^2}\begin{bmatrix}\pi\sqrt{2}-4+2\sqrt{2}\\ 4-\pi-2\sqrt{2}\end{bmatrix} \tag{6.132}$$

6.5.4 Global Force Vector Assembly

After assembling the element contributions and applying the tip force P at Node 3, the global force vector becomes

$$\mathbf{F} = \begin{bmatrix} f_1^{(b,1)} \\ f_2^{(b,1)} + f_1^{(b,2)} \\ f_2^{(b,2)} + P \end{bmatrix} \tag{6.133}$$

Substituting the evaluated element vectors gives

$$\mathbf{F} \begin{bmatrix} \frac{2b_0L}{\pi^2}(\pi-2\sqrt{2}) \\ \frac{8b_0L}{\pi^2}(\sqrt{2}-1) \\ \frac{2b_0L}{\pi^2}(4-\pi-2\sqrt{2})+P \end{bmatrix} \tag{6.134}$$

The following SymPy code is then used to solve for the nodal displacements.

```
xi, L, A, b0, E, P = sp.symbols('xi L A b_0 E P',real=True) # Symbols

# Use linear shape functions in natrual coordinates:
N = sp.Matrix([(1 - xi)/2,
               (1 + xi)/2 ])

J = L/4            # Jacobian for each element (length = L/2 → J = L/4)

# --- Element 1: x   [0, L/2]
x1 = (L/4)*(1 + xi)          # mapping
b1 = b0 * sp.sin(sp.pi * x1 / (2*L))

fe1 = sp.integrate(N * A * b1 * J, (xi, -1, 1))
fe1 = sp.simplify(fe1)

# --- Element 2: x   [L/2, L]
x2 = L/2 + (L/4)*(1 + xi)
b2 = b0 * sp.sin(sp.pi * x2 / (2*L))

fe2 = sp.integrate(N * A * b2 * J, (xi, -1, 1))
```

```
fe2 = sp.simplify(fe2)

display(Math(f"\\text{{Element 1 load}} = {latex(fe1)}"))
display(Math(f"\\text{{Element 2 load}} = {latex(fe2)}"))

# --- Element stiffness matrix (linear element)
ke = (E*A)/(L/2) * sp.Matrix([[1, -1],
                              [-1, 1]])

# --- Global assembly to form K (3 nodes)
K = sp.zeros(3,3)
F = sp.zeros(3,1)

# Element 1 assembly (nodes 1-2)
K[0:2,0:2] += ke
F[0:2,0] += fe1

# Element 2 assembly (nodes 2-3)
K[1:3,1:3] += ke
F[1:3,0] += fe2

# Add end force P at node 3
F[2] += P

display(Math(f"\\text{{Global K}} = {latex(K)}"))
display(Math(f"\\text{{Global F}} = {latex(F)}"))

# --- Apply boundary condition u1 = 0
u1, u2, u3 = sp.symbols('u1 u2 u3')

K_reduced = K[1:3,1:3]
F_reduced = F[1:3,0]

D = sp.Matrix([u2, u3])

solution = sp.solve(K_reduced*D - F_reduced, (u2, u3))

u2_sol = sp.simplify(solution[u2])
u3_sol = sp.simplify(solution[u3])

display(Math(f"\\text{{FEM }} u_2 = {latex(u2_sol)}"))
display(Math(f"\\text{{FEM }} u_3 = {latex(u3_sol)}"))

# --- Exact solution
```

```
X = sp.symbols('X')
u_exact=(4*L**2*b0/(sp.pi**2*E))*sp.sin(sp.pi*X/(2*L)) + (P/(E*A))*X

u2_ex = sp.simplify(u_exact.subs(X, L/2))
u3_ex = sp.simplify(u_exact.subs(X, L))

display(Math(f"\\text{{Exact }} u_2 = {latex(u2_ex)}"))
display(Math(f"\\text{{Exact }} u_3 = {latex(u3_ex)}"))
```

$$\text{Element 1 load} = \begin{bmatrix} \frac{2ALb_0\left(\pi-2\sqrt{2}\right)}{\pi^2} \\ \frac{\sqrt{2}ALb_0(4-\pi)}{\pi^2} \end{bmatrix}$$

$$\text{Element 2 load} = \begin{bmatrix} \frac{ALb_0\left(-8+\sqrt{2}(\pi+4)\right)}{\pi^2} \\ \frac{4ALb_0\left(2-\sqrt{2}\right)}{\pi^2} \end{bmatrix}$$

$$\text{Global K} = \begin{bmatrix} \frac{2AE}{L} & -\frac{2AE}{L} & 0 \\ -\frac{2AE}{L} & \frac{4AE}{L} & -\frac{2AE}{L} \\ 0 & -\frac{2AE}{L} & \frac{2AE}{L} \end{bmatrix}$$

$$\text{Global F} = \begin{bmatrix} \frac{2ALb_0\left(\pi-2\sqrt{2}\right)}{\pi^2} \\ \frac{\sqrt{2}ALb_0(4-\pi)}{\pi^2} + \frac{ALb_0\left(-8+\sqrt{2}(\pi+4)\right)}{\pi^2} \\ \frac{4ALb_0\left(2-\sqrt{2}\right)}{\pi^2} + P \end{bmatrix}$$

$$\text{FEM } u_2 = \frac{2\sqrt{2}L^2b_0}{\pi^2E} + \frac{LP}{2AE}$$

$$\text{FEM } u_3 = \frac{4L^2b_0}{\pi^2E} + \frac{LP}{AE}$$

$$\text{Exact } u_2 = \frac{2\sqrt{2}L^2b_0}{\pi^2E} + \frac{LP}{2AE}$$

$$\text{Exact } u_3 = \frac{4L^2b_0}{\pi^2E} + \frac{LP}{AE}$$

We observe the following:

- Both u_2 and u_3 (end displacement) is **exactly equal** to the analytical solution

6.6 A General Modular Solver for 1D solids

When the load become non-polynomial, no standard 1D elements can produce the exact solution inside the elements. We need to refine the mesh refinement to find the exact nodal displacement solution and to approximate the solution inside each element. The best way to achieve this is to use a Python code.

6.6.1 FEM Code for 1D Solid

The following is a set of codes written in modular formate, which can be used for 1D solid mechanics problems. The generalization is achieved through

1. Using the natural coordinate for constructing shape functions.

2. Using Gaussian quadrature for numerical integrations.

```
# ------------------------------------
# Gaussian quadrature (2-pt and 3-pt)
# ------------------------------------
def gauss(n):
    """
    Returns weights and integration points for Gaussian quadrature.
    Gaussian quadrature is used to approximate the integral of a
    function as a weighted sum of values at specific points.
    """
    if n == 2:
        # Standard points for 2-point integration (exact for
        # polynomials up to degree 3)
        xi = np.array([-1/np.sqrt(3), 1/np.sqrt(3)])
        w  = np.array([1.0, 1.0])
    elif n == 3:
        # Standard points for 3-point integration (exact for
        # polynomials up to degree 5)
        xi = np.array([-np.sqrt(3/5), 0.0, np.sqrt(3/5)])
        w  = np.array([5/9, 8/9, 5/9])
    else:
        raise ValueError("Only 2-point and 3-point Gauss rules are \
                                                        implemented.")
    return xi, w

# -----------------------------------------------------------
# Shape functions in natural coordinates (xi from -1 to 1)
# -----------------------------------------------------------
def shape_p1(xi):
    """
    Linear (P1) shape functions for a 2-node element.
    N: values of shape functions at xi
    dN_dxi: derivative of shape functions with respect to natural
    coordinate xi
    """
    N = np.array([(1 - xi) / 2, (1 + xi) / 2])
    dN_dxi = np.array([-0.5, 0.5])
```

```
    return N, dN_dxi

def shape_p2(xi):
    """
    Quadratic (P2) shape functions for a 3-node element.
    Allows for curved interpolation of displacements.
    """
    N = np.array([
        0.5 * xi * (xi - 1),  # Node 1 (left)
        1 - xi**2,            # Node 2 (middle)
        0.5 * xi * (xi + 1)   # Node 3 (right)
    ])
    dN_dxi = np.array([
        xi - 0.5,             # Derivative at Node 1
        -2 * xi,              # Derivative at Node 2
        xi + 0.5              # Derivative at Node 3
    ])
    return N, dN_dxi

# ------------------------------------------------
# FEM solver for 1D Linear Elasticity numerically
# ------------------------------------------------
def FEM_solver_1D(ne, order, L, E, A, P, b_func):
    """
    Solves a 1D axial bar problem.
    ne: number of elements
    order: polynomial order (1 for linear, 2 for quadratic)
    L: Length, E: Young's modulus, A: Area, P: Tip load,
    b_func: body force
    """
    # 1. Mesh Generation
    if order == 1:
        nn = ne + 1  # Number of nodes
        coords = np.linspace(0, L, nn)
        # Element connectivity: which nodes belong to which element
        conn = [np.array([i, i+1]) for i in range(ne)]
        ngauss = 3
        shape_fn = shape_p1
    elif order == 2:
        nn = 2 * ne + 1
        coords = np.linspace(0, L, nn)
        # Connect nodes (i, i+1, i+2) for quadratic elements
        conn = [np.array([2*i, 2*i+1, 2*i+2]) for i in range(ne)]
        ngauss = 3
```

```
        shape_fn = shape_p2
    else:
        raise ValueError("order must be 1 (linear) or 2 (quad.).")

    # 2. Initialization of Global Matrices
    K = np.zeros((nn, nn))  # Global Stiffness Matrix
    F = np.zeros(nn)        # Global Force Vector
    xi_g, w_g = gauss(ngauss)

    # 3. Element Assembly
    for el in conn:
        x_e = coords[el] #Physical nodal coordinates in this element
        nloc = len(el)
        ke = np.zeros((nloc, nloc))
        fe = np.zeros(nloc)

        # Numerical Integration (Gauss Loop)
        for xi, w in zip(xi_g, w_g):
            N, dN_dxi = shape_fn(xi)

            # Mapping from natural (xi) to physical (x) space
            x = np.dot(N, x_e)
            # Maps derivatives from natural to physical space
            J = np.dot(dN_dxi, x_e) # Jacobian

            # Derivative of shape functions wrt physical coord. x
            dN_dx = dN_dxi / J

            # Here B = dN_dx
            ke += E * A * np.outer(dN_dx, dN_dx) * J * w

            fe += N * A * b_func(x) * J * w

        # Assemble local element matrices into global matrices
        # np.ix_ efficiently maps local indices to global positions
        K[np.ix_(el, el)] += ke
        F[el] += fe

    # 4. Boundary Conditions
    # Neumann BC: Concentrated load at the end (x = L)
    F[-1] += P

    # Dirichlet BC: Fixed at x = 0 (u(0) = 0)
    # Solve the system by partitioning (removing the 1st row/column)
```

```
    K_bc = K[1:, 1:]
    F_bc = F[1:]

    # 5. Solve
    u_sol = np.linalg.solve(K_bc, F_bc)

    # Reconstruct full displacement vector including the fixed one
    u = np.zeros(nn)
    u[1:] = u_sol

    return coords, u
```

6.6.2 Error Measures and Convergence Metrics

Because finite element solutions are generally approximations of the exact continuum fields, quantitative error measures are required to evaluate solution accuracy and convergence behavior. For multidimensional problems, these measures are formulated using vector fields.

6.6.2.1 L^2 Norm for Displacement Errors

The global displacement error is commonly measured using the L^2 norm:

$$|e_{\mathbf{u}}|_{L^2} = \left(\int_{\Omega} (\mathbf{u} - \mathbf{u}_h)^T (\mathbf{u} - \mathbf{u}_h), d\Omega \right)^{1/2} \tag{6.135}$$

where $\mathbf{u}$ is the exact displacement field and $\mathbf{u}_h$ is the finite element approximation.

This norm measures the mean-square error of the displacement field over the entire domain. In practice, the integral is evaluated numerically using element-wise Gauss quadrature.

6.6.2.2 Energy Norm for Stress and Strain Errors

To evaluate the accuracy of derivative quantities such as strains and stresses, the energy norm is used:

$$|e|_E = \left(\int_{\Omega} (\sigma - \sigma_h)^T \mathbf{C}^{-1} (\sigma - \sigma_h), d\Omega \right)^{1/2} \tag{6.136}$$

where:

- σ is the exact stress field,
- σ_h is the finite element stress field,
- and $\mathbf{C}$ is the constitutive matrix.

Using Hooke's law,

$$\sigma = \mathbf{C}\varepsilon \tag{6.137}$$

the energy norm can equivalently be written in terms of strain errors:

$$|e|_E = \left(\int_\Omega (\varepsilon - \varepsilon_h)^T \mathbf{C} (\varepsilon - \varepsilon_h), d\Omega \right)^{1/2} \tag{6.138}$$

This norm is closely related to the H^1 norm because it depends on the first derivatives of the displacement field. Like the L^2 norm, it is evaluated numerically using Gaussian quadrature over the discretized domain.

The preceding definitions are fully general and apply to 1D, 2D, and 3D problems. For the 1D bar problems considered in this chapter, the displacement, strain, and stress fields reduce to scalar quantities.

6.6.3 Numerical Integration (Gauss Quadrature)

Because the Finite Element Method (FEM) is based on weak formulations, evaluating domain and boundary integrals is fundamental. For simple problems, these integrals may be computed analytically. However, for complex integrands, higher-order elements, distorted geometries, or error norm evaluations, analytical integration becomes impractical. In such cases, numerical integration is used.

The most common technique in finite element analysis is **Gauss quadrature**, also called **Gauss-Legendre quadrature**.

Consider the 1D integral

$$I = \int_{-1}^{1} f(\xi)\, d\xi \tag{6.139}$$

defined over the standard natural coordinate domain $[-1, 1]$. Gauss quadrature approximates this integral as

$$I \approx \sum_{i=1}^{n} w_i f(\xi_i) \tag{6.140}$$

where:

- n is the number of Gauss points,
- ξ_i are the Gauss point locations,
- and w_i are the corresponding weights.

A major advantage of Gauss-Legendre quadrature is its high polynomial accuracy: an n-point rule integrates polynomials of degree up to $2n - 1$ exactly.

The most commonly used 1D Gauss rules are listed below.

Number of Points (n)	Gauss Points (ξ_i)	Weights (w_i)	Exact Polynomial Degree
1	0	2	1
2	$\pm\frac{1}{\sqrt{3}}$	1	3
3	$0 \ \pm\sqrt{\frac{3}{5}}$	$\frac{8}{9}$ $\frac{5}{9}$	5
4	± 0.3399810436 ± 0.8611363116	0.6521451549 0.3478548451	7

Symmetry Property: The Gauss points are symmetric about the origin, and symmetric points share identical weights. This symmetry reduces computational cost and simplifies element integration.

6.6.3.1 Two-Dimensional Gauss Quadrature

For 2D elements, such as Q4 quadrilateral elements, integration is performed over the domain $[-1, 1]^2$ using a tensor product of the 1D rules:

$$I = \int_{-1}^{1} \int_{-1}^{1} f(\xi, \eta) \, d\xi, d\eta \approx \sum_{i=1}^{n} \sum_{j=1}^{m} w_i w_j f(\xi_i, \eta_j) \tag{6.141}$$

where ξ and η are the natural coordinates.

6.6.3.2 Three-Dimensional Gauss Quadrature

For 3D elements, such as hexahedral (brick) elements, the integration extends naturally to the domain $[-1, 1]^3$:

$$\begin{aligned} I &= \int_{-1}^{1} \int_{-1}^{1} \int_{-1}^{1} f(\xi, \eta, \zeta) \, d\xi, d\eta, d\zeta \\ &\approx \sum_{i=1}^{n} \sum_{j=1}^{m} \sum_{k=1}^{p} w_i w_j w_k f(\xi_i, \eta_j, \zeta_k) \end{aligned} \tag{6.142}$$

where:

- ξ, η, and ζ are the natural coordinates,
- ξ_i, η_j, and ζ_k are the Gauss points,
- and w_i, w_j, and w_k are the associated weights.

Higher-dimensional Gauss quadrature therefore follows directly from repeated tensor products of the 1D integration rule.

6.7 Case Study: Convergence Analysis of a 1D Bar Under Sinusoidal Loading

Building on the previously established boundary value problem, this case study investigates the spatial convergence behavior of finite element solutions using linear (P_1) and quadratic (P_2) elements. Accuracy is assessed using the global L^2 **displacement error norm** and the **energy norm for stress error**, allowing computation of empirical convergence rates under h-refinement and comparison with theoretical a priori estimates.

6.7.1 FEM Model Setup

The numerical experiment is performed on a sequence of progressively refined meshes. The discretization parameters are:

- **Number of elements:** $N_e = 1, 2, 4, 8, 16, 32$
- **Element types:**
  - **Linear elements (P_1):** 2-node elements, with $N_n = N_e + 1$
  - **Quadratic elements (P_2):** 3-node elements, with $N_n = 2N_e + 1$

The spatial resolution is characterized by the nodal density

$$\rho_n = \frac{N_n}{L} \tag{6.143}$$

6.7.2 Theoretical Background of Convergence

In computational mechanics, convergence is typically studied through two refinement strategies:

1. **h-refinement:** Reducing the element size h by increasing the number of elements while keeping the polynomial order fixed.
2. **p-refinement:** Increasing the polynomial order p within fixed element geometries.

In this study, we focus on h-convergence. For a uniform 1D mesh, the characteristic element size is

$$h = \frac{L}{N_e} \tag{6.144}$$

so that $h \to 0$ corresponds to mesh refinement, and the nodal density scales as $\rho_n \sim 1/h$.

6.7.3 Finite Element Approximation Space

A finite element discretization of order p implies that:

- The displacement field within each element is represented by a complete polynomial basis of degree p.
- The global approximation satisfies

$$u_h(x) \in \mathbb{P}^p \subset H^1(0, L) \tag{6.145}$$

Thus, the finite element solution lies in a finite-dimensional subspace of the Sobolev space $H^1(0, L)$.

6.7.4 Regularity Assumption

To ensure optimal convergence behavior, we assume sufficient smoothness of the exact solution:

$$u(x) \in H^{p+1}(0, L) \tag{6.146}$$

This condition implies that:

- The solution possesses square-integrable derivatives up to order $p + 1$,
- No singularities or discontinuities degrade the approximation quality,
- The polynomial space of degree p can fully realize its approximation capability.

6.7.5 Convergence Interpretation

Because the Galerkin finite element method is an energy projection onto the discrete space, Cea's lemma ensures that the finite element solution is quasi-optimal in the energy norm. Therefore, for a stable formulation (i.e., with a symmetric positive definite reduced stiffness matrix $\mathbf{K}$), the convergence rate is governed primarily by:

- the polynomial degree p, and
- the regularity of the exact solution.

In other words, the asymptotic error decay is determined by the approximation power of the chosen shape functions rather than the specific loading configuration or discretization details.

6.7.6 A Priori Interpolation Error Bounds

Using standard interpolation theory in Sobolev spaces, the finite element approximation error in the primary variable satisfies the bound

$$|u - u_h|_{L^2} \leq C_1 h^{p+1} \tag{6.147}$$

This establishes that the displacement error converges with asymptotic order

$$\boxed{|u - u_h|_{L^2} = \mathcal{O}(h^{p+1})} \tag{6.148}$$

In contrast, errors involving derivatives of the solution are measured in the H^1 seminorm, which corresponds to the energy norm in linear elasticity:

$$|u - u_h|_{H^1} \leq C_2 h^p \tag{6.149}$$

Hence, the energy norm error behaves as

$$\boxed{|u - u_h|_E = \mathcal{O}(h^p)} \tag{6.150}$$

This result reflects a fundamental property of displacement-based finite element methods: **stress and strain fields converge one order slower than the displacement field**. The reason is that stresses are obtained by differentiating the displacement field,

$$\sigma = \mathbf{C}\varepsilon = \mathbf{CBd}_e, \tag{6.151}$$

and differentiation reduces the polynomial order from p to $p-1$, thereby lowering the achievable convergence rate for derivative quantities.

> **Engineering Insight:** Displacement predictions typically achieve high accuracy even on relatively coarse meshes. However, accurate stress evaluation—especially near gradients or concentrations—requires finer meshes or higher-order elements due to the reduced convergence rate of derivative fields.

6.7.7 Asymptotic Convergence Rate Summary

The theoretical convergence behavior under uniform h-refinement is summarized below.

Element Order (p)	Structural Analogues	Displacement Error $\lvert e_u \rvert_{L^2}$	Energy / Stress Error $\lvert e \rvert_E$
$p = 1$	Linear bar element (L_2)	$\mathcal{O}(h^2)$	$\mathcal{O}(h)$
$p = 2$	Quadratic bar element (L_3)	$\mathcal{O}(h^3)$	$\mathcal{O}(h^2)$
$p = 3$	Cubic bar element (L_4)	$\mathcal{O}(h^4)$	$\mathcal{O}(h^3)$

The table highlights the consistent shift of one order between displacement and energy norms across all polynomial degrees.

6.7.8 Empirical Convergence Curve Diagnostics

In numerical analysis, **log–log convergence plots** are used to quantify how discretization errors decay under systematic uniform h-refinement. The slope of the resulting curve provides an empirical convergence rate, which can be directly compared with the a priori theoretical predictions.

1. **Displacement field convergence:** For linear (P_1) elements, the L^2 error exhibits an observed slope of approximately **2.00**, consistent with the theoretical rate $\mathcal{O}(h^2)$. For quadratic (P_2) elements, the slope increases to approximately **3.00**, confirming the expected rate $\mathcal{O}(h^3)$.
2. **Stress (derivative) convergence:** The energy norm error converges one order slower, consistent with interpolation theory. The linear element exhibits a slope of approximately **1.00** ($\mathcal{O}(h)$), while the quadratic element achieves a slope of approximately **2.00** ($\mathcal{O}(h^2)$). This confirms that derivative-based quantities converge at a reduced rate compared to displacements.

These numerical results validate the theoretical error estimates derived in the previous section.

The following section presents the computational implementation used to evaluate these convergence curves for the 1D sinusoidally loaded bar problem.

6.7.9 Utility Code for Error Computation

```
# ----------------------------
# Solution error computation
# ----------------------------
def compute_error(coords, u, order):
    ne = (len(coords)-1) if order==1 else (len(coords)-1)//2

    if order == 1:
        conn = [(i, i+1) for i in range(ne)]
        shape = shape_p1
        ngauss = 3
    else:
        conn = [(2*i,2*i+1,2*i+2) for i in range(ne)]
        shape = shape_p2
        ngauss = 3

    xi_g, w_g = gauss(ngauss)

    err_L2 = 0.0
    err_E  = 0.0
```

```
    for el in conn:
        x_e = coords[list(el)]
        u_e = u[list(el)]

        for xi, w in zip(xi_g, w_g):
            N, dN_dxi = shape(xi)

            x = np.dot(N, x_e)
            J = np.dot(dN_dxi, x_e)
            dN_dx = dN_dxi / J

            uh = np.dot(N, u_e)
            duh = np.dot(dN_dx, u_e)

            ue = u_exact(x)
            due = du_exact(x)

            err_L2 += (ue - uh)**2 * J * w
            err_E  += E*(due - duh)**2 * J * w

    return np.sqrt(err_L2), np.sqrt(err_E)

def compute_rates(x, err):
    """
    Compute convergence rates:
    slope = d(log(err)) / d(log(x))
    """
    rates = []
    for i in range(1, len(x)):
        r = np.log(err[i]/err[i-1]) / np.log(x[i]/x[i-1])
        rates.append(r)
    return np.array(rates)
```

6.7.10 Numerical Convergence Study

The following code snippets compute and plot the convergence curves, and evaluate their slopes, which correspond to the empirical convergence rates.

```
# ---------------------------------------------------
# Problem parameters that defines the 1D bar problem
# ---------------------------------------------------
E  = 1.0                  # material property
A  = 1.0; L = 1.0          # geometry
b0 = 1.0; P = 1.0         # loads
```

```
# The exact solution u(x) for error analysis:
def u_exact(x):
    return (4*L**2*b0/(np.pi**2*E))*np.sin(np.pi*x/(2*L))+(P/(E*A))*x

def du_exact(x):
    return (2*L*b0/(np.pi*E))*np.cos(np.pi*x/(2*L)) + P/(E*A)

# Function of the body force applied on the bar:
def b_func(x):
    return b0*np.sin(np.pi*x/(2*L))
```

6.7.11 Verification of Nodal Exactness

Using the modular finite element framework, we can numerically verify the nodal exactness property associated with one-dimensional Galerkin formulations. Even for a non-polynomial sinusoidal body force, the computed displacement values at element nodes coincide with the exact analytical solution, up to machine precision.

The following demonstration uses a coarse five-element linear mesh. The same script can be extended to arbitrary mesh densities to confirm that this nodal exactness property is preserved under uniform h-refinement.

```
# Compute the nodal displacements of a FEM model, linear elements
Ne = 5        # use Ne elements
coords, u = FEM_solver_1D(Ne, 1, L, E, A, P, b_func) #order 1,linear

np.set_printoptions(precision=8,suppress=True) # Digits in print-out␣
↪
print(f"FEM nodal solution using {Ne} linear elements:\n{u}")
print(f"Exact solution at nodes:\n{u_exact(coords)}")
```

```
FEM nodal solution using 5 linear elements:
[0.         0.32523987 0.63822039 0.92788224 1.18544869 1.40528473]

Exact solution at nodes:
[0.         0.32523987 0.63822039 0.92788224 1.18544869 1.40528473]
```

It is seen that the nodal values are exact to at least 8 digits.

The following snippet performs the same analysis but using quadratic elements.

```
# Compute the nodal displacements of a FEM model, quadratic elements
Ne = 5 # use Ne elements
coords, u = FEM_solver_1D(Ne, 2, L, E, A, P, b_func)          # order 2
```

```
print(f"FEM nodal solution using {Ne} quadratic elements:\n{u}")
print(f"Exact solution at nodes:\n{u_exact(coords)}")
```

```
FEM nodal solution using 5 quadratic elements:
[0.         0.16340018 0.32523987 0.48399449 0.63822039 0.78657813
 0.92788224 1.06110952 1.18544869 1.30029298 1.40528473]

Exact solution at nodes:
[0.         0.1634005  0.32523987 0.48399542 0.63822039 0.78657958
 0.92788224 1.06111134 1.18544869 1.30029501 1.40528473]
```

6.7.12 Analysis of Nodal Exactness and Quadrature Effects

The numerical results show that, for linear (P_1) elements, the computed nodal displacements match the exact analytical solution at element boundaries to at least eight significant digits. For quadratic (P_2) elements, a small deviation appears: endpoint nodes remain exact to machine precision, while internal midside nodes exhibit minor errors on the order of the fifth significant digit under non-polynomial loading.

These observations are consistent with the theoretical results derived earlier.

> **Critical Theoretical Caveat:** The nodal exactness property in one-dimensional problems holds only when the system matrices and load vectors are evaluated exactly. In the present case, the distributed load involves a non-polynomial function ($\sin(\cdot)$), so numerical integration must be used in place of exact quadrature.

Here, a high-order Gauss–Legendre quadrature rule with three integration points is employed. This level of accuracy is sufficient to preserve nodal exactness to near machine precision. However, reducing the number of quadrature points introduces integration error in the consistent load vector, which propagates into the global system and degrades nodal accuracy.

Readers are encouraged to vary the `n_gauss` parameter in the implementation to directly observe how reduced quadrature order affects nodal precision.

6.7.13 Convergence Study Implementation

To assess solution behavior under systematic h-refinement, the following script evaluates both displacement and stress errors across a sequence of progressively refined meshes. The displacement error is measured using the global L^2 norm, while the stress error is evaluated using the energy norm.

```
# Compute the solution error in L2 norm and energy norm
Ne_list = [2, 4, 8, 16, 32]

results_p1 = []
```

```
results_p2 = []

for ne in Ne_list:
    coords, u = FEM_solver_1D(ne, 1, L, E, A, P, b_func) # order 1
    eL2, eE = compute_error(coords, u, order=1)
    results_p1.append((len(coords), eL2, eE))

    coords, u = FEM_solver_1D(ne, 2, L, E, A, P, b_func) # order 2
    eL2, eE = compute_error(coords, u, order=2)
    results_p2.append((len(coords), eL2, eE))

# Convert to arrays
nodes_p1 = np.array([r[0] for r in results_p1])
errL2_p1 = np.array([r[1] for r in results_p1])
errE_p1  = np.array([r[2] for r in results_p1])

nodes_p2 = np.array([r[0] for r in results_p2])
errL2_p2 = np.array([r[1] for r in results_p2])
errE_p2  = np.array([r[2] for r in results_p2])

# Print tables
print("\nP1 Results:")
for n, e1, e2 in results_p1:
    print(f"Nodes={n:3d}, L2={e1:.3e}, Energy={e2:.3e}")

print("\nP2 Results:")
for n, e1, e2 in results_p2:
    print(f"Nodes={n:3d}, L2={e1:.3e}, Energy={e2:.3e}")
```

```
P1 Results:
Nodes=  3, L2=1.592e-02, Energy=1.010e-01
Nodes=  5, L2=4.021e-03, Energy=5.090e-02
Nodes=  9, L2=1.008e-03, Energy=2.550e-02
Nodes= 17, L2=2.521e-04, Energy=1.276e-02
Nodes= 33, L2=6.303e-05, Energy=6.379e-03

P2 Results:
Nodes=  5, L2=6.617e-04, Energy=1.027e-02
Nodes=  9, L2=8.330e-05, Energy=2.582e-03
Nodes= 17, L2=1.043e-05, Energy=6.465e-04
Nodes= 33, L2=1.304e-06, Energy=1.617e-04
Nodes= 65, L2=1.631e-07, Energy=4.042e-05
```

6.7.14 Interpretation of Convergence Trends

The numerical results show that, as the number of nodes increases through systematic mesh refinement, both the L^2 displacement error and the energy norm error decrease monotonically. This consistent decay is observed for both linear (P_1) and quadratic (P_2) elements.

Since the theoretical error estimates scale as a power of the element size h, the convergence behavior becomes linear when plotted on a log–log scale:

$$\log(|e|) = \log(C) + k\log(h) \tag{6.152}$$

In this representation, the error follows a straight-line relationship, where the slope k corresponds to the empirical **convergence rate**. This makes it possible to directly compare numerical results with the theoretical predictions derived earlier.

6.7.15 Visualization of Spatial Convergence Rates

The following script extracts the error data, transforms it into log–log coordinates, and plots the resulting convergence curves for the displacement field under successive mesh refinements.

```
# Rates (note: using nodes as x-axis)
rates_L2_p1 = compute_rates(nodes_p1, errL2_p1)
rates_E_p1  = compute_rates(nodes_p1, errE_p1)

rates_L2_p2 = compute_rates(nodes_p2, errL2_p2)
rates_E_p2  = compute_rates(nodes_p2, errE_p2)

plt.figure(figsize=(7, 5), dpi=300)
plt.loglog(nodes_p1, errL2_p1, 'o-',label='p1, linear elements')
plt.loglog(nodes_p2, errL2_p2, 's-',label='p2, quadratic elemenets')

# Annotate slopes (last segment)
mid_x_p1 = np.sqrt(nodes_p1[-1]*nodes_p1[-2])
mid_y_p1 = np.sqrt(errL2_p1[-1]*errL2_p1[-2])
plt.text(mid_x_p1, mid_y_p1, f"{rates_L2_p1[-1]:.2f}")

mid_x_p2 = np.sqrt(nodes_p2[-1]*nodes_p2[-2])
mid_y_p2 = np.sqrt(errL2_p2[-1]*errL2_p2[-2])
plt.text(mid_x_p2, mid_y_p2, f"{rates_L2_p2[-1]:.2f}")

plt.xlabel('Number of Nodes')
plt.ylabel('L2 Error')
plt.legend()
```

```
plt.grid(True)
```

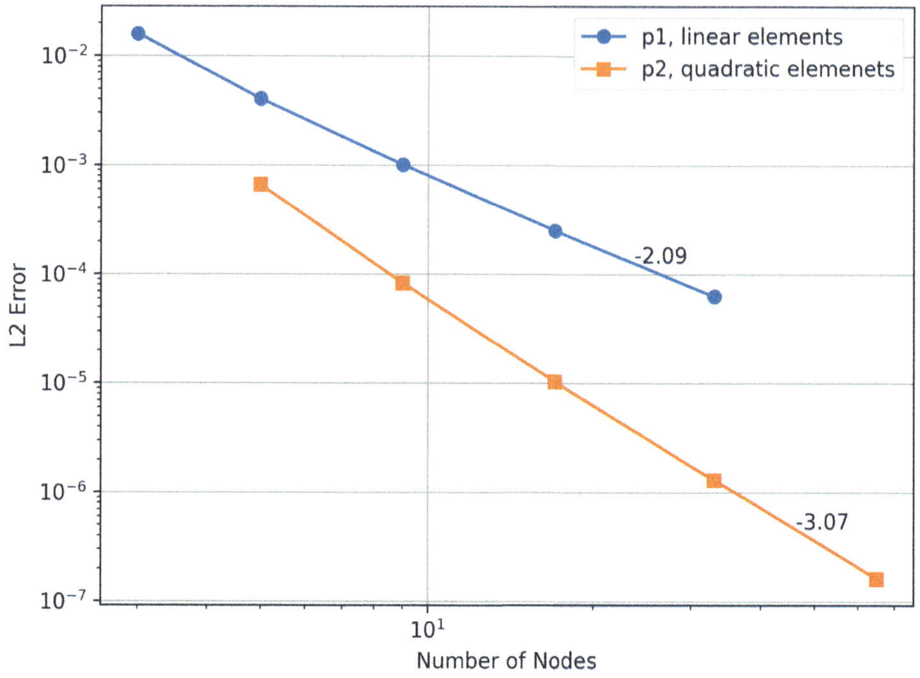

6.7.16 Evaluation of Primary Kinematic Fields

The log–log error curves indicate that the empirical convergence rate for the displacement field is approximately 2.1 for linear (P_1) elements and 3.1 for quadratic (P_2) elements. These values are in close agreement with the optimal theoretical rates predicted by interpolation theory, $\mathcal{O}(h^{p+1})$, confirming that the primary kinematic field achieves its expected convergence behavior.

6.7.17 Gradient Convergence Performance

To assess the accuracy of secondary fields (stresses and strains), the error is evaluated using the energy norm. The following script extracts the energy norm errors for each mesh configuration and plots the corresponding convergence curves on a log–log scale.

```
plt.figure(figsize=(7, 5), dpi=300)
plt.loglog(nodes_p1, errE_p1, 'o-', label='p1 Energy convergence')
plt.loglog(nodes_p2, errE_p2, 's-', label='p2 Energy convergence')

# Annotate slopes
mid_x_p1 = np.sqrt(nodes_p1[-1]*nodes_p1[-2])
mid_y_p1 = np.sqrt(errE_p1[-1]*errE_p1[-2])
plt.text(mid_x_p1, mid_y_p1, f"{rates_E_p1[-1]:.2f}")
```

```
mid_x_p2 = np.sqrt(nodes_p2[-1]*nodes_p2[-2])
mid_y_p2 = np.sqrt(errE_p2[-1]*errE_p2[-2])
plt.text(mid_x_p2, mid_y_p2, f"{rates_E_p2[-1]:.2f}")

plt.xlabel('Number of Nodes')
plt.ylabel('Energy Error')
plt.legend()
plt.grid(True)
```

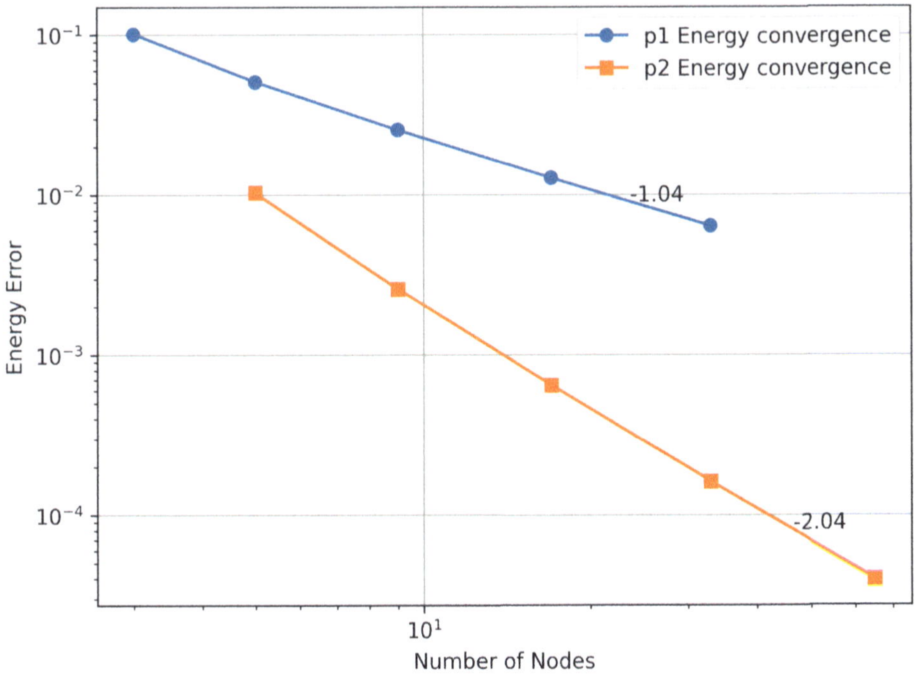

6.7.18 Evaluation of Secondary Derivative Fields

The log–log error curves show that the empirical convergence rate in the energy norm is approximately 1.0 for linear (P_1) elements and 2.0 for quadratic (P_2) elements. These observed slopes agree with the theoretical a priori estimate $\mathcal{O}(h^p)$ derived in the previous section.

This confirms that derivative fields converge one order more slowly than the primary displacement field. The result reflects the fact that spatial differentiation reduces the effective polynomial order of the finite element approximation by one degree, thereby lowering the achievable convergence rate for strains and stresses.

6.7.19 Sobolev Regularity Analysis

6.7.19.1 Loss of Regularity under a Point Force

We consider a one-dimensional uniform bar subjected to a concentrated point load P_c applied at the midpoint $x = L/2$, as shown in Fig. 6.3.

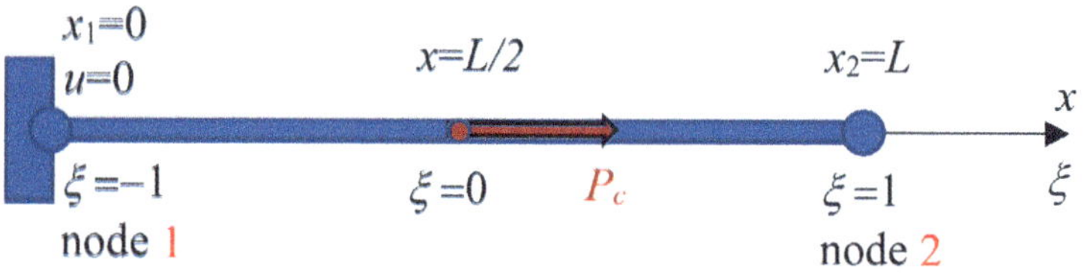

Figure 6.3: A uniform bar subjected to a concentrated point force at its center.

The point load is modeled using the Dirac delta function $\delta(x - L/2)$, leading to the governing equation

$$EA\frac{d^2u}{dx^2} = -P_c\delta\left(x - \frac{L}{2}\right). \tag{6.153}$$

Integrating once gives

$$EA\frac{du}{dx} = -P_cH\left(x - \frac{L}{2}\right) + C_1, \tag{6.154}$$

where $H(\cdot)$ is the Heaviside function [6]. This shows that the strain $\varepsilon(x) = u'(x)$ (and thus the stress $\sigma(x)$) has a jump discontinuity at $x = L/2$.

Although $u'(x)$ remains square-integrable (so $u \in H^1(0, L)$), the second derivative contains a Dirac delta distribution, which is not in $L^2(0, L)$. Hence the solution loses higher regularity. More precisely, the smoothness is limited by [1]:

$$u(x) \in H^m(0, L), \quad \text{with } m < 1.5. \tag{6.155}$$

As a result, the standard FEM assumption $u \in H^{p+1}$ required for optimal convergence is violated for any $p \geq 1$.

6.7.19.2 Effect on Convergence Rates

When the mesh is not aligned such that $x = L/2$ coincides with a node, standard polynomial shape functions cannot resolve the internal slope discontinuity. The approximation error is therefore governed by the reduced Sobolev regularity rather than the polynomial order.

Consequently, the convergence rates deteriorate to

$$|u - u_h|_{L^2} = \mathcal{O}(h^{1.5}), \qquad |u - u_h|_E = \mathcal{O}(h^{0.5}). \tag{6.156}$$

This is significantly lower than the optimal rates for linear elements, $\mathcal{O}(h^2)$ in the L^2 norm and $\mathcal{O}(h)$ in the energy norm. The loss in convergence is purely a consequence of reduced regularity, not element formulation.

Optimal convergence can be recovered if the mesh is aligned such that a node is placed at the singular point $x = L/2$, thereby resolving the discontinuity exactly.

6.7.19.3 Exact Solution for a Point-Loaded Bar

For the governing equation (6.153) with boundary conditions

$$u(0) = 0, \qquad u'(L) = 0, \tag{6.157}$$

the exact solution is obtained by piecewise integration:

$$u(x) = \begin{cases} \dfrac{P_c}{EA}x, & 0 \le x \le \dfrac{L}{2}, \\ \dfrac{P_c L}{2EA}, & \dfrac{L}{2} < x \le L. \end{cases} \tag{6.158}$$

This solution is continuous but not differentiable at $x = L/2$, where the slope is discontinuous. The corresponding stress field is therefore

$$\sigma(x) = \begin{cases} \dfrac{P_c}{A}, & 0 \le x \le \dfrac{L}{2}, \\ 0, & \dfrac{L}{2} < x \le L. \end{cases} \tag{6.159}$$

The following code snippet computes this exact solution for a bar subjected to a concentrated force P_c located at an arbitrary position x_c.

```
def exact_solution(x, L, E, A, xc, Pc):
    '''
    Exact displacement and stress solution
    for a 1D bar subjected to a concentrated load Pc at xc.
    Returns: u, displacement; sigma, stress'''
    x = np.asarray(x)
    u = np.zeros_like(x, dtype=float)   # for displacement

    left = x <= xc
    right = x > xc
    u[left] = (Pc / (E * A)) * x[left]      # displacements
    u[right] = (Pc / (E * A)) * xc
    sigma = np.zeros_like(x, dtype=float)           # stress
    sigma[left] = Pc / A                  # sigma = E * du/dx
    sigma[right] = 0.0
```

```
    return u, sigma
```

This function is continuous and piecewise linear, so it belongs globally to $H^1(0, L)$. However, it does not belong to $H^2(0, L)$ because its first derivative u' contains a jump discontinuity at $x = L/2$.

6.7.20 FEM Modeling

We now construct a finite element model for this problem. The overall formulation is identical to previous examples; the only difference lies in how the global nodal force vector is assembled.

Consider a concentrated force P_c applied at $x = L/2$. Representing it as a body force using the Dirac delta function,

$$b(x) = P_c\, \delta\left(x - \frac{L}{2}\right), \tag{6.160}$$

the consistent nodal force vector is

$$\mathbf{f}^{(e)} = \int_0^L \mathbf{N}^\top(x)\, b(x)\, dx = \int_0^L \mathbf{N}^\top(x)\, P_c \delta\left(x - \frac{L}{2}\right)\, dx. \tag{6.161}$$

Factoring out P_c,

$$\mathbf{f}^{(e)} = P_c \int_0^L \mathbf{N}^\top(x)\, \delta\left(x - \frac{L}{2}\right)\, dx. \tag{6.162}$$

Using the sifting property of the Dirac delta function (see Section 4.7.1 of [6]),

$$\int_a^b f(x)\, \delta(x - x_0), dx = f(x_0), \quad x_0 \in (a, b), \tag{6.163}$$

the integral reduces to a simple evaluation of the shape functions at the load location:

$$\mathbf{f}^{(e)} = P_c\, \mathbf{N}^\top\left(\frac{L}{2}\right). \tag{6.164}$$

For a standard two-node linear element of length L, the shape functions are

$$N_1(x) = 1 - \frac{x}{L}, \qquad N_2(x) = \frac{x}{L}. \tag{6.165}$$

Evaluating at $x = L/2$ gives

$$N_1\left(\frac{L}{2}\right) = \frac{1}{2}, \qquad N_2\left(\frac{L}{2}\right) = \frac{1}{2}. \tag{6.166}$$

Equivalently, in the natural coordinate system ($\xi = 0$ at the element center),

$$\mathbf{f}^{(e)} = P_c \begin{bmatrix} 1/2 \\ 1/2 \end{bmatrix}. \tag{6.167}$$

This shows that a point load applied at the center of a linear element is distributed equally to its two nodes. If the load is applied exactly at an element boundary, it reduces to a single nodal force contribution.

The following Python script computes the nodal displacements for a bar subjected to a concentrated force P_c at an arbitrary location x_c. The solver framework remains unchanged; the only modification is the automatic identification of the host element and the consistent distribution of the point load to its nodes.

```
def FEM_solver_1D_Pc(ne, L, E, A, xc, Pc):
    '''FEM solver for 1D bar under a point load Pc at xc in the
       bar meshed uniformly with ne elements. It computes both
       displacement and element stress.
    '''
    nn = ne + 1                 # Mesh
    coords = np.linspace(0, L, nn)
    conn = [np.array([i, i+1]) for i in range(ne)]

    # Global system
    K = np.zeros((nn, nn))
    F = np.zeros(nn)
    xi_g, w_g = gauss(3)        # for Gaussian quadrature

    # Assembly: stiffness matrix
    for el in conn:
        x_e = coords[el]
        ke = np.zeros((2, 2))
        for xi, w in zip(xi_g, w_g):
            N, dN_dxi = shape_p1(xi)
            J = np.dot(dN_dxi, x_e)
            dN_dx = dN_dxi / J
            ke += E * A * np.outer(dN_dx, dN_dx) * J * w
        K[np.ix_(el, el)] += ke

    # Find host element containing concentrated load
    load_el = None
    for e, el in enumerate(conn):
        xL = coords[el[0]]
        xR = coords[el[1]]
        # left-closed, right-open
```

```
        if (xc >= xL) and (xc < xR):
            load_el = e
            break
    if load_el is None:
        load_el = ne - 1

    # Apply concentrated load in weak form
    el = conn[load_el]
    x_e = coords[el]
    # map physical point -> natural coordinate
    xi_c = 2 * (xc - x_e[0]) / (x_e[1] - x_e[0]) - 1
    # shape functions at load point
    N_c, _ = shape_p1(xi_c)
    # consistent nodal force vector
    F[el] += N_c * Pc

    # Boundary condition: u(0)=0
    K_bc = K[1:, 1:]
    F_bc = F[1:]
    # solve
    u_sol = np.linalg.solve(K_bc, F_bc)
    # reconstruct full displacement vector
    u = np.zeros(nn)
    u[1:] = u_sol

    # Compute element stresses
    stress = np.zeros(ne)
    for e, el in enumerate(conn):
        x_e = coords[el]
        u_e = u[el]
        # linear element: strain/stress constant in element
        dN_dxi = np.array([-0.5, 0.5])
        J = np.dot(dN_dxi, x_e)
        dN_dx = dN_dxi / J
        epsilon = np.dot(dN_dx, u_e)  # strain
        stress[e] = E * epsilon       # stress

    return coords, u, stress
```

The following code snippet computes the nodal displacements of a uniform bar subjected to a point load P_c located at an internal position x_c, using linear finite elements.

In the implementation below, the bar is discretized into a uniform mesh of equal-length elements, with the concentrated force applied at the bar midpoint. The number of elements can be chosen arbitrarily:

1. **Odd number of elements:** The point load coincides with the center of the middle element. In this case, the FEM solution is **exact at all nodal points**, while the displacement field inside the central element is only a linear approximation of the true (locally non-smooth) response.

2. **Even number of elements:** The point load lies exactly on an inter-element node shared by two adjacent elements. Since the singularity is then represented directly at a nodal point, the linear interpolation is fully consistent with the piecewise structure of the exact solution. As a result, the FEM solution reproduces the **exact displacement field over the entire domain**.

```
# Compute the nodal displacements of a bar under Pc at xc:
L  = 1.0; A = 1.0                 # length, area of the bar
E  = 1.0                          # Young's modulus
Ne = 5                            # use Ne elements
xc = 0.5*L; Pc = 1.0
coords, u_fem, σ_fem = FEM_solver_1D_Pc(Ne, L, E, A, xc, Pc)#order 1

np.set_printoptions(precision=5,suppress=True) # Digits in print-out
u_exact, σ_exact = exact_solution(coords, L, E, A, xc, Pc)
print(f"FEM displ. solution using {Ne} linear elements:\n{u_fem}")
print(f"Exact displ. solution at nodes:\n{u_exact}")
print(f"FEM stress solution using {Ne} linear elements:\n{σ_fem}")
print(f"Exact stress solution at nodes:\n{σ_exact}")
```

```
FEM displ. solution using 5 linear elements:
[0.  0.2 0.4 0.5 0.5 0.5]
Exact displ. solution at nodes:
[0.  0.2 0.4 0.5 0.5 0.5]
FEM stress solution using 5 linear elements:
[1.  1.  0.5 0.  0. ]
Exact stress solution at nodes:
[1. 1. 1. 0. 0. 0.]
```

It is observed that the computed nodal displacements match the exact analytical values, illustrating the nodal accuracy property of the finite element method for one-dimensional problems with point loads.

6.7.21 Convergence Study

To assess global accuracy and convergence behavior, we employ the standard L^2 displacement error norm and the energy norm defined in Eqs. (6.135) and (6.136), respectively. The analysis is carried out over a sequence of progressively refined meshes.

From these simulations, we compute the asymptotic convergence rates and generate the corresponding log–log convergence curves, following the same procedure as in the previous case studies. The code implementation is provided below.

```
def compute_L2_error_Pc(coords, conn, uh, exact_func, L, E,A, xc,Pc):
    '''L2 displacement error of FEM solution for a 1D bar under
       a concentrated force Pc at xc.'''
    xi_g, w_g = np.polynomial.legendre.leggauss(5)
    err2 = 0.0
    for el in conn:
        x_e = coords[el]
        u_e = uh[el]
        h = x_e[1] - x_e[0]
        for xi, w in zip(xi_g, w_g):
            N = np.array([0.5*(1-xi), 0.5*(1+xi)])
            x = np.dot(N, x_e)
            uh_x = np.dot(N, u_e)
            uex, _ = exact_func(x, L, E, A, xc, Pc)
            err2 += (uex - uh_x)**2 * (h/2) * w
    return np.sqrt(err2)

def compute_energy_error_Pc(coords, conn, uh, L, E, A, xc, Pc):
    '''L2 energy error of FEM solution for a 1D bar under
       a concentrated force Pc at xc.'''
    xi_g, w_g = np.polynomial.legendre.leggauss(5)
    err2 = 0.0
    for el in conn:
        x_e = coords[el]
        u_e = uh[el]
        h = x_e[1] - x_e[0]
        # constant FE strain/stress in linear element
        dN_dxi = np.array([-0.5, 0.5])
        J = h / 2
        dN_dx = dN_dxi / J
        du_dx_h = np.dot(dN_dx, u_e)
        sigma_h = E * du_dx_h

        for xi, w in zip(xi_g, w_g):
            # shape functions
            N = np.array([0.5 * (1 - xi),
                          0.5 * (1 + xi)])
            x = np.dot(N, x_e) # physical coordinate

            # EXACT stress (GLOBAL definition)
            if x < xc:
                sigma_exact = Pc / A
            else:
                sigma_exact = 0.0
```

```
            sigma_err = sigma_exact - sigma_h   # stress error
            err2 += sigma_err**2 * J * w
    return np.sqrt(err2)

def convergence_study_Pc(L, E, A, xc, Pc):
    #orders = [2, 4, 8, 16, 32, 64]  #even element number,zero errors
    orders = [3, 5, 9, 17, 33, 65]  #odd element number

    h_vals = []
    L2_errors = []
    H1_errors = []
    for ne in orders:
        coords, u_h, _ = FEM_solver_1D_Pc(ne, L, E, A, xc, Pc)
        conn = [np.array([i, i+1]) for i in range(ne)]
        h = L / ne
        h_vals.append(h)

        L2_err = compute_L2_error_Pc(coords, conn, u_h,
            exact_solution, L, E, A, xc, Pc)
        H1_err = compute_energy_error_Pc(coords, conn, u_h,
                L, E, A, xc, Pc )

        L2_errors.append(L2_err)
        H1_errors.append(H1_err)

        print(f"ne={ne:3d}  h={h:.3e}  L2={L2_err:.3e} " \
              f"Energy={H1_err:.3e}")
    return h_vals, L2_errors, H1_errors

# run + plot
h, L2, H1 = convergence_study_Pc(L, E, A, xc, Pc)
# convergence rates:
def rate(h, e):
    return np.polyfit(np.log(h), np.log(e), 1)[0]

r_L2 = rate(h, L2)
r_H1 = rate(h, H1)
print(f"Convergence rate, r_L2 = {r_L2}")
print(f"Convergence rate, r_H1 = {r_H1}")

plt.figure(figsize=(7, 5), dpi=300)
plt.loglog(h, L2, 'o-', label=f"L2 error (rate={r_L2:.2f})")
plt.loglog(h, H1, 's-', label=f"Energy error (rate={r_H1:.2f})")
```

```
plt.gca().invert_xaxis()
plt.xlabel("h")
plt.ylabel("Error")
plt.legend()
plt.grid(True, which="both")
plt.title("Convergence of FEM with point load at center")
plt.show()
```

```
ne=  3  h=3.333e-01  L2=2.999e-02 Energy=2.887e-01
ne=  5  h=2.000e-01  L2=1.394e-02 Energy=2.236e-01
ne=  9  h=1.111e-01  L2=5.771e-03 Energy=1.667e-01
ne= 17  h=5.882e-02  L2=2.223e-03 Energy=1.213e-01
ne= 33  h=3.030e-02  L2=8.220e-04 Energy=8.704e-02
ne= 65  h=1.538e-02  L2=2.973e-04 Energy=6.202e-02
Convergence rate, r_L2 = 1.5000000000008307
Convergence rate, r_H1 = 0.49999999999999856
```

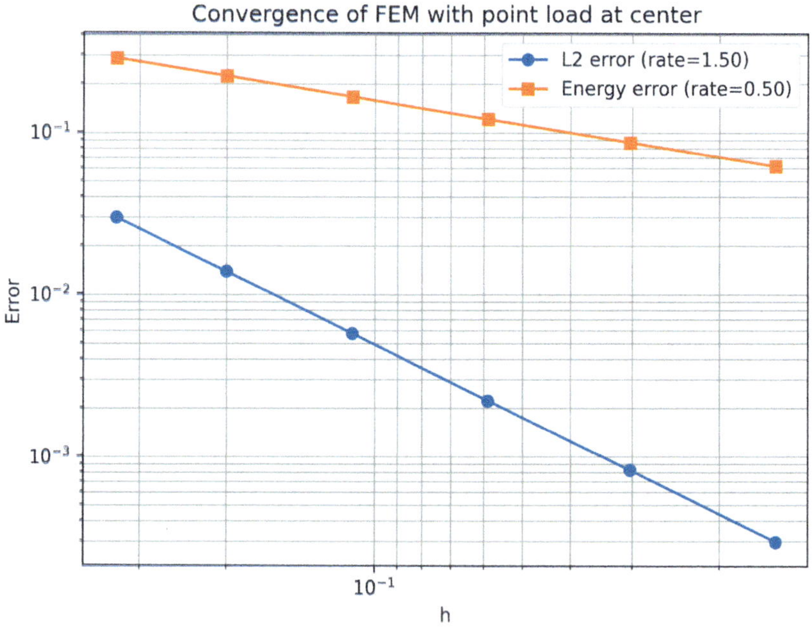

As observed, the FEM solution still converges; however, the convergence rate is reduced by approximately 0.5 relative to the optimal rates in both the displacement and energy norms due to the presence of the singular point load. These results are consistent with Eq. (6.156).

6.8 Final Remarks and Key Takeaways

1. **Global equilibrium vs. field accuracy:** Global static equilibrium is always satisfied exactly in a displacement-based finite element formulation. However, pointwise accuracy of the displacement and stress fields—measured in global L^2 and energy norms—is governed by both the polynomial approximation order and the regularity of the exact solution.

2. **Superconvergent post-processing via local ODE reconstruction:** In one-dimensional problems, the nodal displacements obtained from FEM are exact in a global equilibrium sense. This property allows the continuous solution within each element to be reconstructed by locally solving the element-level ordinary differential equation (ODE) [4], provided the body force $b(x)$ is integrable. In this way, the FEM solution serves as a consistent starting point for recovering piecewise-exact continuum fields in axial systems. The resulting post-processed solution exhibits a form of superconvergence, where the convergence rate exceeds the optimal rate of the original FEM approximation.

3. **Superconvergent stress post-processing:** A different form of superconvergence can also be achieved for stress recovery by computing nodal stresses during post-processing through averaging the stresses from adjacent elements. For one-dimensional problems, however, this averaging approach is generally less effective than reconstructing the solution by solving the local ODE for each bar element. The nodal stress averaging approach will be demonstrated for two-dimensional problems in the next chapter.

4. **Impact of solution regularity:** Higher-order elements (P_2, P_3, etc.) provide improved accuracy per degree of freedom when the exact solution is sufficiently smooth ($u \in H^{p+1}$). When singularities or steep gradients are present, this advantage may be significantly reduced.

5. **Linear element performance (P_1):** Linear elements are robust, simple, and computationally efficient, but exhibit lower asymptotic convergence rates, requiring h-refinement to achieve high accuracy in stress-dominated problems.

6. **Quadratic element performance (P_2):** Quadratic elements offer substantially higher accuracy for smooth solutions and are particularly effective for problems involving smoothly varying loads and fields.

7. **Effect of reduced regularity:** Loss of solution regularity reduces the achievable convergence rates in both displacement and energy norms. In practical applications, aligning singularities such as point loads or geometric discontinuities with mesh nodes is an effective strategy to restore optimal convergence behavior.

The following chapters extend these concepts to two-dimensional (2D) and three-dimensional (3D) continuum formulations and their corresponding finite element implementations.

6.9 Exercises

6.9.1 Part A: Theoretical and Variational Principles

1: Completeness and Constant Strain Conditions

Prove that the 1D linear Lagrangian shape functions $N_1(\xi) = \frac{1}{2}(1-\xi)$ and $N_2(\xi) = \frac{1}{2}(1+\xi)$ satisfy the partition of unity property:

$$\sum_{i=1}^{2} N_i(\xi) = 1$$

Explain the physical significance of this property in the context of rigid-body translation and the reproduction of constant strain states.

2: Alternative Virtual Field Construction

In the proof for nodal exactness, a piecewise-linear virtual field $\mathbf{v}^{(i,k)}$ was constructed to isolate node i. Suppose we choose a continuous, globally differentiable test function $v(x) = \sin(\pi x/L)$ for a single-element bar fixed at $x = 0$ and free at $x = L$.

1. Is this test function kinematically admissible for the weak form?
2. Does the Galerkin orthogonality condition guarantee that the finite element solution will be exact at any point under this virtual field? Justify your answer.

3: Cea's Lemma and Best Approximation

Cea's Lemma states that the finite element solution u_h is the best approximation of the exact solution u in the energy norm:

$$\|u - u_h\|_E = \min_{v_h \in V_h} \|u - v_h\|_E$$

Using the principle of virtual work and Galerkin orthogonality, derive this relation for a 1D elastic bar with varying cross-section $A(x)$ and Young's modulus $E(x)$.

4: Loss of Regularity and Step Loads

The regular convergence rates $\mathcal{O}(h^{p+1})$ for displacement and $\mathcal{O}(h^p)$ for energy assume that the exact solution possesses sufficient smoothness, specifically $u(x) \in H^{p+1}(0, L)$. Suppose a Heaviside step loading $q(x) = H(x - L/2)$ is applied over the right half-span of a uniform axial bar of length L. Discuss the Sobolev regularity of the exact displacement solution $u(x)$.

5: Equivalence of Energy Norm Formulations

Prove that for a 1D bar with constant EA, the stress-based error measure:

$$\|e\|_E = \left(\int_0^L \frac{1}{EA}(\sigma - \sigma_h)^2 \, dx \right)^{1/2}$$

is mathematically identical to the strain-based displacement gradient error measure:

$$\|e\|_E = \left(\int_0^L EA(u' - u_h')^2 \, dx \right)^{1/2}$$

State explicitly where Hooke's law is invoked in your proof.

6.9.2 Part B: Manual Derivations and Hand Calculations

6: Two-Element Assembly under Linear Body Force

Consider a bar of length L, constant EA, fixed at $x = 0$ and free at $x = L$. It is subject to a linearly varying body force $b(x) = b_0 \frac{x}{L}$. Discretize the bar into two equal-length linear (P_1) elements.

1. Derive the consistent element nodal force vectors $\mathbf{f}^{(b,1)}$ and $\mathbf{f}^{(b,2)}$ by performing analytical integration over the natural domain $\xi \in [-1, 1]$.
2. Assemble the unconstrained global system $\mathbf{Ku} = \mathbf{F}$.

7: Manual System Solution and Nodal Exactness

Using the assembled equations derived in **6**:

1. Enforce the essential boundary condition $u_1 = 0$ and solve for the discrete displacements u_2 and u_3.
2. Calculate the exact analytical displacement field $u(x)$ by solving the governing differential equation $EA\frac{d^2u}{dx^2} + b(x) = 0$ with matching boundary conditions.
3. Verify whether the nodal displacements u_2 and u_3 are mathematically exact.

8: Quadratic (P_2) Shape Function Derivation

A quadratic 1D bar element contains three nodes: Node 1 at $\xi = -1$, Node 2 at $\xi = 0$, and Node 3 at $\xi = 1$.

1. Use Lagrange interpolation polynomials to derive the analytical expressions for $N_1(\xi)$, $N_2(\xi)$, and $N_3(\xi)$.
2. Derive the corresponding natural derivatives $\frac{dN_i}{d\xi}$ for all three nodes.

9: Single Quadratic Element Stiffness Matrix

Using the shape functions derived in **8**, evaluate the local stiffness matrix $\mathbf{k}$ for a single quadratic element of physical length l_e and constant structural stiffness EA. Perform the integration analytically over the domain $\xi \in [-1, 1]$ using the relationship:

$$\mathbf{k} = \int_{-1}^{1} \mathbf{B}^T EA\mathbf{B} J \, d\xi$$

10: Gauss Quadrature Verification

Consider the polynomial function $f(\xi) = 3\xi^5 - 2\xi^4 + \xi^2 - 5$.

1. Calculate the exact integral $\int_{-1}^{1} f(\xi)\, d\xi$ analytically.
2. Evaluate the integral numerically using 3-point Gauss-Legendre quadrature.
3. Determine if the 3-point quadrature rule is exact for this polynomial degree and explain why based on orthogonal polynomial theory.

6.9.3 Part C: Computational Extensions and Algorithmic Analysis

11: Implementation of Quadratic (P_2) Assembly Testing

Look at the provided function `assemble_system_1d()`.

1. Trace how the connectivity matrix `mesh['conn']` changes when transitioning from linear elements to quadratic elements for a domain divided into 4 geographic regions. Write out the explicit index layout for `mesh['conn']` when `order = 2`.
2. Explain how the global matrix assembly loop handles the overlapping degrees of freedom between adjacent quadratic elements.

12: Programming a Post-Processing Error Norm Module

Write a Python function stub named `compute_errors_1d(mesh, material, U, exact_u_func, exact_stress_func)` that loops over an existing mesh arrangement and computes the total numerical values for both $\|e_u\|_{L^2}$ and $\|e\|_E$. Ensure your function uses a 5-point Gauss quadrature rule per element to guarantee integration precision regardless of the shape function order.

13: Under-Integration Analysis

In the textbook section, it is noted that reducing the number of Gauss integration points degrades nodal exactness.

1. If a linear element mesh undergoes a sinusoidal body force integration using only **1 Gauss point**, what is the maximum polynomial order of a force distribution that can be integrated exactly?
2. Explain algorithmically how this "under-integration" manifests as an error in the global displacement array U.

14: Non-Uniform Cross-Sections ($A(x)$)

Modify the mathematical structure of the `assemble_system_1d` method to accommodate a bar with a linearly tapering cross-sectional area $A(x) = A_0(1 - \frac{x}{2L})$.

1. Describe how the area parameter inside the nested integration loops must be updated using the natural coordinate shape functions.
2. Will a standard 2-point Gauss quadrature rule still compute the exact stiffness matrix for a linear element with this varying area profile? Explain your reasoning.

15: Extraction of Empirical Convergence Rates

Suppose your finite element pipeline generates the following error profile data sequence under uniform h-refinement:

h (Element Size)	$\lVert e_u \rVert_{L^2}$
0.500	2.510×10^{-2}
0.250	6.275×10^{-3}
0.125	1.569×10^{-3}

1. Calculate the empirical convergence rate k using the data points from the step transitions via the log-log mapping approach.
2. Based on your calculated value of k, determine whether this finite element model is utilizing linear (P_1) or quadratic (P_2) interpolation.

Chapter 7

FEM Solver for 2D Solids

```
# Place cursor here, and press Ctrl+Enter to import dependencies.
import sys                              # for accessing the computer system
import os
import time
import importlib
from sympy import init_printing
import numpy as np
import sympy as sp
import matplotlib.pyplot as plt                                    # For plots
from IPython.display import display, Math
from sympy import symbols, lambdify, latex, Matrix, Function

init_printing(use_unicode=True)     # For latex-like quality printing
np.set_printoptions(precision=4,suppress=True) # Digits in print-out
```

Building on the one-dimensional finite element formulation developed in the previous chapter, we now extend the framework to two-dimensional (2D) continua. This chapter presents a detailed and systematic derivation of the finite element matrices, implementation codes, including:

- geometry and interpolation,
- strain–displacement relations,
- constitutive modeling,
- element stiffness and mass matrices,
- consistent nodal force vectors,
- and a detailed FEM workflow and solver for 2D solid mechanics problems.

7.1 T3 Element Matrices and Codes

The simplest and most widely used 2D continuum element is the three-node linear triangular element (T3), also known as the Constant Strain Triangle (CST).

7.1.1 Problem Setup

Consider a triangular element in a 2D Cartesian coordinate system with nodes ordered counterclockwise to ensure a positive element area (see Fig. 4.2):

- **Node 1:** (x_1, y_1) with displacements (u_1, v_1)
- **Node 2:** (x_2, y_2) with displacements (u_2, v_2)
- **Node 3:** (x_3, y_3) with displacements (u_3, v_3)

The material and geometric parameters are:

- Young's modulus: E
- Poisson's ratio: ν
- Density: ρ
- Thickness: t (assumed constant over the element)

7.1.2 Finite Element Formulation

The element nodal displacement vector collects all translational degrees of freedom:

$$\mathbf{d}_e = \begin{bmatrix} u_1 & v_1 & u_2 & v_2 & u_3 & v_3 \end{bmatrix}^\top \quad (6 \times 1) \tag{7.1}$$

7.1.3 Displacement Approximation

The continuous displacement field $\mathbf{u}(x, y) = [u(x, y), v(x, y)]^\top$ is approximated using linear shape functions:

$$\underset{6\times 1}{\mathbf{d}_e} = \begin{bmatrix} u_1 & v_1 & u_2 & v_2 & u_3 & v_3 \end{bmatrix}^\top \tag{7.2}$$

In compact matrix form:

$$\mathbf{u}(x, y) = \begin{bmatrix} u(x, y) \\ v(x, y) \end{bmatrix} = \mathbf{N}_e(x, y), \mathbf{d}_e \tag{7.3}$$

where the element shape function matrix is

$$\begin{aligned} \mathbf{N}_e &= \begin{bmatrix} N_1 & 0 & N_2 & 0 & N_3 & 0 \\ 0 & N_1 & 0 & N_2 & 0 & N_3 \end{bmatrix}_{2\times 6} \\ &= \begin{bmatrix} \mathbf{N}_1 & \mathbf{N}_2 & \mathbf{N}_3 \end{bmatrix} \end{aligned} \tag{7.4}$$

For each node, the nodal submatrix is

$$\mathbf{N}_i = \begin{bmatrix} N_i & 0 \\ 0 & N_i \end{bmatrix}. \tag{7.5}$$

7.1.4 Shape Function Properties

The scalar shape functions $N_i(x, y)$ satisfy the standard finite element consistency conditions:

$$N_i(x_j, y_j) = \delta_{ij}, \qquad \sum_{i=1}^{3} N_i(x, y) = 1. \tag{7.6}$$

These properties ensure exact reproduction of rigid body motions and guarantee displacement continuity across element boundaries.

7.1.4.1 Strain–Displacement Relation

Under the assumption of small deformations, the infinitesimal strain vector for a 2D continuum is defined as ((see Chapter 2)):

$$\varepsilon = \begin{bmatrix} \varepsilon_x \\ \varepsilon_y \\ \gamma_{xy} \end{bmatrix} = \begin{bmatrix} \frac{\partial u}{\partial x} \\ \frac{\partial v}{\partial y} \\ \frac{\partial u}{\partial y} + \frac{\partial v}{\partial x} \end{bmatrix} \tag{7.7}$$

Substituting the displacement interpolation into the strain definition leads to the standard strain–displacement relation:

$$\underset{3\times1}{\varepsilon} = \underset{3\times6}{\mathbf{B}_e};\ \underset{6\times1}{\mathbf{d}_e} \tag{7.8}$$

where the strain–displacement matrix $\mathbf{B}_e$ is assembled as:

$$\begin{aligned} \mathbf{B}_e &= \begin{bmatrix} b_1 & 0 & b_2 & 0 & b_3 & 0 \\ 0 & c_1 & 0 & c_2 & 0 & c_3 \\ c_1 & b_1 & c_2 & b_2 & c_3 & b_3 \end{bmatrix} \\ &= \begin{bmatrix} \mathbf{B}_1 & \mathbf{B}_2 & \mathbf{B}_3 \end{bmatrix} \end{aligned} \tag{7.9}$$

with the nodal strain operators defined by:

$$\mathbf{B}_i = \begin{bmatrix} b_i & 0 \\ 0 & c_i \\ c_i & b_i \end{bmatrix}_{3\times2} \tag{7.10}$$

The coefficients b_i and c_i are given by the spatial derivatives of the shape functions: $b_i = \frac{\partial N_i}{\partial x}$ and $c_i = \frac{\partial N_i}{\partial y}$.

Key Observation: Since the shape functions $N_i(x, y)$ are linear polynomials, their derivatives b_i and c_i are constants over the element domain. As a result, $\mathbf{B}_e$ is constant within each element, which is why the T3 element is referred to as a **Constant Strain Triangle (CST)**.

7.1.4.2 Constitutive Relation

For an isotropic linear elastic material, the stress–strain relationship is given by:

$$\sigma = \mathbf{C}\varepsilon \tag{7.11}$$

The material stiffness matrix $\mathbf{C}$ has the general form:

$$\mathbf{C} = \begin{bmatrix} C_{11} & C_{12} & 0 \\ C_{12} & C_{22} & 0 \\ 0 & 0 & C_{33} \end{bmatrix}_{3\times 3} \tag{7.12}$$

Depending on the modeling assumption, $\mathbf{C}$ reduces to standard special cases:

- **Plane Stress:** Typically used for thin structures loaded in-plane (see Eq.2.27).
- **Plane Strain:** Typically used for thick or constrained structures in the out-of-plane direction (see Eq.2.29).

The implementation for computing $\mathbf{C}$ is provided in the following code snippet and will be used in subsequent formulations.

```
## Material Matrix C
def get_C_2D(E, nu, plane_stress=True):
    '''Form the matrix of material elatic constants C for 2D solids
    '''
    if plane_stress:
        return (E / (1 - nu**2)) * np.array([
            [1, nu,       0    ],
            [nu, 1,       0    ],
            [0,  0, (1-nu)/2] ])
    else:  # plane strain
        return (E / ((1+nu)*(1-2*nu))) * np.array([
            [1-nu,    nu,       0      ],
            [  nu, 1-nu,       0      ],
            [   0,     0, (1-2*nu)/2] ])
```

7.1.5 T3 Element Stiffness Matrix

The element stiffness matrix $\mathbf{k}_e$ is derived through the domain integral (see Chapter 5):

$$\underset{6\times 6}{\mathbf{k}_e} = \int_{\Omega_e} \mathbf{B}_e^\top \mathbf{C} \mathbf{B}_e \, d\Omega \tag{7.13}$$

Since the strain-displacement matrix $\mathbf{B}_e$ and constitutive matrix $\mathbf{C}$ are constant throughout the integration domain, they can be pulled outside the integral operator:

$$\mathbf{k}_e = \mathbf{B}_e^\top \mathbf{C} \mathbf{B}_e \int_{\Omega_e} d\Omega = \boxed{tA_e \mathbf{B}_e^\top \mathbf{C} \mathbf{B}_e} \tag{7.14}$$

where A_e is the area of the triangle.

7.1.6 Stiffness Matrix in Nodal Submatrix form

Equations (7.14) can be expanded into the following **nodal submatrix** blocks:

$$\mathbf{k}_e = tA_e \begin{bmatrix} \mathbf{B}_{e1}^\top \mathbf{C} \mathbf{B}_{e1} & \mathbf{B}_{e1}^\top \mathbf{C} \mathbf{B}_{e2} & \mathbf{B}_{e1}^\top \mathbf{C} \mathbf{B}_{e3} \\ \mathbf{B}_{e2}^\top \mathbf{C} \mathbf{B}_{e1} & \mathbf{B}_{e2}^\top \mathbf{C} \mathbf{B}_{e2} & \mathbf{B}_{e2}^\top \mathbf{C} \mathbf{B}_{e3} \\ \mathbf{B}_{e3}^\top \mathbf{C} \mathbf{B}_{e1} & \mathbf{B}_{e3}^\top \mathbf{C} \mathbf{B}_{e2} & \mathbf{B}_{e3}^\top \mathbf{C} \mathbf{B}_{e3} \end{bmatrix}_{6\times 6} \tag{7.15}$$

This can be written compactly form:

$$\boxed{\mathbf{k}_e = \begin{bmatrix} \mathbf{k}_{11} & \mathbf{k}_{12} & \mathbf{k}_{13} \\ \mathbf{k}_{21} & \mathbf{k}_{22} & \mathbf{k}_{23} \\ \mathbf{k}_{31} & \mathbf{k}_{32} & \mathbf{k}_{33} \end{bmatrix}} \tag{7.16}$$

Each constituent 2×2 nodal submatrix is defined by:

$$\boxed{\underset{2\times 2}{\mathbf{k}_{ij}} = tA_e \mathbf{B}_{ei}^\top \mathbf{C} \mathbf{B}_{ej} = V_e \bar{\mathbf{k}}_{ij}} \tag{7.17}$$

where $V_e = tA_e$ represents the total physical volume of the element, and

$$\underset{2\times 2}{\bar{\mathbf{k}}_{ij}} = \underset{2\times 3}{\mathbf{B}_{ei}^\top} \underset{3\times 3}{\mathbf{C}} \underset{3\times 2}{\mathbf{B}_{ej}} \tag{7.18}$$

is the nodal submatrix that characterizes the inner energy density distribution of the stiffness matrix.

> Note that the nodal submatrices $\mathbf{k}_{ij}$ are generally **not** symmetric unless they lie on the main diagonal ($i = j$).

7.1.6.1 Standard Code for T3 Stiffness Matrix

The following is Python function that computes the element stiffness matrix for T3 elements, using Eq.(7.18)

```
def element_ke_T3_BCB(coords, C):
    '''
    Compute the stiffness matrix for T3 elements in a FEM model
    with nodal coordinates `coords` given and matrial matrix C.
    It uses the original matrix products form.
    '''
    x1, y1 = coords[0]
    x2, y2 = coords[1]
    x3, y3 = coords[2]

    A = 0.5 * np.linalg.det(np.array([
        [1, x1, y1],
        [1, x2, y2],
        [1, x3, y3] ]))

    b = np.array([y2 - y3, y3 - y1, y1 - y2])
    c = np.array([x3 - x2, x1 - x3, x2 - x1])

    B = np.zeros((3, 6))
    for i in range(3):
        B[0, 2*i]   = b[i]
        B[1, 2*i+1] = c[i]
        B[2, 2*i]   = c[i]
        B[2, 2*i+1] = b[i]

    B /= (2*A)
    ke = A * (B.T @ C @ B)
    return ke, B
```

```
# An example stiffness matrix:
E = 1.0; nu = 0.3
C = get_C_2D(E, nu)

# Define an arbitrary T3 element:
T3_ele = np.array([[0, 0.1], [1, 0.2], [0.5, 0.51]])

ke_standard = element_ke_T3_BCB(T3_ele, C)[0]
print("T3 Stiffness Matrix:")
print(ke_standard)
```

```
T3 Stiffness Matrix:
[[ 0.2802  0.1538 -0.0604 -0.0385 -0.2198 -0.1152]
 [ 0.1538  0.4329 -0.0111  0.3137 -0.1427 -0.7466]
 [-0.0604 -0.0111  0.3901 -0.2034 -0.3297  0.2144]
 [-0.0385  0.3137 -0.2034  0.4714  0.2419 -0.785 ]
 [-0.2198 -0.1427 -0.3297  0.2419  0.5495 -0.0992]
 [-0.1152 -0.7466  0.2144 -0.785  -0.0992  1.5316]]
```

To accelerate finite element computations in production settings, element matrices are often evaluated in batches. This allows most operations to be vectorized, improving performance even when executed on a CPU.

The code below, adapted from Gemini, implements this approach for T3 elements. The mesh is provided as a NumPy array with shape $(N, 3, 2)$, where N is the number of triangular elements, and each element contains the (x, y) coordinates of its three nodes.

The same implementation also supports the single-element case. In that situation, the input array should have shape $(3, 2)$, representing one triangular element defined by its three nodal coordinates.

```python
def batch_ke_T3_BCB(coords, C):
    """
    Vectorized computation of T3 element stiffness matrices.It works
    for both a single element (3,2) and a batch of elements (N,3,2).

    Parameters:
    -----------
    coords : ndarray of shape (3, 2) or (N, 3, 2)
        Nodal coordinates for the element(s).
    C : ndarray of shape (3, 3)
        Material elasticity matrix for 2D plane stress/strain.

    Returns:
    --------
    ke : ndarray of shape (6, 6) or (N, 6, 6)
        Stiffness matrix/matrices.
    B : ndarray of shape (3, 6) or (N, 3, 6)
        Strain-displacement matrix/matrices.
    """
    # 1. Handle dimensionality dynamically to accept single or
    #    multiple elements
    is_single_element = (coords.ndim == 2)
    if is_single_element:
        # Temporarily add a batch dimension: (3, 2) -> (1, 3, 2)
```

```
        all_coords = coords[np.newaxis, :, :]
    else:
        all_coords = coords

    N = all_coords.shape[0]

    # 2. Extract coordinates across the batch
    x1, y1 = all_coords[:, 0, 0], all_coords[:, 0, 1]
    x2, y2 = all_coords[:, 1, 0], all_coords[:, 1, 1]
    x3, y3 = all_coords[:, 2, 0], all_coords[:, 2, 1]

    # 3. Compute Ae via 2D cross product formula (faster than det)
    A = 0.5 * ((x2 - x1) * (y3 - y1) - (x3 - x1) * (y2 - y1))

    # 4. Direct shape function gradient computation (scaled by 2A)
    #    Shape of b_all and c_all: (N, 3)
    b_all = np.column_stack((y2 - y3, y3 - y1, y1 - y2))
    c_all = np.column_stack((x3 - x2, x1 - x3, x2 - x1))

    # 5. Populate the Strain-Displacement Tensor (N, 3, 6) with
    #    stride slicing
    B_all = np.zeros((N, 3, 6))

    B_all[:, 0, 0::2] = b_all
    B_all[:, 1, 1::2] = c_all
    B_all[:, 2, 0::2] = c_all
    B_all[:, 2, 1::2] = b_all

    # Divide the entire tensor block by 2A at once
    B_all /= (2.0 * A[:, np.newaxis, np.newaxis])

    # 6. Batched Triple Matrix Product: A * (B^T @ C @ B)
    # BtC shape: (N, 6, 3)
    BtC = np.einsum('nji,jk->nik', B_all, C)
    # ke_all shape: (N, 6, 6)
    ke_all = np.einsum('nik,nkl->nil', BtC, B_all)

    # Scale by element areas
    ke_all *= A[:, np.newaxis, np.newaxis]

    # 7. Strip out the batch dimension if input was a single element
    if is_single_element:
        return ke_all[0], B_all[0]
```

```
    return ke_all, B_all
```

Let us verify the optimized code against the original standard one:

```
ke_standard = element_ke_T3_BCB(T3_ele, C)[0]          # check only ke
ke_standard_batch = batch_ke_T3_BCB(T3_ele, C)[0]

print(np.isclose(ke_standard, ke_standard_batch))
print(np.allclose(ke_standard, ke_standard_batch))
```

```
[[ True  True  True  True  True  True]
 [ True  True  True  True  True  True]
 [ True  True  True  True  True  True]
 [ True  True  True  True  True  True]
 [ True  True  True  True  True  True]
 [ True  True  True  True  True  True]]
True
```

```
Ne = 100_000                # test on Ne elements
start = time.time()
for i in range(Ne):         # run a number of times for accurate timing
    ke_standard = element_ke_T3_BCB(T3_ele, C)[0]

time_standard = time.time() - start
print(f"Time, the standard approach: {time_standard:.4f}s")

# Create a batch of Ne elements for testing purpose:
all_coords = np.repeat(T3_ele[np.newaxis, :, :], Ne, axis=0)
start = time.time()
ke_batch = batch_ke_T3_BCB(all_coords, C)[0]
time_batch = (time.time() - start)
print(f"Time, standard fast: {time_batch:.4f}s")
print(f"Time ratio: {time_standard/time_batch:.4f}")
```

```
Time, the standard approach: 4.0913s
Time, standard fast: 0.2183s
Time ratio: 18.7436
```

For a batch of $N_e = 100{,}000$ T3 elements, the vectorized implementation is approximately 15 times faster than the standard single-element loop formulation. This improvement is consistent with the expected benefits of interpreter-level vectorization and aligns with general performance trends predicted by Amdahl's Law, where speedups in this class of optimizations typically range from $10\times$ to $50\times$.

7.1.6.2 T3 Nodal Submatrix Analytical Formula

To further reduce computational cost, we now adopt an analytical formulation that minimizes the number of numerical operations required during assembly. The approach begins by using the SymPy script below to symbolically evaluate the nodal stiffness density submatrix $\bar{\mathbf{k}}_{ij}$ defined in Eq. (7.20). From this symbolic expression, closed-form analytical formulas are then derived for direct implementation in the finite element routine.

```
# -------------------------------------------------------------
# SymPy code for deriving stiffness matrix for T3 elements
# -------------------------------------------------------------
bi, ci = sp.symbols('b_i c_i')                        # Symbols
bj, cj = sp.symbols('b_j c_j')
C11, C12, C22, C33 = sp.symbols('C_11 C_12 C_22 C_33')

Bi = sp.Matrix([[bi, 0],                             # Define B_i
                [0,  ci],
                [ci, bi] ])
Bj = sp.Matrix([[bj, 0],                             # Define B_j
                [0,  cj],
                [cj, bj] ])

# Constitutive matrix C for general 2D plane problems:
C = sp.Matrix([[C11, C12, 0],
               [C12, C22, 0],
               [0,   0,   C33] ])

# Stiffness density:
kij_bar = sp.simplify(Bi.T * C * Bj)
#kij_bar  # to display
```

This symbolic derivation leads to an analytical expression for the nodal stiffness density submatrix $\bar{\mathbf{k}}_{ij}$:

$$\bar{\mathbf{k}}_{ij} = \begin{bmatrix} C_{11}b_ib_j + C_{33}c_ic_j & C_{12}b_ic_j + C_{33}c_ib_j \\ C_{33}b_ic_j + C_{12}c_ib_j & C_{22}c_ic_j + C_{33}b_ib_j \end{bmatrix} \tag{7.19}$$

This expression represents the 2×2 contribution of node pair (i, j) to the element stiffness matrix density. It can be further rewritten in a more compact form as:

$$\bar{\mathbf{k}}_{ij} = \begin{bmatrix} C * 11b_{ij} + C_{33}c_{ij} & C_{12}p_{ij} + C_{33}q_{ij} \\ C_{33}p_{ij} + C_{12}q_{ij} & C_{22}c_{ij} + C_{33}b_{ij} \end{bmatrix} \tag{7.20}$$

where the shorthand notations are defined as:

$$b_{ij} = b_ib_j, \qquad c_{ij} = c_ic_j, \qquad p_{ij} = b_ic_j, \qquad q_{ij} = c_ib_j \tag{7.21}$$

This compact formulation significantly reduces computational cost. Compared with the original expanded expression in Eq. (7.18), the number of arithmetic operations is reduced by approximately a factor of four. A simple operation count shows that the original form requires roughly 50 arithmetic operations, whereas the simplified analytical form requires only about 13.

Because this reduction occurs at the elemental kernel level, its impact accumulates across all elements in the mesh, making Eq. (7.20) particularly advantageous for production-scale FEM implementations.

7.1.6.3 Properties of the T3 Element Stiffness Matrix

- Symmetric and positive semi-definite
- Rank-deficient due to rigid-body modes (two translations and one in-plane rotation)
- No numerical integration required
- Fully analytical formulation with minimized operation count

7.1.6.4 NSAF Code for T3 Stiffness Matrix

The function `batch_ke_T3_NSAF` presented below computes the stiffness matrices of T3 elements using the Nodal Submatrix Analytical Formula (NSAF). The implementation is fully vectorized (courtesy of Gemini) and supports both single-element input and batch processing of multiple elements.

```
def batch_ke_T3_NSAF(coords, C):
    """
    Compute stiffness matrices for T3 elements using the Nodal
    Submatrix Analytical Formula (NSAF), with vectorization.It works
    for both a single element (3,2) and a batch of elements (N,3,2).

    Parameters:
    -----------
    coords : ndarray of shape (3,2) or (N,3,2)
        Nodal coordinates.
    C : ndarray of shape (3,3)
        Constitutive matrix.

    Returns:
    --------
    ke : ndarray of shape (6,6) or (N,6,6)
    B  : ndarray of shape (3,6) or (N,3,6)
    """
    # 1. Dynamically manage dimensionality for uniform vectorization
```

```
is_single_element = (coords.ndim == 2)
if is_single_element:
    all_coords = coords[np.newaxis, :, :]  # (3, 2) -> (1, 3, 2)
else:
    all_coords = coords

N = all_coords.shape[0]

# 2. Extract nodal coordinates across the batch
x1, y1 = all_coords[:, 0, 0], all_coords[:, 0, 1]
x2, y2 = all_coords[:, 1, 0], all_coords[:, 1, 1]
x3, y3 = all_coords[:, 2, 0], all_coords[:, 2, 1]

# 3. Fast area calculation (2D Cross-Product Sequence)
A = 0.5 * ((x2 - x1) * (y3 - y1) - (x3 - x1) * (y2 - y1))
scale = 1.0 / (4.0 * A)  # Shape (N,)

# 4. Compute baseline geometric vectors for all elements
b_all = np.column_stack((y2 - y3, y3 - y1, y1 - y2))  # (N, 3)
c_all = np.column_stack((x3 - x2, x1 - x3, x2 - x1))  # (N, 3)

# 5. Compute batched analytical outer products using
#    Einstein Summation
bij_all = np.einsum('ni,nj->nij', b_all, b_all)     # (N, 3, 3)
cij_all = np.einsum('ni,nj->nij', c_all, c_all)     # (N, 3, 3)
bicj_all = np.einsum('ni,nj->nij', b_all, c_all)    # (N, 3, 3)
cibj_all = np.einsum('ni,nj->nij', c_all, b_all)    # (N, 3, 3)

# 6. Extract material parameters
c11, c12, c22, c33 = C[0, 0], C[0, 1], C[1, 1], C[2, 2]

# 7. Construct the sub-component terms scaled by (1 / 4A)
# Shape of each block is (N, 3, 3)
V_scale = scale[:, np.newaxis, np.newaxis]

k_xx = V_scale * (c11 * bij_all + c33 * cij_all)
k_xy = V_scale * (c12 * bicj_all + c33 * cibj_all)
k_yx = V_scale * (c12 * cibj_all + c33 * bicj_all)
k_yy = V_scale * (c22 * cij_all + c33 * bij_all)

# 8. Assemble global (N,6,6) stiffness matrix via stride slicing
ke_all = np.zeros((N, 6, 6), dtype=float)

ke_all[:, 0::2, 0::2] = k_xx
```

```
    ke_all[:, 0::2, 1::2] = k_xy
    ke_all[:, 1::2, 0::2] = k_yx
    ke_all[:, 1::2, 1::2] = k_yy

    # 9. Completely vectorized Strain-Displacement Tensor (N, 3, 6)
    B_all = np.zeros((N, 3, 6), dtype=float)
    B_all[:, 0, 0::2] = b_all
    B_all[:, 1, 1::2] = c_all
    B_all[:, 2, 0::2] = c_all
    B_all[:, 2, 1::2] = b_all
    B_all *= (1.0 / (2.0 * A[:, np.newaxis, np.newaxis]))

    # 10. Strip out batch dimension if input was a single element
    if is_single_element:
        return ke_all[0], B_all[0]

    return ke_all, B_all
```

Let us verify the optimized code against the original standard one:

```
E = 1.0; nu = 0.3
C = get_C_2D(E, nu)
# Define an arbitrary T3 element:
T3_ele = np.array([[0, 0.1], [1, 0.2], [0.5, 0.51]])

ke_standard = element_ke_T3_BCB(T3_ele, C)[0]          # check only ke
ke_NSAF = batch_ke_T3_NSAF(T3_ele, C)[0]

print(np.isclose(ke_standard, ke_NSAF))
print(np.allclose(ke_standard, ke_NSAF))
```

```
[[ True  True  True  True  True  True]
 [ True  True  True  True  True  True]
 [ True  True  True  True  True  True]
 [ True  True  True  True  True  True]
 [ True  True  True  True  True  True]
 [ True  True  True  True  True  True]]
True
```

The results are in agreement to the machine precision. Let's study the performance of the NSAF formulas using these codes.

```
Ne = 10000  # create a batch of Ne elements for testing purpose:
all_coords = np.repeat(T3_ele[np.newaxis, :, :], Ne, axis=0)
```

```
start = time.time()
ke_batch = batch_ke_T3_BCB(all_coords, C)[0]
time_BCBbatch = (time.time() - start)
print(f"Time, time_BCBbatch: {time_BCBbatch:.4f}s")
#print(f"Time ratio: {time_standard/time_batch:.4f}")

start = time.time()
ke_batch = batch_ke_T3_NSAF(all_coords, C)[0]
time_NSAFbatch = (time.time() - start)
print(f"Time, NSAF batch: {time_NSAFbatch:.4f}s")
print(f"Time ratio: {time_BCBbatch/time_NSAFbatch:.4f}")
```

```
Time, time_BCBbatch: 0.0380s
Time, NSAF batch: 0.0146s
Time ratio: 2.5971
```

7.1.6.5 T3 NSAF Performance Evaluation

For a batch of $N_e = 10{,}000$ T3 elements, the vectorized NSAF implementation based on Eq. (7.20) is approximately 2–3 times faster than the corresponding vectorized implementation using the standard formulation in Eq. (7.18).

These performance tests are deliberately carried out in batch mode to expose the true computational differences predicted by the operation-count analysis. When the number of elements is sufficiently large, the fixed overhead associated with Python-level data allocation, array construction, and tensor setup becomes negligible. As a result, the runtime is dominated by low-level floating-point operations, allowing the efficiency of the underlying mathematical formulation to be observed directly.

In this regime, the measured performance reflects the intrinsic cost of the algebraic expressions rather than implementation artifacts. Therefore, we conclude:

> **Key Principle of Computational Efficiency** A formulation with a lower arithmetic operation count will consistently achieve better performance, provided that implementation overhead is sufficiently amortized over a large number of elements.

In the next chapter, similar benchmark studies will be extended to 3D formulations, where the impact of the underlying mathematical structure on computational efficiency becomes even more pronounced.

7.1.7 T3 Element Mass Matrix

7.1.7.1 Mass Matrix Formulation

The consistent element mass matrix $\mathbf{m}_e$ represents the inertial distribution of the element. Using Eq. (5.27) and noting that $d\Omega = t\,dA$ for a uniform thickness t, we obtain:

$$\underset{6\times 6}{\mathbf{m}_e} = \int_{\Omega_e} \rho \mathbf{N}_e^\top \mathbf{N}_e \, d\Omega$$

$$= \rho t \begin{bmatrix} \int_{A_e} \mathbf{N}_1^\top \mathbf{N}_1 \, dA & \int_{A_e} \mathbf{N}_1^\top \mathbf{N}_2 \, dA & \int_{A_e} \mathbf{N}_1^\top \mathbf{N}_3 \, dA \\ \int_{A_e} \mathbf{N}_2^\top \mathbf{N}_1 \, dA & \int_{A_e} \mathbf{N}_2^\top \mathbf{N}_2 \, dA & \int_{A_e} \mathbf{N}_2^\top \mathbf{N}_3 \, dA \\ \int_{A_e} \mathbf{N}_3^\top \mathbf{N}_1 \, dA & \int_{A_e} \mathbf{N}_3^\top \mathbf{N}_2 \, dA & \int_{A_e} \mathbf{N}_3^\top \mathbf{N}_3 \, dA \end{bmatrix} \tag{7.22}$$

This block structure can be written compactly as a symmetric 6×6 matrix composed of 2×2 nodal submatrices:

$$\boxed{\mathbf{m}_e = \begin{bmatrix} \mathbf{m}_{11} & \mathbf{m}_{12} & \mathbf{m}_{13} \\ \mathbf{m}_{21} & \mathbf{m}_{22} & \mathbf{m}_{23} \\ \mathbf{m}_{31} & \mathbf{m}_{32} & \mathbf{m}_{33} \end{bmatrix}} \tag{7.23}$$

where each nodal coupling submatrix is defined as:

$$\underset{2\times 2}{\mathbf{m}_{ij}} = \rho t \int_{A_e} \mathbf{N}_i^\top \mathbf{N}_j \, dA = \rho t \int_{A_e} \begin{bmatrix} N_i N_j & 0 \\ 0 & N_i N_j \end{bmatrix} dA \tag{7.24}$$

Because the nodal shape function matrices $\mathbf{N}_i$ are uncoupled between the u and v displacement components, the problem reduces to evaluating a single scalar quantity:

$$m_{ij} = \int_{A_e} N_i N_j \, dA \tag{7.25}$$

This shows that the mass matrix assembly can be efficiently reduced to scalar integrals of shape function products, making the nodal submatrix formulation equally advantageous for inertia computations.

7.1.7.2 Shape Function Products via Area Coordinates

Direct evaluation of these integrals in Cartesian coordinates (x, y) is cumbersome due to arbitrary triangular geometry. To simplify the formulation, we introduce natural **area (barycentric) coordinates** (L_1, L_2, L_3).

By definition, these coordinates represent normalized sub-areas of the triangle and are directly related to the linear shape functions:

$$L_1 + L_2 + L_3 = 1, \quad L_i \geq 0, \quad L_i = N_i(x, y) \tag{7.26}$$

This identity transforms the geometric interpolation problem into a standardized parameter space.

Coordinate Transformation

We map the physical triangular element to a reference right triangle using two independent variables:

$$L\_1 = \xi, \qquad L\_2 = \eta, \qquad L\_3 = 1 - \xi - \eta \tag{7.27}$$

The corresponding integration domain becomes:

$$0 \leq \xi \leq 1, \qquad 0 \leq \eta \leq 1 - \xi \tag{7.28}$$

Since the mapping from physical coordinates to area coordinates is linear, the Jacobian determinant is constant, and the differential area element becomes:

$$dA = \det(\mathbf{J}), d\xi d\eta = 2A\_e, d\xi d\eta$$

where A_e is the area of the triangular element.

Thus, the mass matrix integrals reduce to:

$$\int_{A_e} N_i N_j \, dA = 2A_e \int_0^1 \int_0^{1-\xi} L_i L_j, d\xi d\eta \tag{7.29}$$

which can now be evaluated analytically in closed form.

```
# ---------------------------------------------------------------
# SymPy code for integrals in the mass matrix for T3 elements
# ---------------------------------------------------------------
xi, eta = sp.symbols('xi eta')   # Natural (area) coordinates
A = sp.symbols('A', positive=True)             # Triangle area

# Shape functions (T3), Ni = Li
N1 = xi
N2 = eta
N3 = 1 - xi - eta
N = [N1, N2, N3]

def Nij_integral(i, j):
    integrand = N[i] * N[j]
    integral = sp.integrate(          # the double integral
        sp.integrate(integrand * 2*A, (eta, 0, 1-xi)), (xi, 0, 1))
    return sp.simplify(integral)

# Compute full 3x3 matrix
M = sp.Matrix(3, 3, lambda i, j: Nij_integral(i, j))
#M  # to display
```

Using the Sympy results:

$$\begin{bmatrix} \int_{A_e} N_1 N_1 \, dA & \int_{A_e} N_1 N_2 \, dA & \int_{A_e} N_1 N_3 \, dA \\ \int_{A_e} N_1 N_2 \, dA & \int_{A_e} N_2 N_2 \, dA & \int_{A_e} N_2 N_3 \, dA \\ \int_{A_e} N_1 N_3 \, dA & \int_{A_e} N_2 N_3 \, dA & \int_{A_e} N_3 N_3 \, dA \end{bmatrix} = \frac{A}{12} \begin{bmatrix} 2 & 1 & 1 \\ 1 & 2 & 1 \\ 1 & 1 & 2 \end{bmatrix}$$

and Eq.(??), the mass matrix for the T3 element is finally found in an analytical form as:

$$\boxed{\underset{6\times 6}{\mathbf{m}_e} = \frac{\rho t A}{12} \begin{bmatrix} 2 & 0 & 1 & 0 & 1 & 0 \\ 0 & 2 & 0 & 1 & 0 & 1 \\ 1 & 0 & 2 & 0 & 1 & 0 \\ 0 & 1 & 0 & 2 & 0 & 1 \\ 1 & 0 & 1 & 0 & 2 & 0 \\ 0 & 1 & 0 & 1 & 0 & 2 \end{bmatrix}} \tag{7.30}$$

No numerical integration is needed. The nodal scalar of mass for a T3 element is $\frac{\rho t A}{6}$, $\frac{\rho t A}{12}$, or 0.

7.1.7.3 Use of the Eisenberg–Malvern Formula

A more direct route to Eq. (7.30) is provided by the Eisenberg–Malvern integration formula, which is closely related to Dirichlet-type simplex integrals. This result gives closed-form expressions for monomials of area coordinates over a triangular domain.

Using the area-coordinate transformation, we obtain:

$$\begin{aligned} \int_{A_e} L_1^n L_2^m L_3^p \, dA &= 2A_e \int_0^1 \int_0^{1-\xi} \xi^n \eta^m (1-\xi-\eta)^p \, d\eta d\xi \\ &= 2A_e \frac{m!, n!, p!}{(m+n+p+2)!} \end{aligned} \tag{7.31}$$

This formula allows any polynomial product of shape functions to be evaluated exactly in closed form. As a result, the consistent mass matrix can be derived without numerical quadrature, making the approach particularly efficient for analytical element formulations. We will reuse this result extensively in later chapters.

For the T3 mass matrix, only two types of integrals are required:

$$\int_{\Omega_e} L_i^2 \, d\Omega \qquad \text{and} \qquad \int_{\Omega_e} L_i L_j \, d\Omega \tag{7.32}$$

7.1.7.4 Diagonal Terms

For L_i^2, we set $n = 2$, $m = 0$, $p = 0$, yielding:

$$\int_{\Omega_e} L_i^2 \, d\Omega = 2A_e \frac{2!}{4!} = 2A_e \frac{2}{24} = \frac{A_e}{6} \tag{7.33}$$

7.1.7.5 Off-Diagonal Terms

For $L_i L_j$ with $i \neq j$, we set $n = 1$, $m = 1$, $p = 0$, giving:

$$\int_{\Omega_e} L_i L_j \, d\Omega = 2A_e \frac{1!, 1!}{4!} = 2A_e \frac{1}{24} = \frac{A_e}{12} \tag{7.34}$$

Using Eqs. (7.33) and (7.34), we directly recover Eq. (7.30). These identities will also be used in the derivation of the T4 element mass matrix for 3D solids in the next chapter.

7.1.7.6 Lumped Mass Matrix

For explicit dynamic algorithms, vibration and wave propagation analyses, and high-velocity impact simulations, the fully populated consistent mass matrix can introduce substantial computational overhead due to repeated matrix inversions or matrix-vector operations. To reduce this cost, the inertial distribution is often diagonalized using a row-sum or proportional scaling procedure, producing the **lumped mass matrix**. For T3 elements, it is computed using:

$$\mathbf{m}_e^{\text{lumped}} = \frac{\rho t A_e}{3} \mathbf{I}_6 \tag{7.35}$$

In this approximation, each node carries one-third of the total element mass.

7.1.8 T3 Nodal Force Vectors

External loads acting on a T3 element are converted into equivalent nodal forces using the Principle of Virtual Work. We consider two primary load types: distributed body forces over the element domain and surface tractions along element boundaries.

7.1.8.1 Body Force Vector

For a distributed body force field $\mathbf{b} = [b_x, b_y]^\top$ (force per unit volume), the equivalent nodal force vector is defined as:

$$\underset{6\times 1}{\mathbf{f}_b} = \int_{\Omega_e} \mathbf{N}_e^\top \mathbf{b} \, d\Omega \tag{7.36}$$

Assuming a constant thickness t, the volume integral reduces to an area integral via $d\Omega = t\,dA$. For a uniform body force field, and using the fact that $N_i = L_i$, we can apply the triangular integration identity (Eq. 7.31) to evaluate the result exactly.

Expanding the expression gives:

$$\mathbf{f}_b = t\int_{A_e} \begin{bmatrix} N_1\mathbf{I} \\ N_2\mathbf{I} \\ N_3\mathbf{I} \end{bmatrix} \begin{bmatrix} b_x \\ b_y \end{bmatrix} dA = t \begin{bmatrix} \int_{A_e} L_1\,dA \cdot b_x \\ \int_{A_e} L_1\,dA \cdot b_y \\ \int_{A_e} L_2\,dA \cdot b_x \\ \int_{A_e} L_2\,dA \cdot b_y \\ \int_{A_e} L_3\,dA \cdot b_x \\ \int_{A_e} L_3\,dA \cdot b_y \end{bmatrix} \tag{7.37}$$

Using the result $\int_{A_e} L_i\,dA = \frac{A_e}{3}$ (from the Eisenberg–Malvern formula), we obtain the exact nodal distribution of the body force:

$$\boxed{\underset{6\times 1}{\mathbf{f}_b} = \frac{tA_e}{3}\begin{bmatrix} b_x & b_y & b_x & b_y & b_x & b_y \end{bmatrix}^\top} \tag{7.38}$$

> **Insight:** For a constant body force field (e.g., gravity or centrifugal loading), the total resultant force tA_eb_x and tA_eb_y is distributed equally among the three nodes of the element. This finding holds as long as the shape functions are linear and the body force is constant over the element.

7.1.8.2 Boundary Traction Vector

Computing nodal forces due to surface tractions requires identifying the element edges that lie on the prescribed Neumann boundary Γ_t. The formulation below considers a single active boundary edge for clarity.

For a prescribed surface traction vector $\bar{\mathbf{t}} = [t_x, t_y]^\top$ (force per unit area) acting along a boundary segment Γ_{edge}, the nodal force vector is defined as:

$$\mathbf{f}_{\text{edge}} = \int_{\Gamma_{\text{edge}}} \mathbf{N}_{\text{edge}}^\top \bar{\mathbf{t}}\, t\, d\Gamma \tag{7.39}$$

On a straight edge, the interpolation matrix $\mathbf{N}_{\text{edge}}$ simplifies significantly because the shape function associated with the opposite (third) node vanishes identically. The remaining two shape functions vary linearly from 1 to 0 along the edge length l_{edge}.

If the traction $\bar{\mathbf{t}}$ is uniform along the boundary segment, the line integral reduces directly to:

$$\underset{4\times 1}{\mathbf{f}_{\text{edge}}} = \frac{t \cdot l_{\text{edge}}}{2} \begin{bmatrix} \bar{\mathbf{t}} \\ \bar{\mathbf{t}} \end{bmatrix} = \frac{t \cdot l_{\text{edge}}}{2} \begin{bmatrix} t_x \\ t_y \\ t_x \\ t_y \end{bmatrix} \tag{7.40}$$

where l_{edge} is the physical length of the edge segment. This result shows that the total traction resultant on the edge is distributed equally between its two end nodes.

Note: The local vector $\mathbf{f}_{\text{edge}}$ has dimension 4, corresponding to the two nodes on the active edge with two degrees of freedom each. During global assembly, this vector is mapped into the full 6-degree-of-freedom element force vector $\mathbf{f}_e$.

7.1.9 Remarks and Insights

7.1.9.1 Algorithmic Simplicity and Analytical Integration

As seen throughout this chapter, the T3 element formulation is remarkably compact and computationally efficient. Because both the strain–displacement matrix $\mathbf{B}_e$ and the constitutive matrix $\mathbf{C}$ remain constant over the element domain, and because all shape function products (all essentially in polynomial form) can be evaluated exactly using area-coordinate integration, the element matrices ($\mathbf{k}_e$, $\mathbf{m}_e$) and nodal force vectors ($\mathbf{f}_b$, $\mathbf{f}_{\text{edge}}$) admit closed-form expressions.

This eliminates the need for numerical quadrature (such as Gauss integration), reducing computational cost and removing quadrature-related approximation errors. Combined with the ability to generate unstructured triangular meshes over complex geometries, this makes the T3 element a robust and practical choice for general-purpose solid mechanics simulations involving irregular domains.

7.1.9.2 Constant Strain Behavior (CST)

The displacement field in the T3 element is linear, which implies that all spatial derivatives are constant within the element. Consequently, strain and stress are uniform over each element.

While this property simplifies implementation and analytical evaluation, it also leads to limitations in regions with strong gradients, such as bending-dominated behavior or stress concentrations. In such cases, mesh refinement or structured h-adaptivity is required to achieve accurate stress predictions.

A particularly effective remedy is the **Smoothed Finite Element Method (S-FEM)**, which constructs smoothed strain fields over local smoothing domains. This reduces the excessive stiffness inherent in standard CST formulations and improves convergence behavior significantly, while preserving the simplicity and mesh flexibility of triangular discretizations [11].

7.1.9.3 Consistency and the Patch Test

Because the T3 element exactly reproduces any linear displacement field, it satisfies the completeness requirement and passes the patch test under linear displacement boundary conditions. This ensures consistency with continuum mechanics and guarantees convergence as the mesh is refined ($h \to 0$).

7.1.9.4 Rigid Body Modes and Rank Deficiency

The unconstrained 6×6 stiffness matrix $\mathbf{k}_e$ has rank 3, reflecting the presence of three rigid body modes that produce zero internal strain energy:

- Translation in the x-direction
- Translation in the y-direction
- In-plane rigid body rotation

These correspond to the three zero eigenvalues of the element stiffness operator prior to the application of boundary conditions.

7.1.9.5 Practical Advantages and Computational Trade-offs

The T3 element remains widely used due to several practical advantages:

- **Robustness:** Works reliably on highly irregular geometries with unstructured meshes.
- **Simplicity:** No numerical integration is required for stiffness, mass, or load computations.
- **Automation:** Naturally compatible with automatic mesh generation and adaptive refinement workflows.

However, its primary limitation is its relatively low convergence rate in stress-sensitive applications. Compared with higher-order elements (such as Q4), it requires significantly more degrees of freedom to achieve comparable stress accuracy.

7.2 Q4 Element Matrices and Codes

Following the formulation of triangular elements, we now consider the four-node bilinear quadrilateral element (Q4). This element is widely used in structured and mapped meshes and generally provides higher accuracy than the T3 element, particularly for bending-dominated problems, provided that geometric distortion remains moderate. This section presents a detailed formulation of the Q4 element in a systematic manner.

7.2.1 Problem Setup

Consider a quadrilateral element defined by four nodes ordered counterclockwise to ensure a positive Jacobian determinant (see Fig.4.6):

- **Node 1:** (x_1, y_1) with displacements (u_1, v_1)
- **Node 2:** (x_2, y_2) with displacements (u_2, v_2)
- **Node 3:** (x_3, y_3) with displacements (u_3, v_3)
- **Node 4:** (x_4, y_4) with displacements (u_4, v_4)

The material parameters (Young's modulus E, Poisson's ratio ν, and density ρ) and the geometric thickness t are identical to those defined for the T3 element.

7.2.2 Element Matrix Formulation

The Q4 element has eight degrees of freedom, and its nodal displacement vector is assembled as:

$$\underset{8\times 1}{\mathbf{d}_e} = \begin{bmatrix} u_1 & v_1 & u_2 & v_2 & u_3 & v_3 & u_4 & v_4 \end{bmatrix}^\top \tag{7.41}$$

7.2.3 Isoparametric Formulation

Unlike the T3 element, which has constant spatial derivatives, the Q4 element is formulated using an **isoparametric mapping** between the physical domain and a reference square domain:

$$(\xi, \eta) \in [-1, 1] \times [-1, 1] \tag{7.42}$$

In this formulation, both geometry and displacement fields are interpolated using the same bilinear shape functions.

The physical coordinates (x, y) are expressed as

$$x(\xi, \eta) = \sum_{i=1}^{4} N_i(\xi, \eta), x_i, \qquad y(\xi, \eta) = \sum_{i=1}^{4} N_i(\xi, \eta), y_i, \tag{7.43}$$

where the bilinear shape functions $N_i(\xi, \eta)$ are given in Eq.4.79). Using this isoparametric formulation, element edges in the parent domain are mapped to element edges in the physical domain, thereby preserving inter-element compatibility and ensuring that the approximated displacement field remains continuous across element boundaries.

This continuity requirement is fundamental because standard displacement-based finite element formulations seek solutions in the Sobolev space $H^1(\Omega)$, whose functions must be continuous across element interfaces. For general quadrilateral elements, the isoparametric mapping provides a natural and consistent framework for satisfying this requirement.

It is worth noting that T3 elements may also be formulated using an isoparametric mapping. However, because the geometry and displacement interpolation are both linear, the mapping reduces to a simple affine transformation, and the element can be formulated directly in physical coordinates without introducing a parent coordinate system. This distinguishes the T3 element from the Q4 element, whose bilinear interpolation generally requires the isoparametric framework for practical implementation.

The isoparametric mapping leads to a the important **Jacobian matrix**.

7.2.3.1 Jacobian Transformation

To relate derivatives in the physical and natural coordinate systems, we need to introduce the Jacobian matrix $\mathbf{J}$:

$$\mathbf{J} = \begin{bmatrix} \frac{\partial x}{\partial \xi} & \frac{\partial y}{\partial \xi} \\ \frac{\partial x}{\partial \eta} & \frac{\partial y}{\partial \eta} \end{bmatrix} = \begin{bmatrix} \sum_{i=1}^{4} \frac{\partial N_i}{\partial \xi} x_i & \sum_{i=1}^{4} \frac{\partial N_i}{\partial \xi} y_i \\ \sum_{i=1}^{4} \frac{\partial N_i}{\partial \eta} x_i & \sum_{i=1}^{4} \frac{\partial N_i}{\partial \eta} y_i \end{bmatrix} \tag{7.44}$$

which is used to transform the spatial derivatives of the shape functions between these to coordinate system, which is a must for Q4 elements. The expanded expressions of the Jacobian components are provided in Eq.(4.96).

7.2.4 Displacement Approximation

The continuous displacement field is approximated using the element shape function matrix $\mathbf{N}_e$:

$$\underset{2\times 1}{\mathbf{u}(x,y)} = \begin{bmatrix} u(x,y) \\ v(x,y) \end{bmatrix} = \underset{2\times 8}{\mathbf{N}_e(\xi,\eta)};\ \underset{8\times 1}{\mathbf{d}_e} \tag{7.45}$$

The shape function matrix is organized in block form using nodal submatrices:

$$\begin{aligned} \mathbf{N}_e &= \begin{bmatrix} N_1 & 0 & N_2 & 0 & N_3 & 0 & N_4 & 0 \\ 0 & N_1 & 0 & N_2 & 0 & N_3 & 0 & N_4 \end{bmatrix} \\ &= \begin{bmatrix} \mathbf{N}_1 & \mathbf{N}_2 & \mathbf{N}_3 & \mathbf{N}_4 \end{bmatrix} \end{aligned} \tag{7.46}$$

For each node i, the nodal submatrix is defined as:

$$\mathbf{N}_i = \begin{bmatrix} N_i & 0 \\ 0 & N_i \end{bmatrix}_{2\times 2} \tag{7.47}$$

7.2.5 Strain–Displacement Relation

The strain vector is expressed in standard form:

$$\varepsilon = \mathbf{B}_e \mathbf{d}_e \tag{7.48}$$

where the strain–displacement matrix is assembled from nodal contributions:

$$\mathbf{B}_e = \begin{bmatrix} \mathbf{B}_1 & \mathbf{B}_2 & \mathbf{B}_3 & \mathbf{B}_4 \end{bmatrix} \tag{7.49}$$

Each nodal strain operator $\mathbf{B}_i$ is given by:

$$\mathbf{B}_i = \begin{bmatrix} \frac{\partial N_i}{\partial x} & 0 \\ 0 & \frac{\partial N_i}{\partial y} \\ \frac{\partial N_i}{\partial y} & \frac{\partial N_i}{\partial x} \end{bmatrix} = \begin{bmatrix} g_i(x,y) & 0 \\ 0 & h_i(x,y) \\ h_i(x,y) & g_i(x,y) \end{bmatrix} \tag{7.50}$$

where $g_i(x,y) = \frac{\partial N_i}{\partial x}$ and $h_i(x,y) = \frac{\partial N_i}{\partial y}$.

Since the shape functions are defined in natural coordinates (ξ, η), their spatial derivatives are obtained through the inverse Jacobian transformation:

$$\begin{bmatrix} g_i(x,y) \\ h_i(x,y) \end{bmatrix} = \begin{bmatrix} \frac{\partial N_i}{\partial x} \\ \frac{\partial N_i}{\partial y} \end{bmatrix} = \mathbf{J}^{-1} \begin{bmatrix} \frac{\partial N_i}{\partial \xi} \\ \frac{\partial N_i}{\partial \eta} \end{bmatrix} \tag{7.51}$$

Observation: Unlike the T3 element, the derivatives $g_i(x,y)$ and $h_i(x,y)$ vary spatially within the element. As a result, the strain field is no longer constant.

7.2.6 Constitutive Relation

For an isotropic linear elastic material, stress and strain are related through:

$$\sigma = \mathbf{C}\varepsilon \tag{7.52}$$

where the constitutive matrix $\mathbf{C}$ is defined in Eq.(7.12) for plane stress and plane strain cases.

7.2.7 Q4 Element Stiffness Matrix

Starting from the weak form of the governing equations, the element stiffness matrix $\mathbf{k}_e$ is computed using a domain integral:

$$\underset{8\times 8}{\mathbf{k}_e} = \int_{\Omega_e} \mathbf{B}_e^\top \mathbf{C} \mathbf{B}_e \, d\Omega \tag{7.53}$$

To evaluate this integral numerically, we transform it to the parent (natural) coordinate system using the isoparametric mapping. This converts the physical domain integral into a standard square domain:

$$\mathbf{k}_e = \int_{-1}^{1}\int_{-1}^{1} \mathbf{B}_e^\top(\xi,\eta)\mathbf{C}\mathbf{B}_e(\xi,\eta)\,\det(\mathbf{J})\,t\,d\xi d\eta \tag{7.54}$$

where:

- $\mathbf{B}_e(\xi, \eta)$ is the strain–displacement matrix expressed in natural coordinates,
- $\mathbf{C}$ is the constitutive (material stiffness) matrix,
- $\det(\mathbf{J})$ is the Jacobian determinant accounting for the mapping $dA = \det(\mathbf{J})\, d\xi d\eta$,
- t is the constant element thickness.

The Jacobian matrix $\mathbf{J}$ and its determinant are evaluated using Eq.(4.96). Matrix $\mathbf{B}_e$ is computed using Eq.(7.50), in which $\mathbf{J}^{-1}$ is involved. Thus, the integrand in Eq.(7.54) is quite a complicated function (a rational function of ξ and η). An exact analytical integration is not possible or practical, and numerical means is needed.

As seen, the major computations for a Q4 element are carried out in the natural coordinates, including numerical integrations.

7.2.8 Numerical Integration for Q4 Elements

As discussed in the previous section, the integrand in Eq.(7.54) is generally non-polynomial due to the presence of the Jacobian and spatially varying strain–displacement terms. As a result, analytical integration is not practical, and numerical quadrature is typically used.

For the bilinear Q4 element, the standard choice is a 2×2 **Gauss quadrature rule**, which provides an efficient balance between accuracy and computational cost. This rule exactly integrates polynomial fields up to the required order for undistorted rectangular elements and remains accurate for moderately distorted meshes.

The Gauss points in natural coordinates are located at:

$$(\xi_i, \eta_j) = \pm \frac{1}{\sqrt{3}} \approx \pm 0.577350269189626$$

with uniform weights:

$$w_i = w_j = 1.0$$

Using this quadrature rule, the stiffness matrix integral becomes:

$$\mathbf{k}_e = \sum_{i=1}^{2} \sum_{j=1}^{2} \mathbf{B}_e^\top(\xi_i, \eta_j)\, \mathbf{C}\, \mathbf{B}_e(\xi_i, \eta_j)\, \det(\mathbf{J}(\xi_i, \eta_j))\, t\, w_i w_j \tag{7.55}$$

In compact form, the integrand may also be interpreted as:

$$\bar{\mathbf{k}}_{ij}(\xi, \eta) \det\left(\mathbf{J}(\xi, \eta)\right)$$

This formulation is robust, straightforward to implement, and remains the standard approach in most commercial finite element codes, even for highly distorted meshes.

7.2.9 Standard Code for Q4 Stiffness Matrix

7.2.9.1 Standard Plain Version

The following code computes the Q4 element stiffness matrix by directly implementing Eq.(7.54).

```python
# Compute ke using the standard formula, BCB
def element_ke_Q4_BCB(coords, C):
    """
    Q4 element stiffness using 2x2 Gauss points for each element
    Using original dense matrix products formulas.
    """
    gp = [-1/np.sqrt(3), 1/np.sqrt(3)]                    # Gauss points
    weights = [1, 1]                                      # Gauss weights

    ke = np.zeros((8, 8))
    for xi in gp:                           # loop over the Gauss points
        for eta in gp:
            # Shape function derivatives wrt natural coords
            dN_dxi = 0.25 * np.array([
                [-(1 - eta),  (1 - eta),  (1 + eta), -(1 + eta)],
                [-(1 - xi),  -(1 + xi),   (1 + xi),   (1 - xi)]
            ])  # 2 x 4

            # Jacobian
            J = dN_dxi @ coords      # (2x4)(4x2) = 2x2
            detJ = np.linalg.det(J)
            invJ = np.linalg.inv(J)

            # Derivatives wrt x, y
            dN_dx = invJ @ dN_dxi    # 2 x 4

            # B matrix (3 x 8)
            B = np.zeros((3, 8))
            for i in range(4):
                B[0, 2*i]     = dN_dx[0, i]
                B[1, 2*i + 1] = dN_dx[1, i]
                B[2, 2*i]     = dN_dx[1, i]
                B[2, 2*i + 1] = dN_dx[0, i]

            ke += B.T @ C @ B * detJ
    return ke
```

Example: Stiffness Matrix of a Q4 Element The following Python script computes and prints the element stiffness matrix for a general Q4 element as an example.

```
# Martrix of material constants:
C = get_C_2D(E=1.0, nu=0.3) #(E=210e9, nu=0.3)

Q4_ele = np.array([[0.0, 0.01], [ 2.0, 0.2 ],      # an arbitrary Q4
                   [2.2,  1.1], [-0.1, 1.02] ])
ke_BCB = element_ke_Q4_BCB(Q4_ele, C)
ke66 = sp.Matrix(ke_BCB)[:6,:6].evalf(4)
display(Math(f"\\text{{Partial }}k_e=\scriptsize{latex(ke66)}"))
```

$$\text{Partial } k_e = \begin{bmatrix} 0.4128 & 0.1401 & -0.03379 & -0.0234 & -0.183 & -0.1603 \\ 0.1401 & 0.8475 & 0.004074 & 0.3241 & -0.1603 & -0.4013 \\ -0.03379 & 0.004074 & 0.5354 & -0.2223 & -0.227 & 0.01884 \\ -0.0234 & 0.3241 & -0.2223 & 0.9939 & 0.04631 & -0.8305 \\ -0.183 & -0.1603 & -0.227 & 0.04631 & 0.4111 & 0.1403 \\ -0.1603 & -0.4013 & 0.01884 & -0.8305 & 0.1403 & 0.8443 \end{bmatrix}$$

Vectorized Batch Version The following code presents a vectorized batch implementation for computing the stiffness matrices of Q4 elements. The formulation is designed to operate efficiently on both single-element inputs and large batches of elements without modification.

In this implementation, all element-level computations (shape function derivatives, Jacobians, and constitutive operations) are evaluated simultaneously across the input set, significantly reducing Python-level looping overhead. This makes the approach particularly suitable for large-scale finite element simulations where the number of elements is high and computational efficiency becomes critical.

The same function can therefore be used in two modes:

- **Single-element mode:** when the input array contains one Q4 element
- **Batch mode:** when the input array contains N_e elements stacked together

This unified structure ensures consistency in formulation while enabling high-performance execution through vectorization.

```
def batch_ke_Q4_BCB(coords, C):
    """
    Vectorized batch computation of Q4 element stiffness matrices
    using 2x2 Gauss quadrature. Courtesy of Gimini.
    Handles a single element (4,2), or a batch of elements (N,4,2).
    Parameters::
    coords : ndarray of shape (4, 2) or (N, 4, 2), nodal coordinates
          for the element(s). Order: [Node1, Node2, Node3, Node4]
    C : ndarray of shape (3, 3)
```

```
        Constitutive elasticity matrix for 2D plane problems.
    Returns::
    ke_all : ndarray of shape (8, 8) or (N, 8, 8)
             Stiffness matrix/matrices.
    """
    # 1. Manage input dimensionality dynamically
    is_single_element = (coords.ndim == 2)
    if is_single_element:
        all_coords = coords[np.newaxis, :, :]  # (4, 2) -> (1, 4, 2)
    else:
        all_coords = coords

    N = all_coords.shape[0]

    # 2. Hardcode the 4 Gauss integration points without loops
    # Layout order matching the combined integration points:(xi,eta)
    g = 0.5773502691896257
    gp_xi  = np.array([-g,  g,  g, -g])
    gp_eta = np.array([-g, -g,  g,  g])
    # Weight for all points in standard 2x2 2D Gauss integration
    # is 1.0 * 1.0 = 1.0

    # 3. Vectorized calculation of natural derivatives dN/dxi and
    #    dN/deta; Shape of arrays: (4, 4) ->
    #    [integration_point, shape_function_node]
    dN_dxi_all  = 0.25 * np.stack((-(1.0 - gp_eta), (1.0 - gp_eta),\
                          (1.0 + gp_eta), -(1.0 + gp_eta)), axis=1)
    dN_deta_all = 0.25 * np.stack((-(1.0 - gp_xi),  -(1.0 + gp_xi),\
                          (1.0 + gp_xi),    (1.0 - gp_xi)),  axis=1)

    # 4. Compute Jacobians for all elements and all Gauss points
    # simultaneously. Using np.einsum allows us to calculate
    # everything without loop iterations. J shape: (N, 4, 2, 2) ->
    # [Element, GaussPoint, Row, Column]; Rows correspond to d/dxi
    # and d/deta. Columns correspond to x and y.
    J = np.zeros((N, 4, 2, 2), dtype=float)
    J[:, :, 0, 0] = np.einsum('qi,ni->nq', dN_dxi_all,\
                                all_coords[:, :, 0]) # J11
    J[:, :, 0, 1] = np.einsum('qi,ni->nq', dN_dxi_all,\
                                all_coords[:, :, 1]) # J12
    J[:, :, 1, 0] = np.einsum('qi,ni->nq', dN_deta_all,\
                                all_coords[:, :, 0]) # J21
    J[:, :, 1, 1] = np.einsum('qi,ni->nq', dN_deta_all,\
                                all_coords[:, :, 1]) # J22
```

```
    # 5. Fast batched analytical matrix inversion for 2x2 Jacobians
    detJ = J[:, :, 0, 0] * J[:, :, 1, 1] - J[:, :, 0, 1] *\
                                           J[:, :, 1, 0]  # (N, 4)
    inv_detJ = 1.0 / detJ  # (N, 4)

    invJ = np.empty((N, 4, 2, 2), dtype=float)
    invJ[:, :, 0, 0] =  J[:, :, 1, 1] * inv_detJ
    invJ[:, :, 0, 1] = -J[:, :, 0, 1] * inv_detJ
    invJ[:, :, 1, 0] = -J[:, :, 1, 0] * inv_detJ
    invJ[:, :, 1, 1] =  J[:, :, 0, 0] * inv_detJ

    # 6. Derivatives dN/dx and dN/dy across all configurations
    # dN_dx_all and dN_dy_all shapes: (N, 4, 4) ->
    # [Element, GaussPoint, Node]
    dN_dx_all=(invJ[:,:,0,0,np.newaxis]*dN_dxi_all[np.newaxis,:,:] +
               invJ[:,:,0,1,np.newaxis]*dN_deta_all[np.newaxis,:,:])

    dN_dy_all=(invJ[:,:,1,0,np.newaxis]*dN_dxi_all[np.newaxis,:,:] +
               invJ[:,:,1,1,np.newaxis]*dN_deta_all[np.newaxis,:,:])

    # 7. Construct the Strain-Displacement Tensor B_all of (N,4,3,8)
    B_all = np.zeros((N, 4, 3, 8), dtype=float)
    B_all[:, :, 0, 0::2] = dN_dx_all
    B_all[:, :, 1, 1::2] = dN_dy_all
    B_all[:, :, 2, 0::2] = dN_dy_all
    B_all[:, :, 2, 1::2] = dN_dx_all

    # 8. Batched matrix products integrated with Gauss point weights
    # Step A: Multiply B matrices by material stiffness matrix C
    # BtC shape: (N, 4, 8, 3)
    BtC = np.einsum('nqji,jk->nqik', B_all, C)

    # Step B: Compute local point stiffness contributions (B.T@C@ B)
    # point_ke shape: (N, 4, 8, 8)
    point_ke = np.einsum('nqik,nqkl->nqil', BtC, B_all)

    # Step C: Scale each point stiffness by its corresponding detJ
    #         and sum over Gauss axis (axis=1)
    ke_all = np.einsum('nqil,nq->nil', point_ke, detJ)

    # 9. Format output shape based on initial input configuration
    if is_single_element:
        return ke_all[0]
```

```
    return ke_all
```

Performance of the Vectorized Code It is a good practice to verify the optimized code against the original standard one:

```
ke_standard = element_ke_Q4_BCB(Q4_ele, C)          # check only ke
ke_standard_batch = batch_ke_Q4_BCB(Q4_ele, C)

#print(np.isclose(ke_standard[:6,:6], ke_standard_batch[:6,:6]))
print(np.allclose(ke_standard, ke_standard_batch))
```

```
True
```

We can now conduct a performance test:

```
Ne = 10_000               # test on Ne elements
start = time.time()
for i in range(Ne):        # run a number of times for accurate timing
    ke_standard = element_ke_Q4_BCB(Q4_ele, C)[0]

time_standard = time.time() - start
print(f"Time, the standard approach: {time_standard:.4f}s")

# Create a batch of Ne elements for testing purpose:
all_coords = np.repeat(Q4_ele[np.newaxis, :, :], Ne, axis=0)
start = time.time()
ke_batch = batch_ke_Q4_BCB(all_coords, C)[0]
time_batch = (time.time() - start)
print(f"Time, standard fast: {time_batch:.4f}s")
print(f"Time ratio: {time_standard/time_batch:.4f}")
```

```
Time, the standard approach: 3.1145s
Time, standard fast: 0.1267s
Time ratio: 24.5755
```

For a batch of $N_e = 10{,}000$ T3 elements, the vectorized implementation is approximately **20 times faster** than the standard single-element loop-based code. This performance gain is consistent with the speedup trends predicted by Amdahl's Law.

7.2.9.2 Q4 Nodal Submatrix Analytical Formula

Expanding $\mathbf{B}_e$ into its nodal submatrix structure yields a partitioned representation of the element stiffness matrix, denoted as $\bar{\mathbf{k}}_e = \mathbf{B}_e^\top \mathbf{C} \mathbf{B}_e$:

$$\underset{8\times 8}{\bar{\mathbf{k}}_e} = \begin{bmatrix} \mathbf{B}_{e1}^\top \mathbf{C}\mathbf{B}_{e1} & \mathbf{B}_{e1}^\top \mathbf{C}\mathbf{B}_{e2} & \mathbf{B}_{e1}^\top \mathbf{C}\mathbf{B}_{e3} & \mathbf{B}_{e1}^\top \mathbf{C}\mathbf{B}_{e4} \\ \mathbf{B}_{e2}^\top \mathbf{C}\mathbf{B}_{e1} & \mathbf{B}_{e2}^\top \mathbf{C}\mathbf{B}_{e2} & \mathbf{B}_{e2}^\top \mathbf{C}\mathbf{B}_{e3} & \mathbf{B}_{e2}^\top \mathbf{C}\mathbf{B}_{e4} \\ \mathbf{B}_{e3}^\top \mathbf{C}\mathbf{B}_{e1} & \mathbf{B}_{e3}^\top \mathbf{C}\mathbf{B}_{e2} & \mathbf{B}_{e3}^\top \mathbf{C}\mathbf{B}_{e3} & \mathbf{B}_{e3}^\top \mathbf{C}\mathbf{B}_{e4} \\ \mathbf{B}_{e4}^\top \mathbf{C}\mathbf{B}_{e1} & \mathbf{B}_{e4}^\top \mathbf{C}\mathbf{B}_{e2} & \mathbf{B}_{e4}^\top \mathbf{C}\mathbf{B}_{e3} & \mathbf{B}_{e4}^\top \mathbf{C}\mathbf{B}_{e4} \end{bmatrix} \tag{7.56}$$

More compactly, the stiffness matrix is structured as an array of 2×2 **nodal submatrix** blocks:

$$\boxed{\underset{8\times 8}{\bar{\mathbf{k}}_e} = \begin{bmatrix} \bar{\mathbf{k}}_{11} & \bar{\mathbf{k}}_{12} & \bar{\mathbf{k}}_{13} & \bar{\mathbf{k}}_{14} \\ \bar{\mathbf{k}}_{21} & \bar{\mathbf{k}}_{22} & \bar{\mathbf{k}}_{23} & \bar{\mathbf{k}}_{24} \\ \bar{\mathbf{k}}_{31} & \bar{\mathbf{k}}_{32} & \bar{\mathbf{k}}_{33} & \bar{\mathbf{k}}_{34} \\ \bar{\mathbf{k}}_{41} & \bar{\mathbf{k}}_{42} & \bar{\mathbf{k}}_{43} & \bar{\mathbf{k}}_{44} \end{bmatrix}} \tag{7.57}$$

where each subblock represents the localized coupling between node i and node j:

$$\underset{2\times 2}{\bar{\mathbf{k}}_{ij}} = \mathbf{B}_{ei}^\top \mathbf{C}\mathbf{B}_{ej} \tag{7.58}$$

For isotropic plane elasticity, substituting the detailed matrices into Eq.(7.58) yields the analytical expansion for each nodal submatrix:

$$\bar{\mathbf{k}}_{ij} = \begin{bmatrix} C_{11}g_{ij} + C_{33}h_{ij} & C_{12}s_{ij} + C_{33}t_{ij} \\ C_{12}t_{ij} + C_{33}s_{ij} & C_{22}h_{ij} + C_{33}g_{ij} \end{bmatrix} \tag{7.59}$$

where the spatial derivative products are defined as:

$$g_{ij} = g_i g_j, \qquad h_{ij} = h_i h_j, \qquad s_{ij} = g_i h_j, \qquad t_{ij} = h_i g_j \tag{7.60}$$

As previously noted, the quantities g_i and h_i are nonlinear functions of the natural coordinates (ξ, η). As a result, the derived parameters g_{ij}, h_{ij}, s_{ij}, and t_{ij} are also spatially varying across the element domain. Consequently, the element stiffness matrix $\mathbf{k}_e$ cannot, in general, be evaluated in closed analytical form for an arbitrary quadrilateral and must instead be computed numerically:

$$\mathbf{k}_{ij} = t \int_{-1}^{1} \int_{-1}^{1} \bar{\mathbf{k}}_{ij}(\xi, \eta) \det(\mathbf{J}(\xi, \eta)), d\xi d\eta \tag{7.61}$$

In practice, this integral is evaluated using Gauss quadrature, most commonly with a 2×2 integration rule.

> **Summary of Distinction:** This requirement represents a fundamental difference from the T3 element. The T3 formulation benefits from a constant strain–displacement matrix, enabling exact analytical integration of

element matrices. In contrast, the Q4 element involves spatially varying derivatives and a nonlinear Jacobian determinant, which necessitates numerical quadrature to accurately capture the integral over the element domain.

7.2.9.3 NSAF Code for Q4 Stiffness Matrix

In this section, we present the Python function `element_ke_Q4_NSAF()`, which implements the nodal submatrix analytical formula (NSAF) introduced in Eq.(7.59). This formulation reduces the number of arithmetic operations significantly and is therefore well-suited for production-scale finite element simulations.

Such computational efficiency becomes particularly important in large-scale FEM analyses, especially in commercial-grade industrial software, where stiffness assembly must be performed repeatedly over thousands or even millions of elements.

```python
def element_ke_Q4_NSAF(coords, C):
    """
    Q4 element stiffness matrix using 2x2 nodal submatrix analytical
    formulation, which is integrated with 2x2 Gauss quadrature.
    """
    # 2x2 Gauss quadrature points
    gp = [-1.0 / np.sqrt(3.0), 1.0 / np.sqrt(3.0)]

    # Force constitutive entries to float
    C11 = float(C[0, 0])
    C12 = float(C[0, 1])
    C22 = float(C[1, 1])
    C33 = float(C[2, 2])

    # Element stiffness matrix
    ke = np.zeros((8, 8), dtype=float)

    # Gauss integration
    for xi in gp:
        for eta in gp:
            # Shape function derivatives wrt natural coordinates
            dN_dxi = 0.25*np.array([[-(1.0 - eta), (1.0 - eta),
                                      (1.0 + eta),-(1.0 + eta)],
                                    [-(1.0 - xi), -(1.0 + xi),
                                      (1.0 + xi),  (1.0 - xi)]
            ], dtype=float)
            J = dN_dxi @ coords          # Jacobian
            detJ = np.linalg.det(J)
            invJ = np.linalg.inv(J)
```

```
        # Shape function derivatives wrt x,y
        dN_dx = invJ @ dN_dxi
        gx = dN_dx[0, :]
        gy = dN_dx[1, :]

        # Assemble analytical submatrices
        for i in range(4):
            gi = gx[i]; hi = gy[i]
            ii = 2 * i
            for j in range(4):
                gj = gx[j]; hj = gy[j]
                jj = 2 * j
                # Common products
                gij = gi * gj; hij = hi * hj
                sij = gi * hj; tij = hi * gj
                # Analytical 2x2 block
                kij = np.array([[C11 * gij + C33 * hij,
                                 C12 * sij + C33 * tij],
                                [C12 * tij + C33 * sij,
                                 C22 * hij + C33 * gij]
                ], dtype=float)
                # Assembly
                ke[ii:ii+2, jj:jj+2] += kij * detJ
    return ke
```

```
def batch_ke_Q4_NSAF(coords, C):
    """
    Vectorized batch computation of Q4 element stiffness matrices
    using the Nodal Submatrix Analytical Formula (NSAF). Handles both
    a single element or a batch of elements. Courtesy of Gemini.
    Parameters::
    coords : ndarray of shape (4, 2) or (N, 4, 2)
             Nodal coordinates for the element(s).
    C :      ndarray of shape (3, 3), Constitutive elasticity matrix.
    Returns::
    ke_all : ndarray of shape (8, 8) or (N, 8, 8)
        Element stiffness matrix/matrices.
    """
    # 1. Manage input dimensionality dynamically for vectorization
    is_single_element = (coords.ndim == 2)
    if is_single_element:
        all_coords = coords[np.newaxis, :, :]  # (4, 2) -> (1, 4, 2)
    else:
```

```
        all_coords = coords

    N = all_coords.shape[0]

    # 2. Extract constitutive components
    c11, c12, c22, c33 = C[0, 0], C[0, 1], C[1, 1], C[2, 2]

    # 3. Unroll 2x2 Gauss points into a flat structural axis of 4
    g = 0.5773502691896257
    gp_xi  = np.array([-g,  g,  g, -g])
    gp_eta = np.array([-g, -g,  g,  g])

    # 4. Vectorized natural derivatives evaluated at 4 Gauss points
    # Shape: (4, 4) -> [GaussPoint, Node]
    dN_dxi_all  = 0.25 * np.stack((-(1.0 - gp_eta),  (1.0 -gp_eta),\
                        (1.0 + gp_eta), -(1.0 + gp_eta)), axis=1)
    dN_deta_all = 0.25 * np.stack((-(1.0 - gp_xi),  -(1.0 + gp_xi),\
                        (1.0 + gp_xi),   (1.0 - gp_xi)),  axis=1)

    # 5. Compute 2x2 Jacobians across the entire batch via tensor
    # tracking; J: (N, 4, 2, 2) -> [Element, GaussPoint, Row, Col]
    J = np.zeros((N, 4, 2, 2), dtype=float)
    J[:, :, 0, 0] = np.einsum('qi,ni->nq', dN_dxi_all, \
                                all_coords[:, :, 0])  # J11
    J[:, :, 0, 1] = np.einsum('qi,ni->nq', dN_dxi_all,  \
                                all_coords[:, :, 1])  # J12
    J[:, :, 1, 0] = np.einsum('qi,ni->nq', dN_deta_all, \
                                all_coords[:, :, 0])  # J21
    J[:, :, 1, 1] = np.einsum('qi,ni->nq', dN_deta_all, \
                                all_coords[:, :, 1])  # J22

    # 6. High-speed manual inversion of 2x2 Jacobians
    detJ = J[:, :, 0, 0] * J[:, :, 1, 1] - J[:, :, 0, 1] *\
                                      J[:, :, 1, 0]  # (N, 4)
    inv_detJ = 1.0 / detJ

    invJ = np.empty((N, 4, 2, 2), dtype=float)
    invJ[:, :, 0, 0] =  J[:, :, 1, 1] * inv_detJ
    invJ[:, :, 0, 1] = -J[:, :, 0, 1] * inv_detJ
    invJ[:, :, 1, 0] = -J[:, :, 1, 0] * inv_detJ
    invJ[:, :, 1, 1] =  J[:, :, 0, 0] * inv_detJ

    # 7. Compute physical derivatives dN/dx & dN/dy via broadcasting
    # Shape of gx and gy: (N, 4, 4) -> [Element, GaussPoint, Node]
```

```
    gx = (invJ[:,:, 0, 0, np.newaxis] * dN_dxi_all[np.newaxis,:,:] +
          invJ[:,:, 0, 1, np.newaxis] * dN_deta_all[np.newaxis,:,:])

    gy = (invJ[:,:, 1, 0, np.newaxis] * dN_dxi_all[np.newaxis,:,:] +
          invJ[:,:, 1, 1, np.newaxis] * dN_deta_all[np.newaxis,:,:])

    # 8. Compute analytical outer products scaled by detJ using
    # Einstein Summation; The 'nq' represents [Element, GaussPoint],
    # and 'i,j' represent the spatial [Node, Node] grids.
    # (N, 4, 4) summed over q -> (N, 4, 4):
    g_g = np.einsum('nqi,nqj,nq->nij', gx, gx, detJ)
    h_h = np.einsum('nqi,nqj,nq->nij', gy, gy, detJ)
    g_h = np.einsum('nqi,nqj,nq->nij', gx, gy, detJ)
    h_g = np.einsum('nqi,nqj,nq->nij', gy, gx, detJ)

    # 9. Direct submatrix mapping to the global (N, 8, 8) array
    #    via stride slicing
    ke_all = np.zeros((N, 8, 8), dtype=float)

    ke_all[:, 0::2, 0::2] = c11 * g_g + c33 * h_h  # k_xx
    ke_all[:, 0::2, 1::2] = c12 * g_h + c33 * h_g  # k_xy
    ke_all[:, 1::2, 0::2] = c12 * h_g + c33 * g_h  # k_yx
    ke_all[:, 1::2, 1::2] = c22 * h_h + c33 * g_g  # k_yy

    # 10. Return appropriate shape based on initial input format
    if is_single_element:
        return ke_all[0]
    return ke_all
```

Let us verify the optimized code against the original standard one:

```
ke_standard = element_ke_Q4_BCB(Q4_ele, C)
ke_NSAF = batch_ke_Q4_NSAF(Q4_ele, C)

#print(np.isclose(ke_standard, ke_NSAF)) # to check all entries
print(np.allclose(ke_standard, ke_NSAF))
```

```
True
```

The results are in agreement to the machine precision. Let's study the performance of the NSAF formulas using the following codes.

```
Ne = 10_000  # create a batch of Ne elements for testing purpose:
all_coords = np.repeat(Q4_ele[np.newaxis, :, :], Ne, axis=0)
```

```
start = time.time()
ke_batch = batch_ke_Q4_BCB(all_coords, C)[0]
time_BCBbatch = (time.time() - start)
print(f"Time, time_BCBbatch: {time_BCBbatch:.4f}s")
#print(f"Time ratio: {time_standard/time_batch:.4f}")

start = time.time()
ke_batch = batch_ke_Q4_NSAF(all_coords, C)[0]
time_NSAFbatch = (time.time() - start)
print(f"Time, NSAF batch: {time_NSAFbatch:.4f}s")
print(f"Time ratio: {time_BCBbatch/time_NSAFbatch:.4f}")
```

```
Time, time_BCBbatch: 0.0898s
Time, NSAF batch: 0.0448s
Time ratio: 2.0041
```

7.2.9.4 Q4 NSAF Performance Evaluation

For a batch of $N_e = 10{,}000$ Q4 elements, the vectorized NSAF implementation based on Eq. (7.59) is approximately **2× faster** than the vectorized implementation based on the standard formulation in Eq. (7.54).

This result is largely consistent with the earlier observation for T3 elements and further supports the **core principle of computational efficiency**: reductions in the underlying arithmetic operation count directly translate into measurable performance gains, when both approaches are fully vectorized.

7.2.10 Q4 Element Mass Matrix

7.2.10.1 Mass Matrix Formulation

The consistent element mass matrix $\mathbf{m}_e$ governs the structural kinetic energy distribution. Formulated in the physical domain, it reads:

$$\mathbf{m}_e = \int_{\Omega_e} \rho \mathbf{N}_e^\top \mathbf{N}_e \, d\Omega \tag{7.62}$$

Mapping this integral to the natural parametric domain yields:

$$\mathbf{m}_e = \int_{-1}^{1} \int_{-1}^{1} \rho \mathbf{N}_e^\top \mathbf{N}_e \, \det(\mathbf{J}) \, t \, d\xi d\eta \tag{7.63}$$

By expanding the global shape function matrix $\mathbf{N}_e$ from Eq.(7.46), the full 8×8 consistent mass matrix partitions into a **nodal submatrix** blocks:

$$\mathbf{m}_e = \begin{bmatrix} \mathbf{m}_{11} & \mathbf{m}_{12} & \mathbf{m}_{13} & \mathbf{m}_{14} \\ \mathbf{m}_{21} & \mathbf{m}_{22} & \mathbf{m}_{23} & \mathbf{m}_{24} \\ \mathbf{m}_{31} & \mathbf{m}_{32} & \mathbf{m}_{33} & \mathbf{m}_{34} \\ \mathbf{m}_{41} & \mathbf{m}_{42} & \mathbf{m}_{43} & \mathbf{m}_{44} \end{bmatrix} \tag{7.64}$$

where each 2×2 nodal submatrix scales the inertial coupling between node i and node j:

$$\begin{aligned} \mathbf{m}_{ij} &= \rho t \int_{-1}^{1}\int_{-1}^{1} \mathbf{N}_i^\top \mathbf{N}_j \det(\mathbf{J})\, d\xi d\eta \\ &= \rho t \int_{-1}^{1}\int_{-1}^{1} \begin{bmatrix} N_i N_j & 0 \\ 0 & N_i N_j \end{bmatrix} \det(\mathbf{J})\, d\xi d\eta \end{aligned} \tag{7.65}$$

Because the nodal displacement interpolations are uncoupled between the orthogonal u and v directions, the cross-axis inertial coupling terms vanish. As a result, the full element mass matrix reduces to evaluating a single **nodal scalar quantity** $\bar{m}_{ij}$ for each node pair:

$$\bar{m}_{ij} = \int_{-1}^{1}\int_{-1}^{1} N_i N_j \det(\mathbf{J})\, d\xi d\eta \tag{7.66}$$

In standard finite element implementations, Eq.(7.66) is typically evaluated using the same 2×2 Gauss quadrature rule employed for the stiffness matrix. However, in this case the integrand is purely polynomial: it consists of bilinear shape functions multiplied by the Jacobian determinant, which is linear in ξ and η. As a result, the integral admits an exact closed-form solution.

7.2.10.2 Analytical Formula for Nodal Scalar $\bar{m}_{ij}$

Because the integrand in Eq.(7.66) is polynomial in nature, the double integral can be evaluated analytically. This eliminates the need for numerical quadrature and reduces computational cost in mass matrix assembly.

SymPy for Analytical Integration of $\bar{m}_{ij}$ The exact expression for $\bar{m}_{ij}$ can be derived symbolically using SymPy. The script below defines the bilinear shape functions, computes the Jacobian determinant, and performs exact integration over the reference square domain $[-1, 1] \times [-1, 1]$, yielding closed-form algebraic expressions for all nodal mass contributions:

```
# ========================================================
# SymPy code for computing mass matrix for Q4 elements
# ========================================================
xi, eta = sp.symbols('xi eta')              # Natural coordinates
rho = sp.symbols('rho')                       # Material density
```

```
x1, x2, x3, x4 = sp.symbols('x1 x2 x3 x4') # Nodal coordinates
y1, y2, y3, y4 = sp.symbols('y1 y2 y3 y4')

N1 = sp.Rational(1,4)*(1-xi)*(1-eta)        # Shape functions
N2 = sp.Rational(1,4)*(1+xi)*(1-eta)
N3 = sp.Rational(1,4)*(1+xi)*(1+eta)
N4 = sp.Rational(1,4)*(1-xi)*(1+eta)
N = [N1, N2, N3, N4]

dN_dxi = [sp.diff(Ni, xi)  for Ni in N] # Shape function derivatives
dN_deta= [sp.diff(Ni, eta) for Ni in N]

# Form the Jacobian matrix:
x = [x1, x2, x3, x4]
y = [y1, y2, y3, y4]
J11 = sum(x[i] * dN_dxi[i]  for i in range(4))
J12 = sum(x[i] * dN_deta[i] for i in range(4))
J21 = sum(y[i] * dN_dxi[i]  for i in range(4))
J22 = sum(y[i] * dN_deta[i] for i in range(4))
J = sp.Matrix([[J11, J12],
               [J21, J22] ])

detJ = sp.simplify(J.det())           # Determinant of Jacobian

# Compute symbolic m_ij
M = sp.zeros(4,4)
for i in range(4):
    for j in range(4):
        integrand = N[i] * N[j] * detJ
        mij = sp.integrate(sp.integrate(integrand, ( xi, -1, 1)),
                                                   (eta, -1, 1))
        M[i,j] = (sp.simplify(mij)).factor()

for i in range(4):
    for j in range(i+1):
        display(Math(f"\n$$\\bar{{m}}_{{{j+1}{i+1}}}$$=\
                        \small{latex(M[i,j])}"))
```

$$\bar{m}_{11}=\frac{3x_1y_2-3x_1y_4-3x_2y_1+x_2y_3+2x_2y_4-x_3y_2+x_3y_4+3x_4y_1-2x_4y_2-x_4y_3}{36}$$

$$\bar{m}_{12}=\frac{3x_1y_2-x_1y_3-2x_1y_4-3x_2y_1+2x_2y_3+x_2y_4+x_3y_1-2x_3y_2+x_3y_4+2x_4y_1-x_4y_2-x_4y_3}{72}$$

$$\bar{m}_{22}=\frac{3x_1y_2-2x_1y_3-x_1y_4-3x_2y_1+3x_2y_3+2x_3y_1-3x_3y_2+x_3y_4+x_4y_1-x_4y_3}{36}$$

$$\bar{m}_{13}=\frac{x_1y_2-x_1y_4-x_2y_1+x_2y_3-x_3y_2+x_3y_4+x_4y_1-x_4y_3}{72}$$

$$\bar{m}_{23}=\frac{2x_1y_2-x_1y_3-x_1y_4-2x_2y_1+3x_2y_3-x_2y_4+x_3y_1-3x_3y_2+2x_3y_4+x_4y_1+x_4y_2-2x_4y_3}{72}$$

$$\bar{m}_{33}=\frac{x_1y_2-x_1y_4-x_2y_1+3x_2y_3-2x_2y_4-3x_3y_2+3x_3y_4+x_4y_1+2x_4y_2-3x_4y_3}{36}$$

$$\bar{m}_{14}=\frac{2x_1y_2+x_1y_3-3x_1y_4-2x_2y_1+x_2y_3+x_2y_4-x_3y_1-x_3y_2+2x_3y_4+3x_4y_1-x_4y_2-2x_4y_3}{72}$$

$$\bar{m}_{24}=\frac{x_1y_2-x_1y_4-x_2y_1+x_2y_3-x_3y_2+x_3y_4+x_4y_1-x_4y_3}{72}$$

$$\bar{m}_{34}=\frac{x_1y_2+x_1y_3-2x_1y_4-x_2y_1+2x_2y_3-x_2y_4-x_3y_1-2x_3y_2+3x_3y_4+2x_4y_1+x_4y_2-3x_4y_3}{72}$$

$$\bar{m}_{44}=\frac{x_1y_2+2x_1y_3-3x_1y_4-x_2y_1+x_2y_3-2x_3y_1-x_3y_2+3x_3y_4+3x_4y_1-3x_4y_3}{36}$$

Due to the symmetry of the consistent mass matrix ($\bar{m}_{ij} = \bar{m}_{ji}$), symbolic computation tools typically return only the independent entries corresponding to the upper triangular portion of the matrix.

Efficient Analytical Formulas for $\bar{m}_{ij}$ A closer inspection of the raw symbolic expressions shows that the resulting integrals can be systematically decomposed into combinations of vector dot products and cross-product-like terms (see Chapter 2 of Ref. [9]). These expressions are directly related to the geometric area of the Q4 element, A_e, which can be computed explicitly using the classical shoelace formula:

$$\begin{aligned} A_e =& \frac{1}{2}(x_1y_2 + x_2y_3 + x_3y_4 + x_4y_1 \\ & - x_2y_1 - x_3y_2 - x_4y_3 - x_1y_4) \end{aligned} \tag{7.67}$$

The quantity A_e is a geometric invariant of the quadrilateral element, meaning it remains unchanged under rigid translations or consistent coordinate transformations.

In modern scientific computing environments such as Python, this expression can be implemented efficiently using vectorized operations:

$$A_{\text{element}} = \frac{1}{2}\left[\mathbf{x} \cdot \text{roll}(\mathbf{y} - 1) - \mathbf{y} \cdot \text{roll}(\mathbf{x} - 1)\right] \tag{7.68}$$

where the nodal coordinate vectors are ordered counterclockwise as $\mathbf{x} = [x_1, x_2, x_3, x_4]^\top$ and $\mathbf{y} = [y_1, y_2, y_3, y_4]^\top$. The operator $\text{roll}(\mathbf{v}, n)$ denotes a circular shift of the vector entries by n positions. This formulation provides a compact and highly efficient implementation compatible with vectorized numerical libraries such as NumPy (`np.roll`).

To further reduce computational cost, the symbolic expressions for $\bar{m}_{ij}$ can be simplified into a minimal set of algebraic operations. The resulting optimized closed-form expressions are presented below.

Fundamental Geometric Invariants We first define the six unique cross-product combinations of nodal coordinates. These quantities represent signed area contributions of local triangular substructures and serve as the fundamental building blocks of the mass matrix formulation:

$$\begin{aligned} A_{12} &= x_1y_2 - x_2y_1 & A_{23} &= x_2y_3 - x_3y_2 \\ A_{13} &= x_1y_3 - x_3y_1 & A_{24} &= x_2y_4 - x_4y_2 \\ A_{14} &= x_1y_4 - x_4y_1 & A_{34} &= x_3y_4 - x_4y_3 \end{aligned} \tag{7.69}$$

Matrix Component Expansions By factoring out a global scaling coefficient of $\frac{1}{72}$, the individual mass matrix entries are expressed as linear combinations of these six geometric invariants.

Diagonal Entries (Self-Inertia):

$$\begin{aligned} \bar{m}_{11} &= \frac{1}{72}\left[6A_{12} - 6A_{14} + 2A_{23} + 4A_{24} + 2A_{34}\right] \\ \bar{m}_{22} &= \frac{1}{72}\left[6A_{12} - 2A_{14} - 4A_{13} + 6A_{23} + 2A_{34}\right] \\ \bar{m}_{33} &= \frac{1}{72}\left[2A_{12} - 2A_{14} + 6A_{23} - 4A_{24} + 6A_{34}\right] \\ \bar{m}_{44} &= \frac{1}{72}\left[2A_{12} - 6A_{14} + 4A_{13} + 2A_{23} + 6A_{34}\right] \end{aligned} \tag{7.70}$$

Symmetric Diagonal Invariants (Opposite Node Coupling): The cross-quadrilateral couplings between diagonally opposite nodes are algebraically identical, sharing a single unique invariant expression:

$$\bar{m}_{13} = \bar{m}_{31} = \bar{m}_{24} = \bar{m}_{42} = \frac{1}{72}\left[A_{12} - A_{14} + A_{23} + A_{34}\right] \tag{7.71}$$

Symmetric Off-Diagonal Entries (Adjacent Edge Coupling):

$$\begin{aligned} \bar{m}_{12} = \bar{m}_{21} &= \frac{1}{72}\left[3A_{12} - 2A_{14} - A_{13} + 2A_{23} + A_{24} + A_{34}\right] \\ \bar{m}_{14} = \bar{m}_{41} &= \frac{1}{72}\left[2A_{12} - 3A_{14} + A_{13} + A_{23} + A_{24} + 2A_{34}\right] \\ \bar{m}_{23} = \bar{m}_{32} &= \frac{1}{72}\left[2A_{12} - A_{14} - A_{13} + 3A_{23} - A_{24} + 2A_{34}\right] \\ \bar{m}_{34} = \bar{m}_{43} &= \frac{1}{72}\left[A_{12} - 2A_{14} + A_{13} + 2A_{23} - A_{24} + 3A_{34}\right] \end{aligned} \tag{7.72}$$

Implementation and Performance Summary

1. **Minimal Flop Count:** The complete element mass matrix construction requires only approximately $\sim$ 35 floating-point multiplications, avoiding the dense loop structures typically associated with numerical quadrature.

2. **Memory Optimization:** Only six scalar invariant quantities A_{ij} need to be stored in local memory to fully reconstruct the 4×4 mass submatrix structure.
3. **Exact Symmetry Preservation:** The intrinsic symmetry property ($\bar{m}_{ij} = \bar{m}_{ji}$) is preserved at the algebraic level, eliminating any risk of numerical asymmetry in the assembled global mass matrix.

Evaluation Methodology	Approximate Multiplications	Computational Cost Ratio
Optimized Analytical Invariants	~ 35	**1.0** (Baseline)
Raw Symbolic Integration Output	~ 170	$\sim 4.9\times$ Slower
Standard 2×2 Gaussian Quadrature	~ 230	$\sim 6.6\times$ Slower

Algorithmic Insight: This formulation achieves a substantial reduction in arithmetic operations while completely eliminating element-level numerical integration. As a result, the optimized analytical mass scheme provides an efficient and stable foundation for large-scale explicit and implicit dynamic finite element solvers.

Once the nodal scalars $\bar{m}_{ij}$ are computed, the full uncoupled 2×2 nodal mass submatrices are assembled as:

$$\mathbf{m}_{ij} = \rho t \begin{bmatrix} \bar{m}_{ij} & 0 \\ 0 & \bar{m}_{ij} \end{bmatrix} \tag{7.73}$$

These nodal blocks are then mapped into the element mass matrix $\mathbf{m}_e$ as defined in Eq.(7.64).

Code for Computing $\bar{m}_{ij}$ The following Python code snippet demonstrates the practical implementation of the optimized analytical expressions for $\bar{m}_{ij}$.

```
def compute_mij_NSAF(coords):
    """
    Compute nodal submatrix mij using the Nodal Submatrix
    Analytic Formulas exactly.
    It takes `coords` of a Q4, and returns `m` which is the m_bar.
    """
    x = coords[:, 0]
    y = coords[:, 1]

    # 6 Unique Cross-products (Signed Areas)
    A12 = x[0]*y[1] - x[1]*y[0]
    A13 = x[0]*y[2] - x[2]*y[0]
```

```
    A14 = x[0]*y[3] - x[3]*y[0]
    A23 = x[1]*y[2] - x[2]*y[1]
    A24 = x[1]*y[3] - x[3]*y[1]
    A34 = x[2]*y[3] - x[3]*y[2]

    inv72 = 1.0 / 72.0
    m = np.zeros((4, 4))

    # Precompute repeated terms
    A12_6 = 6*A12; A23_2 = 2*A23; A24_4 = 4*A24
    A34_2 = 2*A34; A14_6 = 6*A14; A23_6 = 6*A23
    A13_4 = 4*A13; A14_2 = 2*A14; A12_2 = 2*A12
    A34_6 = 6*A34

    # --- Diagonals ---
    m[0, 0] = (A12_6 + A23_2 + A24_4 + A34_2 - A14_6) * inv72
    m[1, 1] = (A12_6 + A23_6 - A13_4 + A34_2 - A14_2) * inv72
    m[2, 2] = (A12_2 + A23_6 + A34_6 - A14_2 - A24_4) * inv72
    m[3, 3] = (A12_2 + A23_2 + A34_6 + A13_4 - A14_6) * inv72

    # --- Symmetric Invariant (m13, m24, m31, m42) ---
    m13_val = (A12 + A23 + A34 - A14) * inv72
    m[0, 2] = m[1, 3] = m[2, 0] = m[3, 1] = m13_val

    # --- Symmetric Pairs ---
    # m12 = m21
    m12_val = (3*A12 + A23_2 + A24 + A34 - A13 - A14_2) * inv72
    m[0, 1] = m[1, 0] = m12_val

    # m14 = m41
    m14_val = (A12_2 + A23 + A24 + A34_2 + A13 - 3*A14) * inv72
    m[0, 3] = m[3, 0] = m14_val

    # m23 = m32
    m23_val = (A12_2 + 3*A23 + A34_2 - A13 - A14 - A24) * inv72
    m[1, 2] = m[2, 1] = m23_val

    # m34 = m43
    m34_val = (A12 + A23_2 + 3*A34 + A13 - A14_2 - A24) * inv72
    m[2, 3] = m[3, 2] = m34_val

    return m      # this is m_bar
```

Validation of the Formula for $\bar{m}_{ij}$ The following code snippet evaluates the element mass matrix using the NSAF formulation and compares it against the standard Gauss quadrature-based implementation. This comparison serves to verify both the correctness of the analytical expressions and their numerical consistency with conventional finite element integration approaches.

```
# ============================================================
# Validation of the analytical formulas for Q4 mass matrix
# ============================================================
def N_Q4(xi, eta):    # Shape functions
    return np.array([0.25*(1-xi)*(1-eta), 0.25*(1+xi)*(1-eta),
                     0.25*(1+xi)*(1+eta), 0.25*(1-xi)*(1+eta) ])
def dN(xi, eta):
    dN_dxi = np.array([-0.25*(1-eta), 0.25*(1-eta),
                        0.25*(1+eta),-0.25*(1+eta) ])

    dN_deta = np.array([-0.25*(1-xi),-0.25*(1+xi),
                         0.25*(1+xi), 0.25*(1-xi) ])
    return dN_dxi, dN_deta

def Jmat(coords, xi, eta):
    '''Form Jacobian'''
    x = coords[:,0]
    y = coords[:,1]
    dxi, deta = dN(xi, eta)

    J = np.zeros((2,2))
    J[0,0] = np.dot(x, dxi)
    J[0,1] = np.dot(x, deta)
    J[1,0] = np.dot(y, dxi)
    J[1,1] = np.dot(y, deta)

    return J

# Gauss quadrature (2x2 exact for Q4 mass)
gp = 1.0 / np.sqrt(3)
gauss_pts = [(-gp, -gp), ( gp, -gp), ( gp,  gp), (-gp,  gp)]

def mass_gauss(coords, rho=1.0, t=1.0):
    M = np.zeros((4,4))
    for xi, eta in gauss_pts:
        N = N_Q4(xi, eta)
        detJ = np.linalg.det(Jmat(coords, xi, eta))
        for i in range(4):
```

```
            for j in range(4):
                M[i,j] += rho * t * N[i]*N[j]*detJ
    return M

# Test: analytical formula = Gaussian quadrature?
coords = np.array([[0.0, 0.01], [ 2.0, 0.2 ],        # an arbitrary Q4
                   [2.2,  1.1], [-0.1, 1.02] ])

M_Gauss = mass_gauss(coords)
M_optimized = compute_mij_NSAF(coords)

#print("\nGauss Quadrature Matrix:\n",M_g) # if checking the details
#print("\nAnalytical Matrix:\n", M_0)
#print("\nDifference:\n", M_g - M_0)
print("Max error:", np.max(np.abs(M_Gauss - M_optimized)))
```

```
Max error: 1.1102230246251565e-16
```

Both approaches produce identical results to machine precision, confirming the correctness and numerical consistency of the NSAF formulation.

Performance of NSAF for Mass Matrix The following snippet examines the performance of NSAF formulation against the standard approach.

```
start = time.time()
for i in range(10000):      # run a number of times for accurate timing
    M_Gauss = mass_gauss(coords)

time_Gauss = time.time() - start
print(f"Time, the standard approach: {time_Gauss:.4f}s")

start = time.time()
for i in range(10000):     # run a number of times for accurate timing
    M_optimized = compute_mij_NSAF(coords)

time_optimized = (time.time() - start)
print(f"Time, nodal submatrix analytical: {time_optimized:.4f}s")
print(f"Time ratio: {time_Gauss/time_optimized:.4f}")
```

```
Time, the standard approach: 1.1003s
Time, nodal submatrix analytical: 0.0999s
Time ratio: 11.0157
```

It is found that the NSAF approach for element mass matrix computation provides an approximate **10× speedup** compared with the standard numerical integration approach.

7.2.10.3 Lumped Mass Matrix Approximation

For Q4 elements, the **lumped mass matrix** has the form of:

$$\mathbf{m}_e^{\text{lumped}} = \frac{\rho t A_e}{4} \mathbf{I}_8 \tag{7.74}$$

where A_e is the physical area of the quadrilateral element and $\mathbf{I}_8$ is the 8×8 identity matrix.

Under this approximation, one-fourth of the total element mass is assigned directly to each corner node, fully decoupling the nodal inertial response. Although the resulting matrix is no longer dynamically consistent, the diagonal structure dramatically improves computational efficiency and eliminates the need for numerical integration or coordinate transformations during mass assembly.

7.2.11 Nodal Force Vectors

In the finite element method, the external virtual work terms are converted into equivalent nodal force vectors. These external loads are typically separated into:

- distributed body forces acting throughout the element domain, and
- surface tractions applied along element boundaries.

7.2.11.1 Body Force Vector

For a distributed body force field $\mathbf{b} = [b_x, b_y]^\top$ (force per unit volume), the equivalent element nodal force vector is

$$\mathbf{f}_b = \int_{\Omega_e} \mathbf{N}_e^\top \mathbf{b} \, d\Omega \tag{7.75}$$

Transforming the integral to the natural coordinate system (ξ, η) of the parent element gives

$$\underset{8\times 1}{\mathbf{f}_b} = \int_{-1}^{1} \int_{-1}^{1} \mathbf{N}_e^\top(\xi, \eta) \, \mathbf{b} \, \det(\mathbf{J}(\xi, \eta)) \, t \, d\xi d\eta \tag{7.76}$$

where t denotes the uniform element thickness.

In standard finite element implementations, Eq.(7.76) is usually evaluated numerically using 2×2 Gauss quadrature.

7.2.11.2 Analytical Integration for Uniform Body Forces

If the body force vector $\mathbf{b}$ is uniform over the element domain—such as gravity or constant rotational loading—the integral in Eq.(7.76) can be evaluated analytically.

For a Q4 element, the shape functions $N_i(\xi, \eta)$ are bilinear polynomials, while the Jacobian determinant $\det(\mathbf{J})$ is linear in ξ and η. Their product therefore remains a polynomial, allowing the integral to be computed exactly in closed form.

The following SymPy implementation automatically derives the exact algebraic expressions for the equivalent nodal body force vector under a uniform body force field.

```
# SymPy code for Q4 element body force vector
xi, eta = sp.symbols('xi eta') # Natural coordinates
t = sp.symbols('t') # Thickness

# Nodal coordinates:
x1, y1 = sp.symbols('x1 y1')
x2, y2 = sp.symbols('x2 y2')
x3, y3 = sp.symbols('x3 y3')
x4, y4 = sp.symbols('x4 y4')

# Body force components
bx, by = sp.symbols('b_x b_y')
b = sp.Matrix([bx, by])                 # Constant body force vector

N1 = (1 - xi)*(1 - eta)/4               # Q4 shape functions
N2 = (1 + xi)*(1 - eta)/4
N3 = (1 + xi)*(1 + eta)/4
N4 = (1 - xi)*(1 + eta)/4

N = sp.Matrix([[N1, 0,  N2, 0,  N3,  0, N4, 0 ],       # Matrix form
               [0, N1,  0, N2,  0, N3, 0,  N4]])

# Geometry mapping
x = N1*x1 + N2*x2 + N3*x3 + N4*x4
y = N1*y1 + N2*y2 + N3*y3 + N4*y4

# Jacobian matrix
J = sp.Matrix([[sp.diff(x, xi),  sp.diff(y, xi)],
               [sp.diff(x, eta), sp.diff(y, eta)]])

detJ = sp.simplify(J.det())
integrand = sp.expand(N.T * b * detJ * t)

# Element body force vector
# ---------------------------------------------------------------
fe = sp.integrate(sp.integrate(integrand, (xi, -1,1)), (eta, -1,1))
fe = sp.simplify(fe)

print("\nFormula for computing the body force vector, Q4 elements:")
display(Math(f"f_e=\\small{latex(fe)}"))
```

`Formula for computing the body force vector, Q4 elements:`

$$f_e = \begin{bmatrix} \frac{b_x t(2x_1y_2-2x_1y_4-2x_2y_1+x_2y_3+x_2y_4-x_3y_2+x_3y_4+2x_4y_1-x_4y_2-x_4y_3)}{12} \\ \frac{b_y t(2x_1y_2-2x_1y_4-2x_2y_1+x_2y_3+x_2y_4-x_3y_2+x_3y_4+2x_4y_1-x_4y_2-x_4y_3)}{12} \\ \frac{b_x t(2x_1y_2-x_1y_3-x_1y_4-2x_2y_1+2x_2y_3+x_3y_1-2x_3y_2+x_3y_4+x_4y_1-x_4y_3)}{12} \\ \frac{b_y t(2x_1y_2-x_1y_3-x_1y_4-2x_2y_1+2x_2y_3+x_3y_1-2x_3y_2+x_3y_4+x_4y_1-x_4y_3)}{12} \\ \frac{b_x t(x_1y_2-x_1y_4-x_2y_1+2x_2y_3-x_2y_4-2x_3y_2+2x_3y_4+x_4y_1+x_4y_2-2x_4y_3)}{12} \\ \frac{b_y t(x_1y_2-x_1y_4-x_2y_1+2x_2y_3-x_2y_4-2x_3y_2+2x_3y_4+x_4y_1+x_4y_2-2x_4y_3)}{12} \\ \frac{b_x t(x_1y_2+x_1y_3-2x_1y_4-x_2y_1+x_2y_3-x_3y_1-x_3y_2+2x_3y_4+2x_4y_1-2x_4y_3)}{12} \\ \frac{b_y t(x_1y_2+x_1y_3-2x_1y_4-x_2y_1+x_2y_3-x_3y_1-x_3y_2+2x_3y_4+2x_4y_1-2x_4y_3)}{12} \end{bmatrix}$$

7.2.12 Analytical Interpretation and the Center of Gravity

Symbolic integration produces exact closed-form expressions for the nodal body force vector of a general distorted Q4 element. Although the resulting formulas are algebraically lengthy, they depend only on the physical corner coordinates of the element and can therefore be evaluated efficiently.

More importantly, the analytical expressions reveal an elegant geometric interpretation. The total body force acting on the element,

$$A_e t b_x \quad \text{and} \quad A_e t b_y, \tag{7.77}$$

is distributed among the four corner nodes according to the areas of four associated sub-Q4s. These sub-regions are formed by connecting the corner nodes and edge midpoints to the element's center of gravity, i.e., the true geometric centroid of the physical element rather than the simple average of the vertex coordinates.

As a result, the x-component of the equivalent nodal body force at node i can be written compactly as

$$f_{ex-i} = \frac{A_{\text{subQ4-}i}}{A_e}\,(tA_e b_x) \tag{7.78}$$

where $A_{\text{subQ4-}i}$ denotes the area associated with node i.

For example, the sub-area associated with node 1 is

$$A_{\text{subQ4-}i} = \frac{2x_1y_2 - 2x_1y_4 - 2x_2y_1 + x_2y_3 + x_2y_4 - x_3y_2 + x_3y_4 + 2x_4y_1 - x_4y_2 - x_4y_3}{12} \tag{7.79}$$

which appears directly in the first component of $\mathbf{f}_b$.

This leads to the following interpretation:

> **The bilinear shape functions partition the uniform body load through the element geometric centroid.**

For an undistorted rectangular Q4 element subjected to a uniform body force field, the geometric centroid coincides exactly with the topological center obtained from the vertex average. In this special case, the analytical expressions simplify to the classical uniform load distribution

$$\mathbf{f}_b = \frac{A_e t}{4} \begin{bmatrix} b_x & b_y & b_x & b_y & b_x & b_y & b_x & b_y \end{bmatrix}^\top \tag{7.80}$$

where A_e is the physical element area.

Thus, the total body force is shared equally among the four corner nodes, with each node receiving exactly one-quarter of the total load.

> **Extension to Non-Uniform Body Forces:** The same symbolic procedure can also handle spatially varying body force fields. If the body force is defined as an integrable function, such as a linear hydrostatic pressure distribution or a quadratic centrifugal load, exact closed-form expressions can still be derived. This is accomplished by explicitly defining $\mathbf{b}(\xi, \eta)$ within the integrand before applying SymPy's `integrate` operation, which is similar to 1D case discussed in the previous chapter.

7.2.12.1 Boundary Traction Vector

As with the T3 element, traction forces are evaluated only along element edges that lie on the Neumann boundary Γ_t. The total boundary contribution is therefore assembled edge by edge.

For a prescribed traction vector

$$\bar{\mathbf{t}} = [t_x, t_y]^\top \tag{7.81}$$

acting along an element edge Γ_{edge}, the equivalent nodal traction vector is

$$\mathbf{f}_{\text{edge}} = \int_{\Gamma_{\text{edge}}} \mathbf{N}_{\text{edge}}^\top \bar{\mathbf{t}}\, t\, d\Gamma \tag{7.82}$$

Along a straight edge of a Q4 element, the two shape functions associated with the opposite nodes vanish identically. The remaining two edge shape functions vary linearly from 1 to 0 over the edge length l_{edge}.

If the applied traction is uniform along the edge, the line integral evaluates exactly to

$$\mathbf{f}_{\text{edge}} = \frac{t\, l_{\text{edge}}}{2} \begin{bmatrix} \bar{\mathbf{t}} \\ \bar{\mathbf{t}} \end{bmatrix} = \frac{t\, l_{\text{edge}}}{2} \begin{bmatrix} t_x \\ t_y \\ t_x \\ t_y \end{bmatrix} \tag{7.83}$$

Thus, the total line load is shared equally between the two nodes defining the loaded edge.

Note: The vector $\mathbf{f}_{\text{edge}}$ contains only the four active degrees of freedom associated with the loaded edge (two nodes with two displacement components each). During global assembly, this local edge vector is expanded into the full 8-degree-of-freedom element force vector according to the element connectivity.

7.2.13 Remarks and Insights

1. **Computational Efficiency of NSAF**

The Nodal Submatrix Analytical Formula (NSAF) framework provides a highly efficient approach for evaluating element matrices and vectors. Compared with conventional formulations, NSAF typically achieves a 2–3× speedup for stiffness matrix computations and up to an order-of-magnitude improvement for mass matrix evaluations.

2. **Bilinear Interpolation and Spatial Field Variation**

Unlike the linear T3 element, the Q4 element uses bilinear shape functions, allowing it to represent a richer displacement field within the element domain. As a result, the corresponding strain and stress fields vary spatially. In standard finite element formulations, these varying fields generally require numerical integration, most commonly through Gauss quadrature. An excellent alternative is the Smoothed Finite Element Method (S-FEM) [11] that avoid the numerical integration entirely, known as a domain integral free method.

3. **Improved Accuracy Relative to T3 Elements**

The Q4 element typically delivers substantially higher accuracy per degree of freedom than the linear T3 element, especially in structured or mapped meshes. It captures membrane, shear, and primary tension states more effectively, making it particularly attractive for regular geometries. The use of the S-FEM [11] can further improve the accuracy.

4. **Isoparametric Flexibility and Distortion Sensitivity**

The isoparametric formulation enables Q4 elements to model irregular geometries and curved boundaries through distorted quadrilateral shapes. However, the accuracy of bilinear interpolation is sensitive to element distortion. As the element becomes increasingly skewed, warped, or nearly triangular, the Jacobian determinant $\det(\mathbf{J})$ can vary rapidly across the domain. This behavior degrades the accuracy of both numerical quadrature and analytical Taylor-series approximations. The solution to element distortion is the S-FEM, which is a Jacobian free formulation, and it even allows the use of concave Q4 elements.

5. **Convergence and the Patch Test**

When integrated correctly using at least a standard 2×2 Gauss quadrature rule, the Q4 element reproduces linear displacement fields exactly, as demonstrated in the following section. Consequently, the element passes the patch test, which provides a fundamental consistency check and guarantees monotonic convergence under mesh refinement. Further improvement is the S-FEM which possesses supper-convergence property.

6. **Numerical Locking**

Despite its advantages, the standard Q4 element is susceptible to several locking phenomena under restrictive kinematic conditions. In bending-dominated problems, it may exhibit *shear locking*, where parasitic shear strains artificially increase the element stiffness. For nearly incompressible materials ($\nu \to 0.5$), the element can also suffer from *volumetric locking*. Common remedies include Selective Reduced Integration (SRI), Uniform Reduced Integration (URI) with hourglass stabilization, and advanced formulations such as the S-FEM where the softness effects can be properly made use to overcome locking issues.

7. **Practical Importance**

The Q4 element remains one of the most widely used finite elements in computational solid mechanics. It is especially effective for structured meshes and provides an excellent balance between computational efficiency, implementation simplicity, and solution accuracy across many engineering applications. It is the author's opinion that the use of S-FEM is a promising direction beyond the standard FEM.

7.3 FEM Solver for 2D Solids

Using the theoretical foundations and finite element formulations developed in the preceding sections, we now construct a finite element solver for two-dimensional solid mechanics problems. To illustrate both simplicity and geometric versatility, the solver is designed to support two of the most widely used element types: the 3-node linear triangular (T3) element and the 4-node bilinear quadrilateral (Q4) element.

The implementation is intentionally modular, making it straightforward to extend the solver to other element formulations. Most components of the code, including mesh handling, assembly procedures, boundary condition enforcement, and solution routines, remain unchanged. In general, only the element-specific routines, such as shape functions, numerical integration schemes, and element stiffness matrix calculations, need to be modified when introducing a new element type.

To verify the implementation, we first consider a classical benchmark problem: a square plate subjected to uniform uniaxial tension. This example problem has an exact solution and provides a transparent setting for illustrating the full FEM workflow, including mesh generation, matrix assembly, boundary condition enforcement, and displacement computation.

7.3.1 Problem Setup

Consider the unit square domain $\Omega = (0,1) \times (0,1)$, whose boundary Γ consists of four edges:

- Left edge: $x = 0$
- Right edge: $x = 1$
- Bottom edge: $y = 0$
- Top edge: $y = 1$

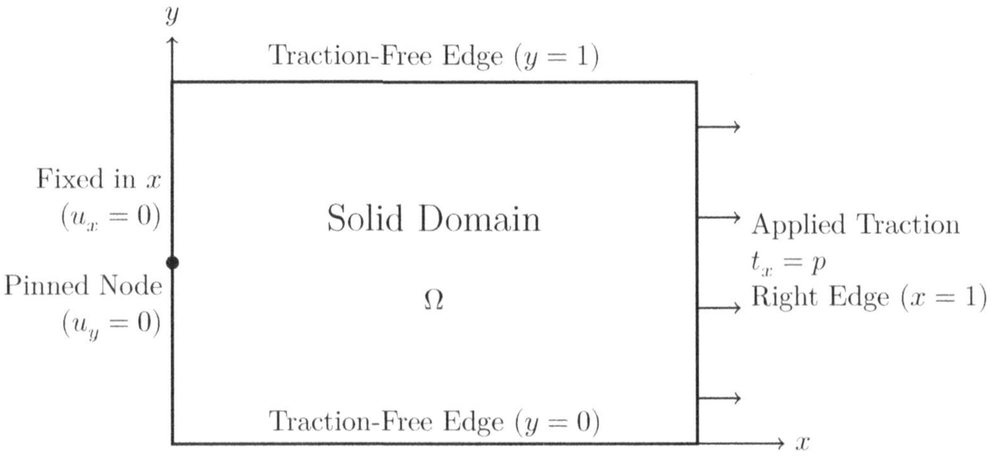

The plate is subjected to a uniform horizontal tensile traction applied along the right boundary. The upper and lower boundaries are traction-free, allowing the body to contract vertically due to Poisson effects. The setting is illustrated in the foregoing figure.

7.3.2 Boundary Conditions

To ensure a well-posed linear system, the problem is prescribed using both Dirichlet (essential) and Neumann (natural) boundary conditions.

7.3.2.1 Dirichlet Boundary Conditions (DBCs)

- **Left Boundary** ($x = 0$)

  The horizontal displacement is constrained along the left edge to provide lateral support and suppress rigid-body motion in the x-direction:

$$u_x(0, y) = 0 \tag{7.84}$$

- **Rigid-Body Motion Suppression**

  To eliminate rigid translation in the vertical direction without over-constraining the structure, the vertical displacement is fixed at a single reference node on the left boundary. This reference point is commonly chosen at the corner node or at the mid-height location:

$$u_y(0, y_{\text{ref}}) = 0 \tag{7.85}$$

7.3.2.2 Neumann Boundary Conditions (NBCs)

- **Top and Bottom Boundaries** $(y = 0,\ y = 1)$

  The upper and lower edges are traction-free:

$$\mathbf{t} = \begin{bmatrix} t_x \\ t_y \end{bmatrix} = \begin{bmatrix} 0 \\ 0 \end{bmatrix} \tag{7.86}$$

- **Left Boundary** $(x = 0)$

  The left boundary acts as a sliding support. Since the horizontal displacement is prescribed, only the natural boundary condition remains active:

$$t_y = 0 \tag{7.87}$$

- **Right Boundary** $(x = 1)$

  A uniform tensile traction is applied along the positive x-direction:

$$\mathbf{t} = \begin{bmatrix} t_x \\ t_y \end{bmatrix} = \begin{bmatrix} p \\ 0 \end{bmatrix} \tag{7.88}$$

This configuration defines a purely force-driven system, where all structural deformation is induced by external traction forces. Since the vertical traction component vanishes everywhere and the horizontal traction is applied only along the right boundary $(x = 1)$, the external force assembly process becomes especially efficient. Only the boundary line integrals associated with the right-edge segments must be evaluated [11].

7.3.3 Exact Analytical Solution

Under the prescribed boundary conditions, the plate is subjected to a homogeneous state of uniaxial tension. This idealized configuration results in a uniform stress and strain field throughout the domain, providing an exact reference solution for validating numerical convergence and discretization accuracy.

7.3.3.1 Stress Field

Assuming plane stress conditions for a thin plate $(\sigma_{zz} = \sigma_{xz} = \sigma_{yz} = 0)$, the stress state reduces to a single non-zero component:

$$\sigma_{xx} = p, \qquad \sigma_{yy} = 0, \qquad \sigma_{xy} = 0 \tag{7.89}$$

7.3.3.2 Strain Field

Using Hooke's law under plane stress conditions, the corresponding in-plane strain components are

$$\varepsilon_{xx} = \frac{p}{E}, \qquad \varepsilon_{yy} = -\frac{\nu p}{E}, \qquad \gamma_{xy} = 2\varepsilon_{xy} = 0 \tag{7.90}$$

where E is Young's modulus and ν is Poisson's ratio.

The out-of-plane strain follows directly from Poisson contraction:

$$\varepsilon_{zz} = -\frac{\nu}{1-\nu}(\varepsilon_{xx} + \varepsilon_{yy}) = -\frac{\nu p}{E}, \qquad \varepsilon_{xz} = \varepsilon_{yz} = 0 \tag{7.91}$$

Since ε_{zz} is fully determined by the in-plane deformation, the finite element formulation only solves for the in-plane strain vector

$$\varepsilon = \begin{bmatrix} \varepsilon_{xx} & \varepsilon_{yy} & \gamma_{xy} \end{bmatrix}^{\top}. \tag{7.92}$$

7.3.3.3 Displacement Field

Integrating the strain–displacement relations and enforcing the rigid-body constraints yields a linear displacement field:

$$u_x(x, y) = \frac{p}{E}x, \qquad u_y(x, y) = -\frac{\nu p}{E}y \tag{7.93}$$

The maximum horizontal displacement occurs at the right boundary ($x = 1$) and equals p/E.

Plane Strain Remark

For thick-plate or constrained configurations, the plane strain assumption imposes $\varepsilon_{zz} = 0$, resulting in a non-zero out-of-plane stress

$$\sigma_{zz} = \nu(\sigma_{xx} + \sigma_{yy}). \tag{7.94}$$

In this case, the structure of the 2D finite element formulation remains unchanged; only the constitutive matrix **C** is modified through the effective material constants E^* and ν^* derived for plane strain conditions, as discussed in Chapter 2.

7.3.4 Typical FEM Analysis Pipeline

The computational workflow for the benchmark problem follows the standard finite element procedure for linear elasticity. The overall pipeline is summarized as follows:

1. **Constitutive Initialization**

   Construct the material elasticity matrix $\mathbf{C}$ for the plane stress condition using Young's modulus E and Poisson's ratio ν.

2. **Global Assembly Loop**

   Loop over all elements in the mesh. For each element, compute the strain-displacement matrix $\mathbf{B}_e$ and assemble the corresponding element stiffness matrix $\mathbf{k}_e$ into the global stiffness matrix $\mathbf{K}$.

3. **Load Vector Formulation**

   Assemble the global force vector $\mathbf{F}$ from element-level contributions. For this benchmark problem, external nodal forces are generated exclusively from the right boundary ($x = 1$) under a uniform traction p.

4. **Essential Boundary Condition Enforcement**

   Identify all nodes along the left boundary ($x = 0$) and apply Dirichlet constraints:

   $$u_x = 0$$

   In addition, a single vertical displacement constraint

   $$u_y = 0$$

   is imposed at one reference node to eliminate rigid-body motion.

5. **Linear System Solution**

   Solve the resulting reduced linear system to obtain the unknown nodal displacement vector. For this we simply use the existing linear algebra solver readily available in Numpy.

6. **Stress and Strain Evaluation**

   After solving for nodal displacements, perform post-processing at the element level. The displacement field is evaluated as

   $$\mathbf{u}_e = \mathbf{N}_e \mathbf{d}_e,$$

   the strain field is computed as

   $$\varepsilon_e = \mathbf{B}_e \mathbf{d}_e,$$

   and stresses are obtained via Hooke's law

   $$\sigma_e = \mathbf{C}\varepsilon_e.$$

   This procedure is carried out for all elements in the mesh.

7. **Nodal Stress Recovery**

   Because stresses are discontinuous across elements, an unweighted averaging scheme is applied to recover smooth nodal stress values. The post-processed stress solution exhibits a form of superconvergence, where the convergence rate exceeds the optimal rate of the original FEM approximation [11].

7.3.5 A Typical Flowchart for FEM Solver

The following is a seven-step flowchart, which provides the overall finite element solution procedure.

PRE-PROCESSING STAGE

1. Mesh Generation and Material Definition
(Nodes, Elements, Material $\mathbf{C}$)

⇓

FEM SOLUTION STAGE

2. Global Assembly Loop
(Compute $\mathbf{B}_e$, $\mathbf{k}_e \rightarrow$ Assemble $\mathbf{K}$)

⇓

3. Load Vector Formulation
(Compute $\mathbf{f}_e \rightarrow$ Assemble Global $\mathbf{F}$)

⇓

4. Essential Boundary Conditions
(Apply DBCs via Static Condensation)

⇓

5. Linear System Inversion
(Solve Condensed Equations)

POST-PROCESSING STAGE ⇓

6. Stress & Strain Evaluation
(Compute Element-Wise Strain/Stress)

⇓

7. Nodal Stress Recovery
(Average Element Stresses at Nodes)

The entire computational pipeline can be compactly implemented in Python using scientific libraries such as NumPy.

7.3.6 Key Functions for the 2D FEM Solver

The mathematical formulations derived in the previous sections are directly translated into Python functions. These implementations are intended as a proof of concept, prioritizing transparency and direct correspondence with the governing equations rather than computational optimization or production-level robustness. Each function is written to be self-contained and extensively documented to clarify the underlying finite element operations.

7.3.6.1 T3 Mesh Generation Function

```
# Modular FEM Code (2D Solids, Static)
def generate_mesh_T3(nx, ny, Lx=1.0, Ly=1.0):
    '''
    Generate regular T3 Mesh for a unit sqaure domain: [0,1]×[0,1].
    nx: number of grid cells in the x-direction
    ny: number of grid cells in the y-direction
    number of T3 elements generated: 2 nx × ny
    '''
    x = np.linspace(0, Lx, nx+1)
    y = np.linspace(0, Ly, ny+1)

    nodes = np.array([[xi, yi] for yi in y for xi in x])

    def node_id(i, j):
        return j*(nx+1) + i

    elements = []
    for j in range(ny):
        for i in range(nx):
            n1 = node_id(i, j)
            n2 = node_id(i+1, j)
```

```
            n3 = node_id(i, j+1)
            n4 = node_id(i+1, j+1)

            elements.append([n1, n2, n4])
            elements.append([n1, n4, n3])
    return nodes, np.array(elements)
```

7.3.6.2 Function for Global K Assembly

```
def assemble_system(nodes, elements, element_stiffness_func, C):
    """
    General FEM assembly supporting both individual and batched
    element stiffness functions.
    Parameters::
    nodes : (n_nodes, 2) array, Nodal coordinates
    elements : (n_elem, n_node_per_elem) array
               Element connectivity matrix
    element_stiffness_func : function, to compute element stiffness
        matrices. Can accept a single element or a batch elements.
    C : array, Material matrix

    Returns::
    K : (ndof, ndof) array, Global stiffness matrix
    F : (ndof,) array, Global force vector
    """
    n_nodes = len(nodes)
    ndof = 2 * n_nodes
    n_elem = len(elements)

    K = np.zeros((ndof, ndof))
    F = np.zeros(ndof)

    # Gather coordinates for all elements
    # Shape: (n_elem, n_node_per_elem, 2)
    all_coords = nodes[elements]

    # Check if the function supports batched evaluation or needs
    # a loop fallback
    try:
        # Attempt to compute all element matrices in a single batch
        ke_data = element_stiffness_func(all_coords, C)
        all_ke=ke_data[0] if isinstance(ke_data, tuple) else ke_data
```

```
        # If batched, output shape should be (n_elem,ndof_e,ndof_e)
        # If it returns a single element matrix instead, it doesn't
        # support batching
        if all_ke.ndim == 2 and n_elem > 1:
            raise TypeError("Function returned a single matrix for\
            a batch input.")

    except (ValueError, TypeError, IndexError):
        # Fallback: compute element stiffnesses one by one if
        # function isn't batched
        all_ke = []
        for coords in all_coords:
            ke_data = element_stiffness_func(coords, C)
            ke=ke_data[0] if isinstance(ke_data, tuple) else ke_data
            all_ke.append(ke)
        all_ke = np.array(all_ke)

    # Loop through elements to map local DOFs to the global K
    for e, elem in enumerate(elements):
        ke = all_ke[e]

        n_en = len(elem)            # nodes per element
        ndof_e = 2 * n_en           # DOFs per element

        # Build DOF map efficiently using numpy vectorization
        dofs = np.zeros(ndof_e, dtype=int)
        dofs[0::2] = 2 * elem
        dofs[1::2] = 2 * elem + 1

        # Assembly using advanced numpy indexing:
        K[np.ix_(dofs, dofs)] += ke
    return K, F
```

7.3.6.3 Function for Nodal Force Vector

As shown in Eqs. (7.40) and (7.83), the consistent nodal force vector resulted from the externally applied force is computed by:

1. Identifying boundary edges on the right side,
2. Computing the length of each edge of the element on right side,
3. Assembling the contribution $\bar{\mathbf{t}}l_e/2$ to the two nodes of each edge.

The following utility functions implement this procedure in a form applicable to both T3 and Q4 elements.

```
def find_boundary_edges(nodes, elements):
    """
    Return all boundary edges (node index pairs) for T3/Q4 mesh.
    Works for unstructured meshes.
    """
    edge_count = {}

    for elem in elements:
        nen = len(elem)
        for i in range(nen):
            a = elem[i]
            b = elem[(i + 1) symmetric positive definite nen]
            edge = tuple(sorted((a, b)))
            edge_count[edge] = edge_count.get(edge, 0) + 1

    # edges appearing once → boundary
    return [e for e, c in edge_count.items() if c == 1]

def find_edges_at_x(nodes, elements, x_target, tol=1e-12):
    """
    Return boundary edges located at x = x_target.
    """
    boundary_edges = find_boundary_edges(nodes, elements)

    edges = []
    for n1, n2 in boundary_edges:
        if (abs(nodes[n1, 0] - x_target) < tol and
            abs(nodes[n2, 0] - x_target) < tol):
            edges.append((n1, n2))

    return edges

def apply_uniform_traction(F, nodes, elements, traction, x_target):
    """
    Apply uniform traction on boundary at x = x_target.

    Parameters
    ----------
    F : global force vector (size 2*n_nodes)
    nodes : nodal coordinates
    elements : connectivity (T3 or Q4)
    traction : [tx, ty]
    x_target : location of boundary (e.g., max or min x)
    """
```

```
    traction = np.asarray(traction)
    edges = find_edges_at_x(nodes, elements, x_target)

    for n1, n2 in edges:
        x1, y1 = nodes[n1]
        x2, y2 = nodes[n2]

        le = np.hypot(x2 - x1, y2 - y1)
        f_edge = traction * le / 2.0

        F[2*n1:2*n1+2] += f_edge
        F[2*n2:2*n2+2] += f_edge

    return edges
```

7.3.7 Imposing Dirichlet Boundary Conditions (General Form)

As demonstrated in Chapter 6, when the prescribed values on Dirichlet boundaries are zero, the constraint enforcement reduces to simply removing the corresponding rows and columns associated with the fixed degrees of freedom (DOFs). However, for non-zero prescribed displacements (e.g., settlements, prestrains, or controlled boundary motion), a more general formulation is required to preserve accuracy and consistency.

7.3.7.1 Mathematical Partitioning Framework

We begin with the global finite element equilibrium system:

$$\mathbf{K}\mathbf{D} = \mathbf{F}. \tag{7.95}$$

The global DOFs are partitioned into two disjoint sets:

- f: free (unconstrained) DOFs,
- c: constrained (Dirichlet) DOFs.

Accordingly, the displacement and force vectors are partitioned as

$$\mathbf{D} = \begin{bmatrix} \mathbf{D}_f \\ \mathbf{D}_c \end{bmatrix}, \qquad \mathbf{F} = \begin{bmatrix} \mathbf{F}_f \\ \mathbf{F}_c \end{bmatrix}. \tag{7.96}$$

The stiffness matrix is correspondingly block-partitioned:

$$\mathbf{K} = \begin{bmatrix} \mathbf{K}_{ff} & \mathbf{K}_{fc} \\ \mathbf{K}_{cf} & \mathbf{K}_{cc} \end{bmatrix}. \tag{7.97}$$

Submatrix	Dimensions	Description
$\mathbf{K}_{ff}$	$N_f \times N_f$	Internal stiffness coupling between active free degrees of freedom (DOFs).
$\mathbf{K}_{fc}$	$N_f \times N_c$	Cross-boundary stiffness coupling free DOFs to the constrained DOFs.
$\mathbf{K}_{cf}$	$N_c \times N_f$	Transpose of $\mathbf{K}_{fc}$; structural feedback from free domains to boundaries.
$\mathbf{K}_{cc}$	$N_c \times N_c$	Localized boundary-to-boundary stiffness matrix.

7.3.7.2 Derivation of the Reduced System

Let the prescribed displacement vector on constrained DOFs be

$$\mathbf{D}_c = \bar{\mathbf{D}}_c. \tag{7.98}$$

Substituting into the first block row yields

$$\mathbf{K}_{ff}\mathbf{D}_f + \mathbf{K}_{fc}\bar{\mathbf{D}}_c = \mathbf{F}_f. \tag{7.99}$$

Rearranging,

$$\mathbf{K}_{ff}\mathbf{D}_f = \mathbf{F}_f - \mathbf{K}_{fc}\bar{\mathbf{D}}_c. \tag{7.100}$$

Homogeneous Case. When $\bar{\mathbf{D}}_c = \mathbf{0}$, the system reduces to

$$\mathbf{K}_{ff}\mathbf{D}_f = \mathbf{F}_f, \tag{7.101}$$

which justifies the common practice of simply removing constrained rows and columns in homogeneous Dirichlet problems.

7.3.7.3 Modified Force Vector

Define the effective load vector:

$$\mathbf{F}_f^* = \mathbf{F}_f - \mathbf{K}_{fc}\bar{\mathbf{D}}_c, \tag{7.102}$$

so that the final reduced system becomes

$$\mathbf{K}_{ff}\mathbf{D}_f = \mathbf{F}_f^*. \tag{7.103}$$

Provided sufficient constraints are applied, $\mathbf{K}_{ff}$ remains symmetric positive definite (SPD), enabling efficient solution via direct solvers (e.g., Cholesky factorization [10]) or iterative methods (e.g., Conjugate Gradient).

After solving for $\mathbf{D}_f$, the full displacement field is reconstructed as

$$\mathbf{D} = \begin{bmatrix} \mathbf{D}_f \\ \bar{\mathbf{D}}_c \end{bmatrix}. \tag{7.104}$$

7.3.7.4 Algorithmic Mapping to Python Implementation

The mathematical components map directly to implementation variables:

Mathematical Notation	Code Variable	Description
f	free_dofs	Indices of unconstrained DOFs
c	fixed_dofs	Indices of constrained DOFs
$\mathbf{K}_{ff}$	K_ff	Reduced stiffness matrix
$\mathbf{K}_{fc}$	K_fc	Coupling stiffness matrix
$\bar{\mathbf{D}}_c$	fixed_vals	Prescribed displacement values
$\mathbf{F}_f^*$	F_mod	Modified force vector

This formulation provides several advantages:

- **Exact enforcement of Dirichlet conditions** at the algebraic level.
- **Preservation of symmetry**, ensuring compatibility with efficient solvers.
- **Equivalence to static condensation**, reducing the system size prior to inversion.

The following Python function `apply_dirichlet` implements this general procedure.

```python
## Apply Dirichlet BCs (General Form)
def apply_dirichlet(K, F, fixed_dofs, fixed_values):
    '''
       Impose the DBC at `fixed_dofs` with values in `fixed_values`.
       This results in a SPD `K_ff` and `F_mod` with reduced size
       of `free_dofs`.
    '''
    ndof = K.shape[0]
    free_dofs = np.setdiff1d(np.arange(ndof), fixed_dofs)

    K_ff = K[np.ix_(free_dofs, free_dofs)]
    K_fc = K[np.ix_(free_dofs, fixed_dofs)]
    F_f = F[free_dofs]

    F_mod = F_f - K_fc @ fixed_values

    return K_ff, F_mod, free_dofs
```

7.3.7.5 Solve System for Nodal Displacements

The following function solves Eq.(7.103) using the standard linear algebraic solver `np.linalg.solve`.

```
def solve_system(K, F, fixed_dofs, fixed_values):
    '''
       Solve the FEM system equation
    '''
    K_ff, F_mod, free_dofs = apply_dirichlet(K, F, fixed_dofs,\
                                                    fixed_values)
    D = np.zeros(len(F))
    D[fixed_dofs] = fixed_values
    D[free_dofs] = np.linalg.solve(K_ff, F_mod)

    return D
```

7.4 Case Study: FEM Solution of Plates under Pure Tension

7.4.1 FEM Solution Using T3 Elements

We begin by generating a T3 mesh for the problem domain, which defines the nodal coordinates and element connectivity for the entire discretized system. The element stiffness matrices are then computed using previously developed routines, such as `element_ke_T3_BCB`, `batch_ke_T3_NSAF`, or `batch_ke_T3_BCB`. The detailed implementation procedure and corresponding code are provided below.

7.4.1.1 FEM Model Creation

```
# FEM Solution for the example problem:

# Mesh generation (regular)
ncx = 1 # number of square cells in the x-direction
ncy = 1 # number of square cells in the y-direction

nodes, elements = generate_mesh_T3(ncx, ncy)
print(f"Number of nodes = {nodes.shape[0]}, "
      f"Number of T3 elements = {elements.shape[0]}")

# Martrix of material constants:
C = get_C_2D(E=1.0, nu=0.3, plane_stress=True) #(E=210e9, nu=0.3)

# Assembly of stiffness matrix and force vector
K, _ = assemble_system(nodes, elements, batch_ke_T3_NSAF, C)
```

```
Number of nodes = 4, Number of T3 elements = 2
```

We have now generated the mesh and assembled the global stiffness matrix **K**. Since the present problem is relatively simple and serves primarily as a verification benchmark, only a small number of elements is used.

At this stage, the assembled stiffness matrix **K** is singular because Dirichlet boundary conditions have not yet been enforced. The system therefore lacks the constraints required to eliminate rigid-body modes.

The following step imposes these boundary conditions.

7.4.1.2 Set Dirichlet Boundary Conditions

The function below applies Dirichlet boundary conditions for rectangular plates discretized using either Q4 or T3 elements. It identifies the constrained degrees of freedom and modifies the global system accordingly, ensuring a well-posed and solvable linear system.

```
def set_DBCs(nodes, elements, delta, dtype=1, tol=1e-10):
    """
    Dirichlet BC setter for rectangular plates with Q4/T3 elements.
    Supports four types of DBCs on the left and right edges.
    """
    fixed_dofs = []
    fixed_vals = []
    x = nodes[:, 0]
    y = nodes[:, 1]

    xmin = np.min(x)
    xmax = np.max(x)
    left_nodes  = np.where(np.abs(x - xmin) < tol)[0]
    right_nodes = np.where(np.abs(x - xmax) < tol)[0]

    # TYPE 0: left ux = 0, pure tension at the right edge
    if dtype == 0:
        for n in left_nodes:
            fixed_dofs.append(2 * n)
            fixed_vals.append(0.0)
        # Fix one uy DOF to remove right motion in y-direction
        y_left = y[left_nodes]
        n_mid = left_nodes[np.argmin(np.abs(y_left-np.mean(y_left)))]
        fixed_dofs.append(2 * n_mid + 1)
        fixed_vals.append(0.0)

    # TYPE 1: Left edge: ux = 0, displacement-pulling at right edge
    elif dtype == 1:
        for n in left_nodes:
```

```
            fixed_dofs.append(2 * n)
            fixed_vals.append(0.0)
        # Fix one uy DOF to remove right motion in y-direction
        y_left = y[left_nodes]
        n_mid = left_nodes[np.argmin(np.abs(y_left-np.mean(y_left)))]
        fixed_dofs.append(2 * n_mid + 1)
        fixed_vals.append(0.0)
        # Right edge: ux = delta
        for n in right_nodes:
            fixed_dofs.append(2 * n)
            fixed_vals.append(delta)

    # TYPE 2: clamped left, displaced right
    elif dtype == 2:
        # Left fully fixed
        for n in left_nodes:
            fixed_dofs.extend([2*n, 2*n+1])
            fixed_vals.extend([0.0, 0.0])
        # Right ux = delta
        for n in right_nodes:
            fixed_dofs.append(2*n)
            fixed_vals.append(delta)

    # TYPE 3: fully constrained tension
    elif dtype == 3:
        # Left fully fixed
        for n in left_nodes:
            fixed_dofs.extend([2*n, 2*n+1])
            fixed_vals.extend([0.0, 0.0])
        # Right: ux = delta, uy = 0
        for n in right_nodes:
            fixed_dofs.extend([2*n, 2*n+1])
            fixed_vals.extend([delta, 0.0])

    else:
        raise ValueError("dtype must be 0, 1, 2, or 3.")

    # Convert to arrays
    fixed_dofs = np.array(fixed_dofs, dtype=int)
    fixed_vals = np.array(fixed_vals, dtype=float)

    # Safe duplicate removal
    unique_map = {}
    for dof, val in zip(fixed_dofs, fixed_vals):
```

```
        unique_map[dof] = val    # last assignment wins (safe)

    fixed_dofs = np.array(sorted(unique_map.keys()), dtype=int)
    fixed_vals = np.array([unique_map[d] for d in fixed_dofs],\
                                                    dtype=float)
    return fixed_dofs, fixed_vals
```

Let us now use `set_DBCs` to set the DBCs on the left edge of the plate:

```
delta = None  # not used for type-0 DBCs
fixed_dofs, fixed_vals = set_DBCs(nodes, elements, delta, dtype=0,\
                                                    tol=1e-10)
```

7.4.1.3 Computation of the Nodal Force Vector

Using the Neumann boundary conditions and the prescribed traction $\bar{\mathbf{t}} = p$, representing the uniform pulling force applied along the right edge, we compute the equivalent nodal force vector.

The following code snippet performs the numerical integration (or exact edge evaluation where applicable) and assembles the global force vector $\mathbf{F}$ consistent with the finite element discretization.

```
# Set NBC (computing the traction at right edge):
F = np.zeros(2 * len(nodes))                # initialization
x_right = np.max(nodes[:, 0])               # find the right edge
p = 0.1                                     # the given pulling force
tx = p; ty =.0                              # convert to traction vector
t_edges = apply_uniform_traction(F, nodes, elements,[tx,ty],x_right)
```

The execution of the above snippet updates the global nodal force vector $\mathbf{F}$ accordingly.

At this stage, all essential components of the finite element system are available: the global stiffness matrix $\mathbf{K}$, the force vector $\mathbf{F}$, and the boundary condition definitions `fixed_dofs` and `fixed_vals`. The system is now ready to be solved.

We solve the linear system

$$\mathbf{KD} = \mathbf{F} \tag{7.105}$$

for the unknown nodal displacement vector $\mathbf{D}$ using the function `solve_system()`, which incorporates the previously defined Dirichlet boundary treatment and returns the complete displacement field.

```
D = solve_system(K, F, fixed_dofs, fixed_vals)
```

We have obtained the nodal displacement field for all nodes in the mesh, stored in the global displacement vector

$$\mathbf{D} = \begin{bmatrix} u_{1x} & u_{1y} & u_{2x} & u_{2y} & \cdots & u_{nx} & u_{ny} \end{bmatrix}^{\mathsf{T}}. \tag{7.106}$$

7.4.1.4 Post-processing

With the full nodal displacement solution available, we can now evaluate derived quantities such as displacement fields, strain fields, and stress distributions over the entire domain.

7.4.1.5 Solution Extraction at Specific Locations

To extract the horizontal displacement component u_x at the nodes located on the right boundary (where the traction is applied), we use the following code snippet:

```python
# use uniqueness feature of set() to remove duplicated nodes:
nodes_on_edge = set()
for e in t_edges:
    nodes_on_edge.update(e)

print(f"Nodes on the right dege: {nodes_on_edge}")

# get the displacements ux at the nodes on the right edge:
edge_ux = [(n, D[2*n:2*n+2][0]) for n in nodes_on_edge]
for i in range(0, len(edge_ux), 2):
    row = edge_ux[i:i+2]
    for key, val in row:
        print(f"Displacement at node {key:<4}: {val!s:<6}", end="")
    print()
```

```
Nodes on the right dege: {1, 3}
Displacement at node 1: 0.1    Displacement at node 3: 0.10
```

It is observed that the finite element solution matches the exact analytical solution up to machine precision, confirming the correctness of the implementation for this benchmark problem.

7.4.1.6 Visualization of the Full Displacement Field

The following function is used to visualize the displacement field over the entire domain using contour plots based on the computed nodal displacement values. This provides a global view of the deformation pattern and allows direct comparison with the expected linear displacement distribution.

```
def plot_disp_FEM(nodes, elements, U, scale=1.0, deformed=True):
    """
    Plot displacement fields and mesh (supports T3 and Q4 elements).
    Uses triangulation internally for visualization.
    """
    ux = U[0::2]
    uy = U[1::2]
    umag = np.sqrt(ux**2 + uy**2)

    # --- coordinates ---
    if deformed:
        x_plot = nodes[:, 0] + scale * ux
        y_plot = nodes[:, 1] + scale * uy
        xlabel = 'x (deformed)'
        ylabel = 'y (deformed)'
    else:
        x_plot = nodes[:, 0]
        y_plot = nodes[:, 1]
        xlabel = 'x'
        ylabel = 'y'

    # Convert elements → triangles if needed
    if elements.shape[1] == 3:
        # Already T3
        triangles = elements

    elif elements.shape[1] == 4:
        # Q4 → split into 2 triangles
        triangles = []
        for e in elements:
            n1, n2, n3, n4 = e
            triangles.append([n1, n2, n3])
            triangles.append([n1, n3, n4])
        triangles = np.array(triangles)
    else:
        raise ValueError("Only T3 and Q4 elements are supported.")

    fig, axs = plt.subplots(2, 2, figsize=(10, 8), dpi=300)
    fields = [ux, uy, umag]
    titles = [r'$u_x$', r'$u_y$', r'$|\mathbf{u}|$']

    for ax, field, title in zip(axs.flat[:3], fields, titles):
        tcf = ax.tricontourf(
            x_plot, y_plot,
```

```
            triangles,
            field,
            levels=20,
            cmap='jet' )
        fig.colorbar(tcf, ax=ax, shrink=0.5, aspect=20)
        ax.set_title(title)
        ax.set_xlabel(xlabel)
        ax.set_ylabel(ylabel)
        ax.set_aspect('equal')
    # --- mesh plot ---
    ax = axs[1, 1]
    ax.triplot(x_plot, y_plot, triangles, color='k', linewidth=0.5)

    ax.set_title('Deformed mesh' if deformed else 'Original mesh')
    ax.set_xlabel(xlabel)
    ax.set_ylabel(ylabel)
    ax.set_aspect('equal')
    plt.tight_layout()
    plt.show()
```

```
# Plot the Entire Displacement Field
plot_disp_FEM(nodes, elements, D, scale=1.0, deformed=True)
```

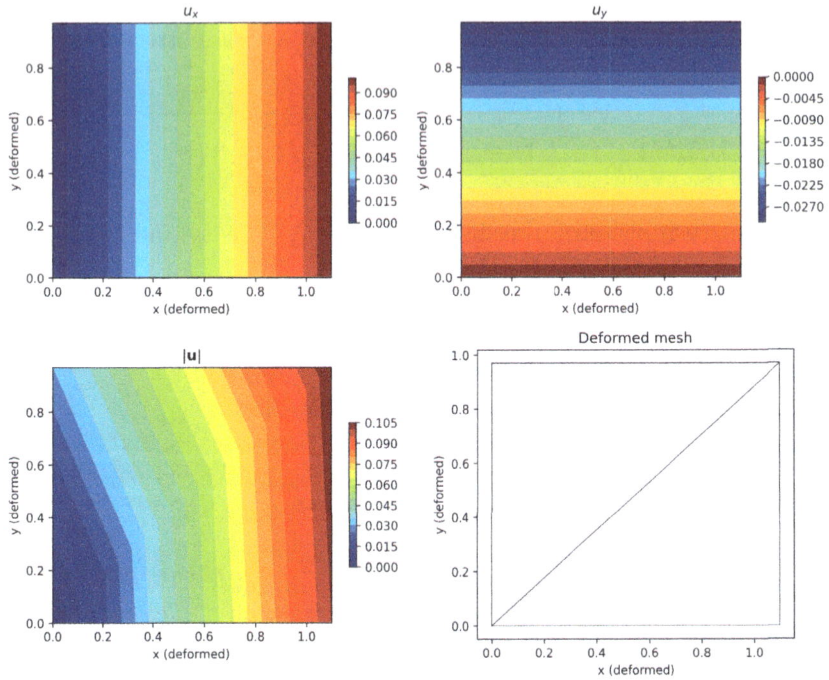

As shown, the linear displacement field is reproduced exactly by the FEM model using only two linear T3 elements.

7.4.1.7 Nodal Stress Recovery for T3 Meshes

As discussed earlier, linear T3 elements produce element-wise constant strain and stress fields due to the linear interpolation of the displacement field. While this is consistent within each element, it introduces two practical issues in post-processing:

1. **Inter-element discontinuity:** Stress values are discontinuous across element boundaries, which is non-physical for most continuum problems.
2. **Nodal indeterminacy:** A single node shared by multiple elements may have multiple conflicting stress values.

To address this, a common post-processing technique is **nodal stress recovery** (or simple nodal averaging). In this approach, the constant stress state from all elements connected to a node is averaged to obtain a single representative nodal stress value.

The recovered nodal stress components are computed as

$$\sigma_{ij}^{(n)} = \frac{1}{n_e} \sum_{e \in \mathcal{E}_n} \sigma_{ij}^{(e)}, \tag{7.107}$$

where $\mathcal{E}_n$ is the set of elements connected to node n, and n_e is the number of such elements. The post-processed stress solution is found much accurate and exhibits a form of superconvergence [11].

Using the recovered nodal stresses, the equivalent von Mises stress under plane stress conditions is given by

$$\sigma_{\text{vm}}^{(n)} = \sqrt{\left(\sigma_{xx}^{(n)}\right)^2 - \sigma_{xx}^{(n)}\sigma_{yy}^{(n)} + \left(\sigma_{yy}^{(n)}\right)^2 + 3\left(\sigma_{xy}^{(n)}\right)^2}. \tag{7.108}$$

This scalar quantity represents the distortion energy density and is widely used as a failure indicator for ductile materials [4].

The following function `recover_nodal_stress_T3` implements this procedure using the global displacement vector as input.

7.4.1.8 Code for Stress Computation

```
def recover_nodal_stress_T3(nodes, elements, U, C):
    '''
    Compute the recovered stresses at `nodes` of the FEM model
    via area-weighted average of the surranding `elements`.
    '''
    n_nodes = len(nodes)
    # Initialize accumulators
```

```
    sxx_n = np.zeros(n_nodes)
    syy_n = np.zeros(n_nodes)
    sxy_n = np.zeros(n_nodes)
    counts = np.zeros(n_nodes)

    for elem in elements:
        coords = nodes[elem]
        dofs = []          # DOFs
        for n in elem:
            dofs.extend([2*n, 2*n+1])
        Ue = U[dofs]

        # Element B matrix
        _, B = batch_ke_T3_NSAF(coords, C)

        # Stress (constant in T3)
        strain = B @ Ue
        stress = C @ strain
        sx, sy, txy = stress

        # Distribute to nodes
        for n in elem:
            sxx_n[n] += sx
            syy_n[n] += sy
            sxy_n[n] += txy
            counts[n] += 1
    # Average:
    sxx_n /= counts
    syy_n /= counts
    sxy_n /= counts

    # Von Mises (nodal)
    svm_n = np.sqrt(sxx_n**2 - sxx_n*syy_n + syy_n**2 + 3*sxy_n**2)

    return sxx_n, syy_n, sxy_n, svm_n
```

7.4.1.9 Visualization of the Stress Field

The following function `plot_nodal_stress_FEM` is used to visualize the nodal stress distribution over the deformed configuration of the domain.

This function maps the recovered nodal stresses onto the finite element mesh and generates contour plots, providing a clear representation of the stress distribution and highlighting regions of stress concentration.

```
import matplotlib.tri as tri

def plot_nodal_stress_FEM(nodes, elements, U,
                          sxx, syy, sxy, svm, scale=1.0):
    """
    Plot nodal stresses on deformed configuration.
    Supports:
        - T3 triangular elements
        - Q4 quadrilateral elements (plit it to 2 T3 for plotting)
    """
    ux = U[0::2]
    uy = U[1::2]
    # Deformed coordinates
    x_def = nodes[:, 0] + scale * ux
    y_def = nodes[:, 1] + scale * uy

    # Convert connectivity to triangles if needed
    nen = elements.shape[1]
    if nen == 3:
        triangles = elements
    elif nen == 4:
        # Split each quad into two triangles
        #   3 ------- 2
        #   |       / |
        #   |     /   |
        #   |   /     |
        #   0 ------- 1
        triangles = []
        for elem in elements:
            n1, n2, n3, n4 = elem
            triangles.append([n1, n2, n3])
            triangles.append([n1, n3, n4])
        triangles = np.array(triangles)
    else:
        raise ValueError("Only T3 and Q4 elements are supported.")

    triang = tri.Triangulation(x_def, y_def, triangles)
    # Plot fields
    fig, axs = plt.subplots(2, 2, figsize=(10, 8), dpi=300)
    fields = [sxx, syy, sxy, svm]
    titles = [r'$\sigma_{xx}$', r'$\sigma_{yy}$',
              r'$\sigma_{xy}$',  'Von Mises' ]

    for ax, field, title in zip(axs.flat, fields, titles):
```

```
        tcf = ax.tricontourf(triang, field, levels=20, cmap='jet')
        fig.colorbar(tcf, ax=ax)
        ax.set_title(title)
        ax.set_xlabel('x (deformed)')
        ax.set_ylabel('y (deformed)')
        ax.set_aspect('equal')
    plt.tight_layout()
    plt.show()
```

```
# Stress at nodes:
sxx_n, syy_n, sxy_n, svm_n=recover_nodal_stress_T3(nodes, elements,\
                                                   D, C )
print(f" Print stress values,  xx={sxx_n}")
print(f" Print stress values,  yy={syy_n}")
print(f" Print stress values,  xy={sxy_n}")
print(f" Print stress values,  vm={svm_n}")
# Plot stress at nodes:
plot_nodal_stress_FEM(nodes, elements, D, sxx_n, syy_n, sxy_n,\
                                        svm_n, scale=1.0)
```

```
Print stress values,  xx=[0.1 0.1 0.1 0.1]
Print stress values,  yy=[0. 0. 0. 0.]
Print stress values,  xy=[0. 0. 0. 0.]
Print stress values,  vm=[0.1 0.1 0.1 0.1]
```

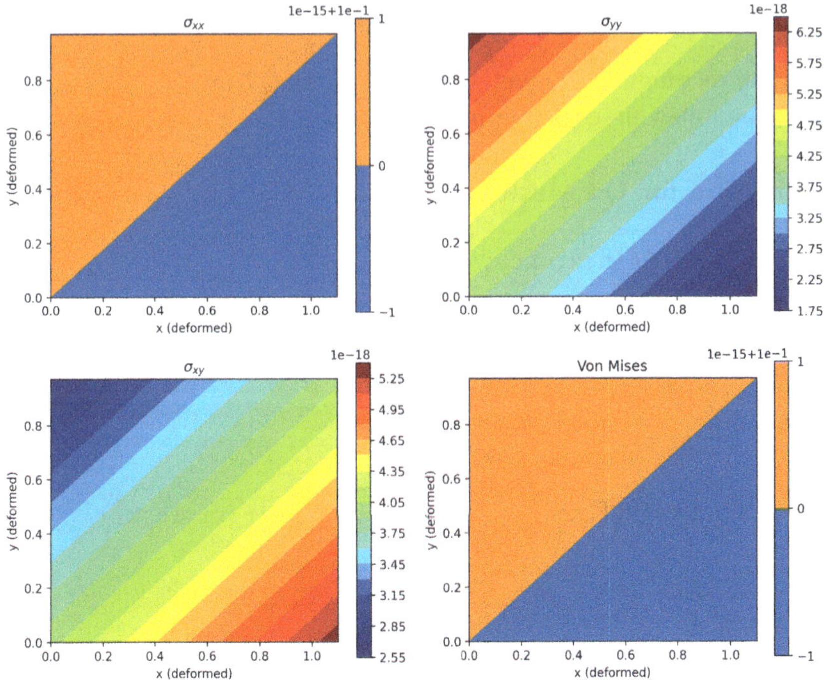

As shown, the stress field is also reproduced exactly by the FEM solution.

It should be noted that the plotted σ_{xx} field may appear visually inaccurate. This is not due to an error in the finite element solution, but rather a limitation of the visualization procedure. In particular, the use of a coarse triangular plotting grid with only two elements can lead to poor contour interpolation and misleading graphical artifacts.

The following snippet evaluates the averaged nodal stress values by aggregating contributions from all adjacent elements connected to each node.

```
print("FEM mean sigma_xx:", np.mean(sxx_n))
#print("Exact:", E*delta/Lx)                # for displacement driven
print("Exact:", p )
```

```
FEM mean sigma_xx: 0.1
Exact: 0.1
```

Since the applied traction is set to $p = 0.1$, the numerical results can be directly compared with the exact analytical solution in Eq.(7.89). The FEM solution obtained using only two T3 elements reproduces the exact solution up to machine precision. This is expected, since the exact displacement field given in Eq.(7.93) is linear, and therefore lies exactly within the interpolation space of the T3 shape functions.

7.4.2 FEM Solver for 2D Solids with Q4 Elements

We now repeat the same verification study using Q4 elements. The overall computational workflow remains unchanged; the only modification is the discretization of the domain using quadrilateral elements instead of triangular ones.

For the present rectangular geometry, a structured Q4 mesh can be generated efficiently using the following code snippet.

7.4.2.1 Function for Q4 Mesh Generation

```
def generate_mesh_Q4(ncx, ncy, Lx=1.0, Ly=1.0):
    """
    Generate structured Q4 mesh on [0,Lx] x [0,Ly]
    """
    x = np.linspace(0, Lx, ncx + 1)
    y = np.linspace(0, Ly, ncy + 1)

    nodes = np.array([[xi, yi] for yi in y for xi in x])

    def node_id(i, j):
        return j * (ncx + 1) + i
```

```
    elements = []
    for j in range(ncy):
        for i in range(ncx):
            n1 = node_id(i, j)
            n2 = node_id(i + 1, j)
            n3 = node_id(i + 1, j + 1)
            n4 = node_id(i, j + 1)

            elements.append([n1, n2, n3, n4])  # counter-clockwise

    return nodes, np.array(elements)
```

The element stiffness matrix for the Q4 formulation is computed using the functions `element_ke_Q4_BCB`, `batch_ke_Q4_NSAF`, or `batch_ke_Q4_BCB`, as established in the previous sections.

For post-processing, the evaluation of nodal stresses requires a dedicated implementation for Q4 elements, which is introduced below.

7.4.2.2 Nodal Stress Recovery for Q4 Meshes

As in the T3 formulation, the raw stress field obtained from Q4 elements is generally discontinuous across element boundaries. Although Q4 elements provide a spatially varying strain field within each element, the resulting stress field is still evaluated independently per element, leading to discontinuities along shared edges.

This behavior arises from the fact that the derivatives of the bilinear shape functions are element-wise defined and are not continuous across element interfaces.

To obtain a smooth and physically meaningful stress field, a post-processing step is required. The following function `recover_nodal_stress_Q4` performs nodal stress recovery by evaluating stress components at element contributions, accumulating them at shared global nodes, and applying an unweighted averaging procedure based on nodal connectivity.

The resulting nodal stresses provide a continuous approximation suitable for contour visualization and engineering interpretation.

```
def recover_nodal_stress_Q4(nodes, elements, U, C):
    n_nodes = len(nodes)

    sxx = np.zeros(n_nodes)
    syy = np.zeros(n_nodes)
    sxy = np.zeros(n_nodes)
    counts = np.zeros(n_nodes)
```

```
    for elem in elements:
        coords = nodes[elem]

        dofs = []
        for n in elem:
            dofs.extend([2*n, 2*n+1])
        Ue = U[dofs]

        # Evaluate at center (xi=0, eta=0)
        xi, eta = 0.0, 0.0

        dN_dxi = 0.25 * np.array([
            [-(1 - eta),  (1 - eta),   (1 + eta), -(1 + eta)],
            [-(1 - xi),  -(1 + xi),    (1 + xi),   (1 - xi)]  ])

        J = dN_dxi @ coords
        dN_dx = np.linalg.inv(J) @ dN_dxi

        B = np.zeros((3, 8))
        for i in range(4):
            B[0, 2*i]     = dN_dx[0, i]
            B[1, 2*i + 1] = dN_dx[1, i]
            B[2, 2*i]     = dN_dx[1, i]
            B[2, 2*i + 1] = dN_dx[0, i]

        stress = C @ (B @ Ue)

        # stress at a node is the average of that of elements at it
        for n in elem:
            sxx[n] += stress[0]
            syy[n] += stress[1]
            sxy[n] += stress[2]
            counts[n] += 1

    sxx /= counts
    syy /= counts
    sxy /= counts

    svm = np.sqrt(sxx**2 - sxx*syy + syy**2 + 3*sxy**2)

    return sxx, syy, sxy, svm
```

7.4.2.3 Solve Using One Q4 Element

With the element-level operators fully defined, the remainder of the solution pipeline—including global stiffness matrix assembly, enforcement of Dirichlet boundary conditions via static condensation, and solution of the resulting linear system—remains structurally identical to the T3 formulation.

The primary difference lies only in the element interpolation scheme and mesh topology. All global-level operations are unchanged.

We are now ready to solve the benchmark uniaxial tension problem using a Q4 mesh.

```
# Mesh generation:
ncx = 1; ncy = 1 # This generates one element
nodes, elements = generate_mesh_Q4(ncx, ncy)

# Martrix of material constants:
C = get_C_2D(E=1.0, nu=0.3) #(E=210e9, nu=0.3)

# Assembly assemble_system_Q4
K, F = assemble_system(nodes, elements, batch_ke_Q4_NSAF, C)

# Set DBC:
delta = 0.1
fixed_dofs, fixed_vals = set_DBCs(nodes, elements, delta, dtype=1,\
                                                    tol=1e-10)

# Set NBC:
F = np.zeros(2 * len(nodes))
x_right = np.max(nodes[:, 0])  # Right edge
p = 0.1
tx = p; ty =.0
t_edges = apply_uniform_traction(F, nodes, elements,[tx,ty],x_right)

# Solve (same as T3)
D = solve_system(K, F, fixed_dofs, fixed_vals)

# Compute Stress
sxx_n, syy_n, sxy_n, svm_n = recover_nodal_stress_Q4(nodes, \
                                                elements, D, C)
print(f" Print stress values, xx={sxx_n}")
```

```
Print stress values, xx=[0.1 0.1 0.1 0.1]
```

```
## Reader may produce diplacement and stress plot using:
# plot_disp_FEM(nodes, elements, D, deformed=True)
```

```
# plot_nodal_stress_FEM(nodes, elements, D, sxx_n, syy_n, sxy_n,\
#                                           svm_n, scale=1.0)
```

The following snippet computes the mean nodal stresses:

```
print("FEM mean sigma_xx:", np.mean(sxx_n))
#print("Exact:", E*delta/Lx)           # for displacement driven
print("Exact:", p )
```

```
FEM mean sigma_xx: 0.09999999999999999
Exact: 0.1
```

The FEM numerical stress solution obtained using a single Q4 element is again exact to machine precision. This is expected because the exact displacement field is linear over the domain, and such fields are exactly representable within the interpolation space of the bilinear Q4 element under uniform pure tension applied on the right edge.

7.5 Case Study: FEM Modeling of Irregular Domains

The true strength of the finite element method lies in its ability to handle boundary value problems defined on complex and irregular geometries. Provided that a suitable mesh can be generated, the same underlying algebraic formulation remains fully applicable.

To demonstrate this capability, we consider a classical benchmark problem: a rectangular plate containing a central elliptical hole subjected to uniaxial tension.

7.5.1 Problem Specification

Consider a rectangular plate of uniform thickness t containing a centrally located elliptical void, as illustrated in Fig. 7.1.

The boundary of the hole may be subjected to prescribed surface tractions, while all other material parameters and external loading conditions remain identical to the uniaxial tension case studied previously.

The governing equations and constitutive model remain unchanged; only the geometry and boundary discretization are modified to accommodate the presence of the internal boundary.

The introduction of the central hole transforms the previously homogeneous stress field into an irregular domain problem. The geometric discontinuity alters internal load paths and induces localized stress gradients, a phenomenon commonly referred to as **stress concentration** [8]. Because the boundary is no longer aligned with a structured Cartesian grid, generating a conforming finite element mesh requires specialized discretization techniques.

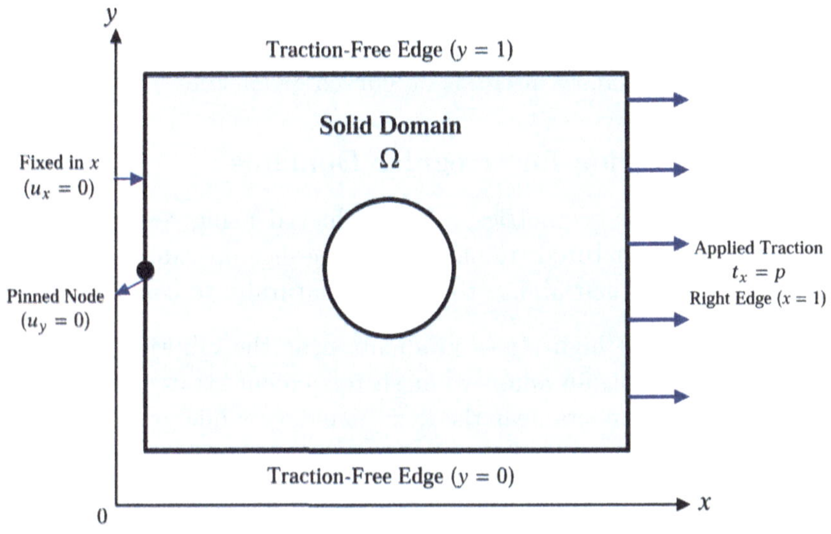

Figure 7.1: A plate with a hole subjected to tension.

7.5.2 Unit Normal Vector of an Elliptical Boundary

Since the boundary of the elliptical hole is curved, its outward unit normal vector $\mathbf{n}$ varies continuously along the surface. For problems involving Neumann boundary conditions (e.g., internal pressure applied on the hole boundary), the traction equilibrium condition must be satisfied:

$$\sigma\mathbf{n} = \bar{\mathbf{t}}. \tag{7.109}$$

To evaluate boundary integrals on the elliptical edge, the outward normal vector must be explicitly computed. Let the ellipse be defined implicitly by the level-set function

$$\phi(x, y) = \frac{(x - x_c)^2}{a^2} + \frac{(y - y_c)^2}{b^2} - 1 = 0, \tag{7.110}$$

where (x_c, y_c) is the center of the ellipse, a is the semi-major axis, and b is the semi-minor axis.

The gradient of this function provides a vector normal to the boundary:

$$\mathbf{g} = \nabla\phi = \begin{bmatrix} \frac{\partial\phi}{\partial x} \\ \frac{\partial\phi}{\partial y} \end{bmatrix} = \begin{bmatrix} \frac{2(x-x_c)}{a^2} \\ \frac{2(y-y_c)}{b^2} \end{bmatrix}. \tag{7.111}$$

The corresponding outward unit normal vector is obtained by normalization:

$$\mathbf{n} = \frac{\mathbf{g}}{\|\mathbf{g}\|} = \frac{1}{\sqrt{\left(\frac{2(x-x_c)}{a^2}\right)^2 + \left(\frac{2(y-y_c)}{b^2}\right)^2}} \begin{bmatrix} \frac{2(x-x_c)}{a^2} \\ \frac{2(y-y_c)}{b^2} \end{bmatrix}. \tag{7.112}$$

Although Eq.(7.112) is not explicitly required in the present benchmark (since no traction is applied on the hole boundary), it provides a useful general formulation for evaluating Neumann boundary integrals on curved geometries.

7.5.3 Mesh Generation for Irregular Domains

Discretization of irregular geometries can be achieved using several meshing strategies. Among them, unstructured triangular (T3) meshes provide the most robust and flexible approach due to their ability to conform naturally to complex boundaries.

To accurately resolve the high stress gradients near the elliptical hole without excessive computational cost, an adaptive mesh refinement strategy is employed. This places smaller, denser elements near the hole boundary while gradually transitioning to coarser elements in regions far from the stress concentration zone.

In this implementation, the Python package `meshpy`, which interfaces with Jonathan Shewchuk's `Triangle` library, is used to generate a high-quality, boundary-conforming T3 mesh with local refinement control.

The following Python script implements the adaptive mesh generation procedure.

```python
import meshpy.triangle as triangle
# Mesh generation using meshpy
def generate_meshpy_T3(nx_boundary=120,
    max_area=5e-4, refine_factor=0.3, a=0.2, b=0.2, h=0.5, k=0.5):
    """
    Generate triangular mesh T3 for a square plate with an
    elliptical hole.
    Returns
    nodes : (nnodes, 2) ndarray, node coordinates.
    elements : (nelems, 3) ndarray Triangle connectivity.
    """
    # Outer square boundary
    outer_pts = [(0, 0), (1, 0), (1, 1), (0, 1) ]
    outer_facets = [(0, 1), (1, 2), (2, 3),(3, 0) ]
    # Elliptical hole boundary
    theta = np.linspace(0, 2*np.pi, nx_boundary, endpoint=False )

    ellipse_pts = [(h+a*np.cos(t), k+b*np.sin(t)) for t in theta]

    offset = len(outer_pts)
    ellipse_facets = [(offset + i, offset + (i + 1) nx_boundary)
                      for i in range(nx_boundary)]
    # Combine geometry
    points = outer_pts + ellipse_pts
    facets = outer_facets + ellipse_facets
```

```
    holes = [(h, k)]
    # Adaptive refinement
    def needs_refinement(vertices, area):
        cx = sum(v[0] for v in vertices) / 3
        cy = sum(v[1] for v in vertices) / 3
        phi = ((cx - h)**2 / a**2 + (cy - k)**2 / b**2)
        target_area = (max_area*(1+5*np.exp(-((phi - 1)/0.15)**2)))

        return area > refine_factor * target_area
    # Build mesh
    mesh_info = triangle.MeshInfo()
    mesh_info.set_points(points)
    mesh_info.set_facets(facets)
    mesh_info.set_holes(holes)
    mesh = triangle.build(mesh_info,
        refinement_func=needs_refinement, min_angle=30 )
    nodes = np.array(mesh.points)
    elements = np.array(mesh.elements)

    return nodes, elements
```

```
# =====================
#   Mesh Visualization
# =====================
def plot_mesh_T3(pts, tris, a=0.2, b=0.2, h=0.5, k=0.5):
    plt.figure(figsize=(4,4), dpi=300)
    for tri in tris:
        poly = pts[tri]
        poly = np.vstack([poly, poly[0]])
        plt.plot(poly[:,0], poly[:,1], linewidth=0.5)

    theta = np.linspace(0, 2*np.pi, 300)
    x = h + a*np.cos(theta)
    y = k + b*np.sin(theta)
    plt.plot(x, y, linewidth=2)
    plt.gca().set_aspect('equal')
    plt.title("Adaptive Mesh (Ellipse Hole)")
    plt.tight_layout()
    plt.show()
```

Let us use `generate_meshpy_T3` to generate an mesh:

```
nodes, elements = generate_meshpy_T3(nx_boundary=200,
       max_area=0.01, refine_factor=0.05, a=0.2, b=0.2, h=0.5, k=0.5)

print("Nodes:", nodes.shape[0])
print("Elements:", elements.shape[0])
plot_mesh_T3(nodes, elements)
```

```
Nodes: 1993;  Elements: 3658
```

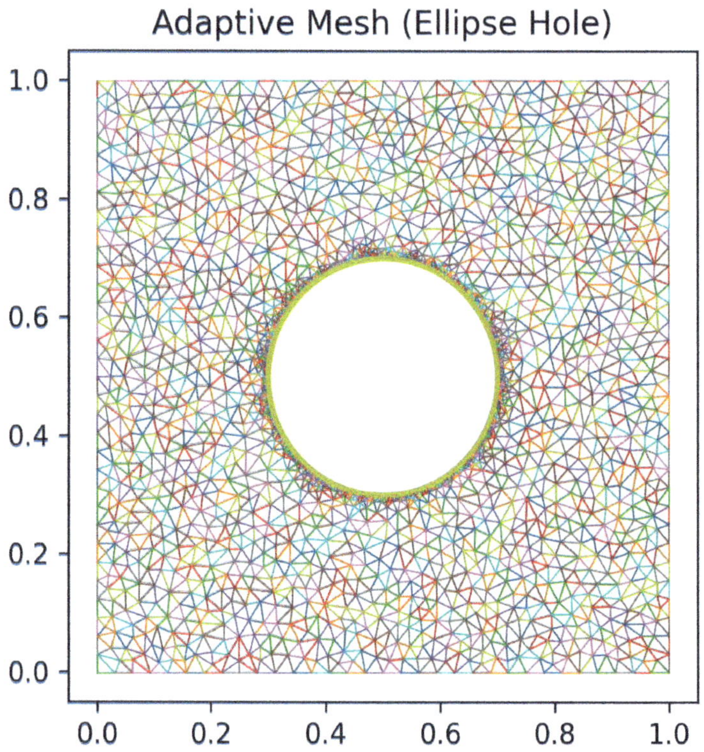

With the boundary-conforming adaptive T3 mesh successfully generated using `meshpy`, we proceed to compute the deformation and stress fields in the perforated plate. Although the geometric complexity has increased due to the presence of the central elliptical hole, the underlying finite element workflow remains identical to the previously studied rectangular benchmark.

The computational pipeline—from global stiffness matrix assembly to final visualization of stress concentration patterns—is implemented in the following section.

7.5.4 Solution Using T3 Elements

The following code snippets solve the problem using linear T3 elements:

```
# Assembly assemble_system_T3
K, F = assemble_system(nodes, elements, batch_ke_T3_NSAF, C)

# Set DBC:
delta = 0.1
fixed_dofs, fixed_vals = set_DBCs(nodes, elements, delta, dtype=0,\
                                                          tol=1e-10)
# Set NBC:
F = np.zeros(2 * len(nodes))
x_right = np.max(nodes[:, 0])                          # on the right edge
p = 0.1                                              # the pulling line load
tx = p; ty =.0                                       # converted to traction
# compute the force vector, and the record nodes on the right edge
t_edges=apply_uniform_traction(F, nodes, elements, [tx,ty], x_right)
# We assumed the hole edge is free of tranction

D = solve_system(K, F, fixed_dofs, fixed_vals)              # Solve KD=F

# Compute Stress from computed D:
sxx, syy, sxy, svm = recover_nodal_stress_T3(nodes, elements, D, C)
# executed in 1.32s, finished 15:06:18 2026-05-22
```

This two-dimensional FEM model consists of 3658 elements and 1993 nodes. Despite the increased geometric complexity introduced by the elliptical hole, the system is solved in less than 2 seconds on a standard Dell laptop. This demonstrates that even with a Python-based implementation on consumer-grade hardware, reasonably complex solid mechanics problems can be solved efficiently.

7.5.4.1 Contour Plots of the Displacement Field

The displacement field can now be visualized using the same plotting routine introduced in the previous benchmark case:

```
plot_disp_FEM(nodes, elements, D, scale=1.0, deformed=True)
```

7.5.4.2 Contour Plots of the Stress Field

The stress field is plotted using:

```
plot_nodal_stress_FEM(nodes, elements, D, sxx,syy,sxy,svm, scale=1.)
```

Both plots are shown in the following, from which we observe a clear and well-defined stress concentration pattern surrounding the central cutout. The localized amplification and spatial distribution of stress components closely resemble the classical analytical solutions for infinite plates containing circular or elliptical holes [8].

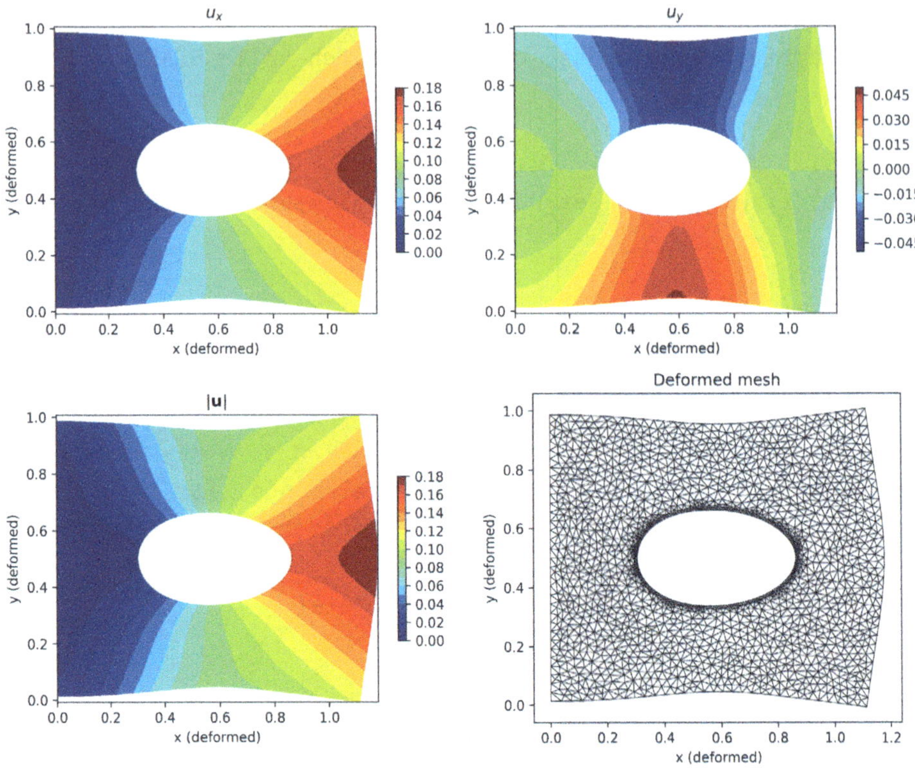

u_x
u_y
$|\mathbf{u}|$
Deformed mesh
x (deformed)
y (deformed)

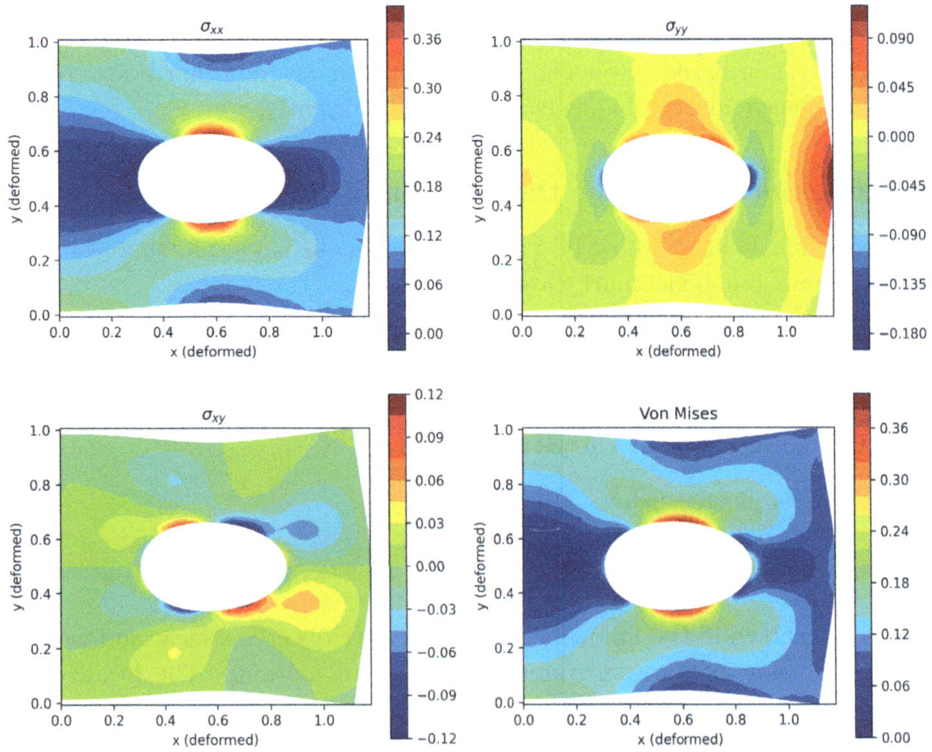

σ_{xx}
σ_{yy}
σ_{xy}
Von Mises
x (deformed)
y (deformed)

7.5.5 Solution Using Q4 Elements

Although unstructured triangular (T3) meshes can be generated efficiently for arbitrary geometries, constructing high-quality quadrilateral (Q4) meshes for irregular domains is significantly more challenging. Unlike triangular elements, quadrilateral elements lack intrinsic geometric flexibility, and standard unstructured meshing procedures often produce highly skewed or distorted elements near curved boundaries.

To address these limitations, advanced meshing workflows employ specialized geometric and topological operations, including transfinite mappings, recombination strategies, and iterative smoothing procedures. In addition, distance-based mesh size fields are typically introduced to enforce local refinement near geometric discontinuities where stress gradients are expected to be high.

In this implementation, we use **Gmsh**, an open-source finite element mesh generator with an integrated CAD kernel and advanced automatic meshing capabilities. Its Python API enables the construction of complex geometries, the enforcement of quad-dominant recombination, and the definition of spatially varying mesh density fields. While the underlying theory of automatic quadrilateral meshing is beyond the scope of this text, further details can be found in the official Gmsh documentation: `https://gmsh.info`.

7.5.5.1 Quadrilateral Mesh Generation Protocol

The Python script below uses the Gmsh API to generate a high-quality Q4 mesh for the perforated plate problem. The procedure performs the following steps:

- **Geometric parameterization (OCC kernel):** The rectangular plate and central elliptical (or circular) hole are constructed using the OpenCASCADE geometry engine, with Boolean subtraction used to define the perforation.
- **Quad-dominant recombination:** A recombination constraint is applied to convert triangular elements into structured quadrilateral elements, producing a predominantly Q4 mesh.
- **Distance-based mesh sizing:** A spatial mesh size field is defined around the hole boundary, enforcing fine discretization near the stress concentration zone and gradually coarsening toward the far-field region.
- **Conforming mesh topology:** The final mesh is guaranteed to be conforming, with no hanging nodes or non-matching interfaces between refinement zones.

7.5.5.2 Code for Generating Q4 Mesh

The following Python snippet demonstrates the full Gmsh-based mesh generation workflow for constructing the Q4 discretization:

```
import gmsh
# Initialize
gmsh.initialize()
gmsh.model.add("plate_hole_clean_fem")

occ = gmsh.model.occ   # use Gmsh OCC kernel

# Parameters for the geometry (one may change)
L = 1.0             # the length of the rectangle
R = 0.2             # Radius of the circular hole
cx = L / 2; cy = L / 2    # center of the hole

# Building geometry:
plate = occ.addRectangle(0, 0, 0, L, L)
hole  = occ.addDisk(cx, cy, 0, R, R)

domain, _ = occ.cut([(2, plate)], [(2, hole)])  # form the domain
occ.synchronize()

# Mesh refinement near hole
h_min = 0.002      # for adaptation control
h_max = 0.03

gmsh.model.mesh.field.add("MathEval", 1)
gmsh.model.mesh.field.setString(1, "F",
    f"sqrt((x-{cx})*(x-{cx}) + (y-{cy})*(y-{cy}))" )

gmsh.model.mesh.field.add("Threshold", 2)
gmsh.model.mesh.field.setNumber(2, "InField", 1)
gmsh.model.mesh.field.setNumber(2, "SizeMin", h_min)
gmsh.model.mesh.field.setNumber(2, "SizeMax", h_max)
gmsh.model.mesh.field.setNumber(2, "DistMin", 0.0)
gmsh.model.mesh.field.setNumber(2, "DistMax", 0.35)

gmsh.model.mesh.field.setAsBackgroundMesh(2)

# Mesh options
gmsh.option.setNumber("Mesh.Algorithm", 6)
gmsh.option.setNumber("Mesh.RecombineAll", 1)

gmsh.model.mesh.generate(2)

# Extract nodes for our Python FEM model
nodeTags, nodeCoords, _ = gmsh.model.mesh.getNodes()
```

```
nodeTags = np.array(nodeTags, dtype=int)
coords = np.array(nodeCoords).reshape(-1, 3)
nodes = coords[:, :2].copy()    # Build pure 0-based node array

# Extract Q4 elements (0-based connectivity)
elemTypes, elemTags, elemNodeTags = gmsh.model.mesh.getElements(2)

Q4 = 3
elements = None

for etype, _, enodes in zip(elemTypes, elemTags, elemNodeTags):

    if etype == Q4:
        enodes = np.array(enodes, dtype=int).reshape(-1, 4)
        # Convert via local indexing (consistent with coords order)
        tag_to_index = {tag: i for i, tag in enumerate(nodeTags)}

        elements = np.array(
            [[tag_to_index[n] for n in elem] for elem in enodes],
            dtype=int )

# OPTIONAL: compact renumbering (recommended FEM practice)
unique_nodes = np.unique(elements)
remap = {old: new for new, old in enumerate(unique_nodes)}

nodes = nodes[unique_nodes]
elements = np.array([[remap[n] for n in e] for e in elements],\
                                                    dtype=int)
# FINAL CHECK: everything should be 0-based
assert nodes.shape[0] == np.max(elements) + 1

# Plot the generated mesh with labelling (STRICT 0-based labeling)
fig, ax = plt.subplots(figsize=(6, 6), dpi=300)

# Elements
for e, elem in enumerate(elements):
    Xe = nodes[elem]
    Xe = np.vstack([Xe, Xe[0]])
    ax.plot(Xe[:, 0], Xe[:, 1], 'k-', linewidth=0.6)

    xc = Xe[:, 0].mean()
    yc = Xe[:, 1].mean()
#      ax.text(xc, yc, str(e),   # 0-based element ID
```

```
#               color='blue', fontsize=6, ha='center', va='center')
# Nodes
for i, (x, y) in enumerate(nodes):
    ax.plot(x, y, 'ro', markersize=1.5)
#     ax.text(x, y, str(i),      # 0-based node ID
#               color='red', fontsize=6, ha='left', va='bottom')

ax.set_aspect('equal')
ax.set_title("Q4 Mesh (Fully Consistent 0-based labeling)")
ax.set_xlabel("x")
ax.set_ylabel("y")
plt.tight_layout()
plt.show()

gmsh.finalize()
```

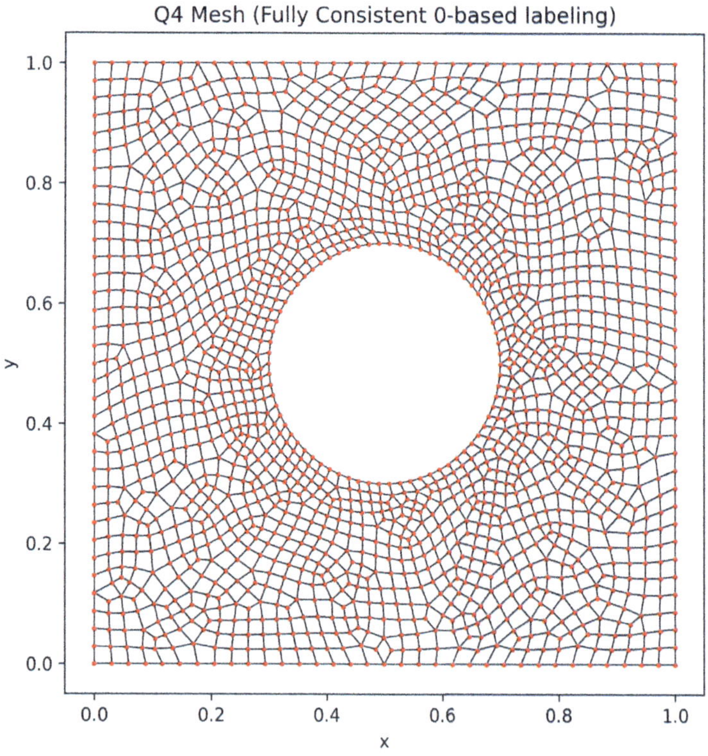

We generate a reasonably high-quality Q4 mesh characterized by:

- refined elements in the vicinity of the hole,
- smooth mesh size grading toward the far field,
- absence of severe distortion or abrupt transitions between element sizes.

7.5.5.3 Solving Using Q4 Mesh

With the mesh successfully constructed, we now proceed to perform the finite element analysis of the perforated plate. The overall solution procedure follows the same workflow as in the previous examples, including global stiffness assembly, application of boundary conditions, solution of the linear system, and post-processing of the resulting displacement and stress fields.

```
# Assembly assemble_system_Q4
K, F = assemble_system(nodes, elements, batch_ke_Q4_NSAF, C)
print("Nodes:", nodes.shape[0])
print("Elements:", elements.shape[0])

# Set DBC:
delta = 0.1
fixed_dofs, fixed_vals = set_DBCs(nodes, elements, delta, dtype=0,\
        tol=1e-10)   # one may use different DBCs by changing dtype
# Set NBC:
F = np.zeros(2 * len(nodes))
x_right = np.max(nodes[:, 0])  # Right edge
p = 0.1
tx = p; ty =.0
t_edges=apply_uniform_traction(F, nodes, elements, [tx,ty], x_right)

# Solve for nodal displacements:
D = solve_system(K, F, fixed_dofs, fixed_vals)
# executed in 782ms, finished 16:10:43 2026-05-26
```

```
Nodes: 1797
Elements: 1694
```

This two-dimensional FEM model consists of 1694 Q4 elements and 1797 nodes (slightly fewer than the corresponding T3 mesh, which contains 1993 nodes). The system is solved in approximately 1 second on a standard Dell laptop, demonstrating the efficiency of the implementation even for geometrically complex domains.

7.5.5.4 Contour Plots of the Displacement Field

We now visualize the numerical solution by plotting the displacement field using the same post-processing routine introduced previously.

```
plot_disp_FEM(nodes, elements, D, scale=1.0, deformed=True)
```

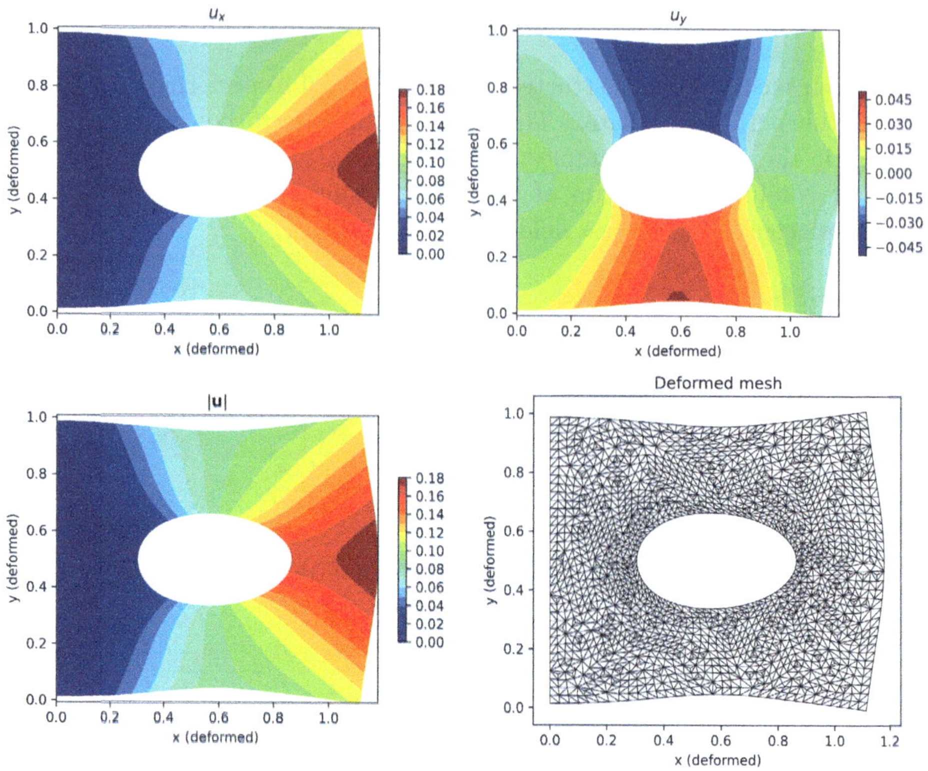

As shown, the field responses obtained using the quadrilateral discretization are in close agreement with the results produced by the unstructured T3 mesh. This consistency verifies the correctness and robustness of the unified finite element implementation across different element types.

> **A Note on Deformed Mesh Visualization:** Although the structural analysis is performed using a Q4 discretization, the deformed configuration may appear visually as a triangular mesh. This is a deliberate choice in the plotting routine `plot_disp_FEM`. For computational efficiency and rendering convenience, each quadrilateral element is internally decomposed into two triangles for visualization purposes. This triangulation is used solely for plotting and does not affect the underlying Q4 solution data or numerical accuracy.

7.5.5.5 Contour Plots of the Stress Field

To visualize the internal stress distribution over the irregular domain, we now apply the Q4 stress recovery procedure. The following code performs unweighted nodal averaging to construct a continuous stress field, which is then used to generate smooth contour plots highlighting the stress concentration around the cutout.

```python
# Compute Stress
sxx, syy, sxy, svm = recover_nodal_stress_Q4(nodes, elements, D, C)
plot_nodal_stress_FEM(nodes, elements, D, sxx, syy, sxy, svm,\
                                                        scale=1.0)
```

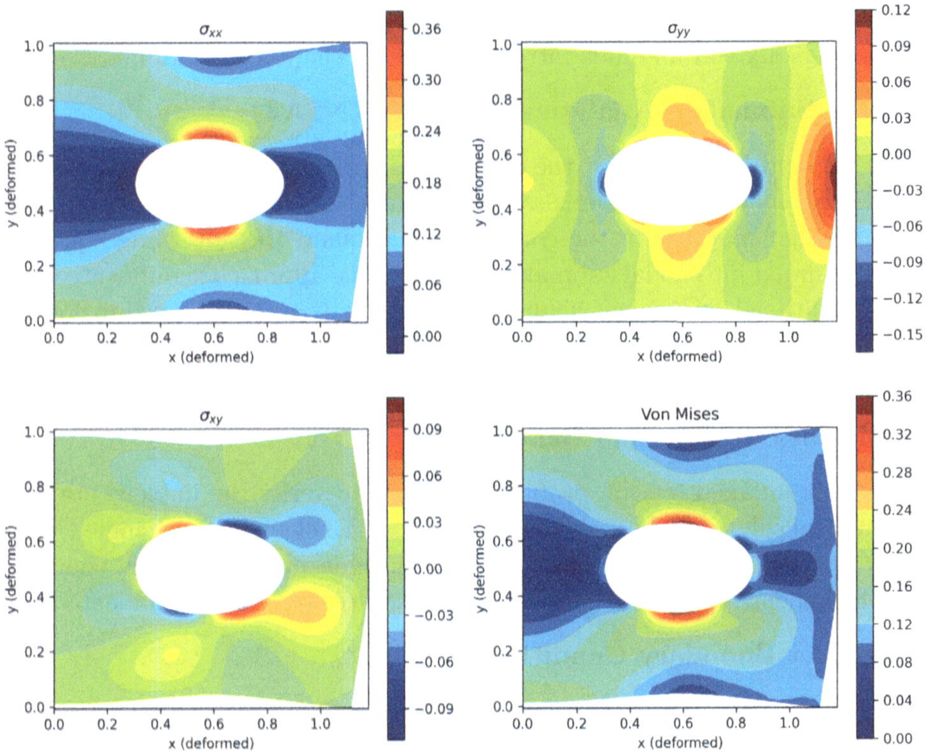

7.5.5.6 Stress Quality and Convergence Analysis

It is observed that the stress contours obtained using a mesh of 1,694 Q4 elements appear noticeably smoother than those produced by a finer T3 mesh containing 3,658 elements. This qualitative difference suggests that Q4 elements provide a more effective representation of spatial stress variation and may achieve higher accuracy per degree of freedom.

For a more rigorous assessment, a quantitative error analysis can be performed using the convergence verification procedures introduced in Chapter 6. Since exact analytical solutions are generally unavailable for complex two-dimensional solid mechanics problems, a high-fidelity numerical solution is typically used as a reference. This reference solution is obtained from an independently generated, highly refined finite element mesh and is treated as a quasi-exact benchmark for error estimation.

Some validation practices and discussions on such error estimation strategies can be found in Ref. [11].

7.6 Final Remarks and Key Takeaways

This chapter bridges the gap between the fundamental theory of the finite element method and the practical implementation of a two-dimensional linear elastic solid mechanics solver. A detailed computational framework was developed for both **3-node linear triangular (T3)** and **4-node bilinear quadrilateral (Q4)** elements.

The main concepts and methods introduced in this chapter are summarized below.

1. **Nodal Submatrix Analytical Formulas (NSAF)**

   This chapter introduced the **Nodal Submatrix Analytical Formulas (NSAF)** for two-dimensional solid mechanics problems. Instead of constructing element matrices in the conventional manner, the governing equations are reformulated into nodal submatrix blocks. This approach significantly improves computational efficiency, yielding approximately a 2–3× speedup for element stiffness matrices and up to a 10× improvement for element mass matrix evaluations.

2. **Uniaxial Pure Tension Benchmark**

   A simple uniaxial tension problem was used to verify the implementation. The numerical solutions obtained using either two T3 elements or a single Q4 element match the exact analytical solution to machine precision. This confirms that both element types pass the patch test and are correctly implemented.

3. **Dirichlet Boundary Conditions**

   A general strategy was implemented for enforcing prescribed displacements, including non-homogeneous Dirichlet boundary conditions. The global system is partitioned into free and constrained degrees of freedom. This procedure enforces boundary conditions exactly while preserving the symmetric positive-definite (SPD) structure of the reduced stiffness matrix.

4. **Element Performance and Computational Efficiency**

   - **T3 simplicity:** T3 elements exhibit constant strain within each element. As a result, no numerical integration is required for internal field evaluation, leading to a simple and computationally efficient formulation that integrates naturally with the NSAF framework.

   - **Q4 optimization:** Using NSAF, the solver avoids explicitly forming the full 3×8 strain-displacement matrix $\mathbf{B}_e$ and the associated product $\mathbf{B}_e^\top \mathbf{C}\mathbf{B}_e$. This reduces computational cost by approximately a factor of four in element-level operations.

   - **Alternative modeling approaches:** Although Q4 elements generally provide higher accuracy than T3 elements, they are also more sensitive to geometric distortion due to the non-constant Jacobian mapping. The **Smoothed Finite Element Method (S-FEM)** provides an alternative finite element formulation in which both T3 and Q4 elements can be employed effectively. In S-FEM, strain smoothing is performed over local smoothing domains, eliminating the need for natural coordinates and the associated Jacobian matrix. As a result, T3 meshes can achieve an accuracy comparable to that of conventional Q4 elements. When Q4 elements are used within the S-FEM framework, the method can further improve solution accuracy, significantly enhance numerical robustness, and greatly eliminate the sensitivity to mesh distortion that often affects conventional FEM formulations. In addition, S-FEM can accommodate highly distorted and even concave Q4 elements without the difficulties typically encountered in standard isoparametric finite element formulations [11].

5. **Post-Processing and Nodal Stress Recovery**

   Standard displacement-based finite element formulations produce discontinuous stress fields across element boundaries. To obtain smoother and more physically interpretable results, a **Nodal Stress Recovery** (or **Simple Nodal Averaging**) procedure was implemented. This approach averages element-level stresses over adjacent elements to construct continuous nodal stress fields and improved von Mises stress visualizations.

6. **Irregular Domains and Stress Concentrations**

   This chapter also demonstrated the capability of the finite element method in handling complex geometries using a tensile plate with a central elliptical cutout. External meshing tools such as `meshpy` and `Gmsh` were employed to generate T3 and Q4 meshes. Local mesh refinement successfully captured the sharp stress gradients near the cutout boundary. These results illustrate that relatively complex two-dimensional solid mechanics problems can be solved using the developed Python framework without relying on commercial finite element software.

The next chapter extends these formulations and computational procedures to three-dimensional (3D) solid mechanics problems.

7.7 Exercises

1: Uniaxial Tension Analytical Verification

Prove analytically why a single 4-node bilinear quadrilateral (Q4) element aligned with the coordinate axes yields a solution exact to machine precision when subjected to uniform uniaxial pure tension $t_x = p$ along its right edge ($x = L$). 1. Write down the exact displacement field solutions $u_x(x, y)$ and $u_y(x, y)$ for this plane stress problem. 2. Show that these exact fields fall completely within the span of the bilinear shape functions of a Q4 element.

2: Symmetry of the Condensed Stiffness Matrix

During the general enforcement of non-zero Dirichlet boundary conditions, the unconstrained global system is partitioned into free (f) and constrained (c) blocks. Prove mathematically that if the initial unconstrained global stiffness matrix $\mathbf{K}$ is symmetric ($\mathbf{K} = \mathbf{K}^\top$), the reduced structural stiffness matrix $\mathbf{K}_{ff}$ used in the condensed equation $\mathbf{K}_{ff}\mathbf{D}_f = \mathbf{F}_f^*$ remains strictly symmetric.

3: Hand Calculation of Non-Zero DBC Condensation

Consider a simplified structural system with three degrees of freedom governed by the unconstrained system equations:

$$\begin{bmatrix} 4 & -2 & 0 \\ -2 & 5 & -3 \\ 0 & -3 & 6 \end{bmatrix} \begin{Bmatrix} u_1 \\ u_2 \\ u_3 \end{Bmatrix} = \begin{Bmatrix} 10 \\ 0 \\ 0 \end{Bmatrix}$$

Assume a non-zero Dirichlet boundary condition is applied at the third degree of freedom such that $u_3 = 2.0$. 1. Partition the system manually into free indices $f = \{1, 2\}$ and constrained index $c = \{3\}$. 2. Calculate the modified right-hand side force vector $\mathbf{F}_f^*$. 3. Solve for the unknown displacement vector $\mathbf{D}_f = \{u_1, u_2\}^\top$.

4: Computational Complexity Matrix Operations

In the textbook, it is stated that the optimized function `element_stiffness_Q4` reduces the floating-point operation (FLOP) count by approximately a factor of four compared to the classic implementation. 1. Quantify the exact number of multiplications and additions required to compute the dense matrix product $\mathbf{B}^\top\mathbf{C}\mathbf{B}$ at a single Gauss point, given that $\mathbf{B}$ is a 3×8 matrix and $\mathbf{C}$ is a 3×3 matrix. 2. Contrast this with the operation count needed to evaluate the analytical block form in Eq.(7.59).

5: The Geometry of the Jacobian in Q4 Elements

For a standard, unskewed rectangular Q4 element of width $2a$ and height $2b$, the mapping Jacobian matrix $\mathbf{J}$ between the physical coordinates (x, y) and natural coordinates (ξ, η) is constant. 1. Derive the algebraic expressions for the components of $\mathbf{J}$ for a rectangular element. 2. Mathematically demonstrate how skewing one of the nodes (e.g., shifting node 3 by a distance Δx) forces the components of $\mathbf{J}$ to become functions of ξ and η, thereby requiring numerical Gauss quadrature rather than simple analytical integration.

6: Simple Nodal Stress Averaging Limitations

Consider two adjacent T3 elements sharing a vertical edge along $x = 0$. Element 1 has an internal constant stress component $\sigma_{xx}^{(1)} = 150$ MPa, and Element 2 has $\sigma_{xx}^{(2)} = 50$ MPa. 1. Calculate the recovered nodal stress at a shared node n assuming both elements meet symmetrically at that node ($c_n = 2$). 2. Discuss a physical scenario (e.g., a bimaterial interface or a step change in plate thickness) where this simple unweighted averaging technique yields a physically incorrect stress distribution, and suggest how the recovery algorithm could be modified to handle material discontinuities.

7: Outward Unit Normal Vectors for a Circular Boundary

A circular cutout centered at $(0, 0)$ with a radius $R = 2.0$ is subjected to an internal uniform hydrostatic pressure p. 1. Define the implicit level-set function $\phi(x, y)$ representing this boundary. 2. Calculate the symbolic outward unit normal vector components $\mathbf{n} = \{n_x, n_y\}^\top$ at the boundary points $P_1 = (2, 0)$, $P_2 = (0, 2)$, and $P_3 = (\sqrt{2}, \sqrt{2})$. 3. Write down the corresponding traction vector $\bar{\mathbf{t}}$ applied at point P_3.

8: Code Implementation: Stress Recovery Tracking

Modify the `recover_nodal_stress_T3` function provided in Section 7.4 to compute and return the global *maximum* principal stress σ_1 at each node, rather than just the Von Mises stress. Recall that the in-plane maximum principal stress is given by the formula:

$$\sigma_1 = \frac{\sigma_{xx} + \sigma_{yy}}{2} + \sqrt{\left(\frac{\sigma_{xx} - \sigma_{yy}}{2}\right)^2 + \tau_{xy}^2} \tag{7.113}$$

Provide the updated Python code block for the function.

9: Mesh Convergence Study near a Stress Concentration

Using the `meshpy` script configuration provided for the plate with an elliptical hole: 1. Run the FEM simulation using three different levels of mesh refinement near the hole boundary (controlled by changing the maximum element area parameter near the cutout). 2. Plot the maximum recovered Von Mises stress $\sigma_{\mathrm{vm,max}}$ against the total number of degrees of freedom (DOFs) in the system. 3. Estimate the converged value of the stress concentration factor $K_t = \sigma_{\mathrm{vm,max}}/p$ and compare it to the analytical Inglis solution.

10: T3-to-Q4 Element Splitting in Post-Processing

The textbook notes that the visualization routine `plot_disp_FEM` splits each Q4 element into two virtual T3 elements for convenience in plotting. 1. Given a quadrilateral element defined by global node indices $[N_1, N_2, N_3, N_4]$ in counter-clockwise order, list the two valid node combinations that form the two sub-triangles without altering the topological volume. 2. Explain how linear interpolation across these two sub-triangles differs from the true bilinear interpolation of the Q4 element, and comment on whether this visual simplification introduces errors into the structural engineering calculations.

Chapter 8

FEM Solver for 3D Solids

```
# Place cursor here, and press Ctrl+Enter to import dependencies.
import time
import importlib
from sympy import init_printing
import numpy as np
import sympy as sp
import matplotlib.pyplot as plt                                 # For plots
from IPython.display import display, Math
from sympy import symbols, lambdify, latex, Matrix, Function

init_printing(use_unicode=True)     # For latex-like quality printing
np.set_printoptions(precision=5,suppress=True) # Digits in print-out
```

We now extend the finite element formulation from two-dimensional problems to three-dimensional (3D) solid continua. In 3D elasticity, the displacement field depends on the spatial coordinates (x, y, z), and each node has three translational degrees of freedom.

This chapter develops the finite element matrix formulation and code implementation for commonly used 3D solid elements, including:

- Geometric interpolation and shape functions
- Displacement approximation
- Strain–displacement relations
- Constitutive equations
- Element stiffness matrices
- Consistent mass matrices
- Equivalent nodal force vectors

These formulations provide the foundation for a 3D FEM solver presented in the final section of this chapter.

8.1 T4 Element Matrices and Codes

One of the most fundamental 3D continuum elements is the four-node linear tetrahedral element, commonly called the T4 element or linear tetrahedron. Due to its simple geometry and strong flexibility for meshing complex domains, the T4 element is widely used in FEM solvers with automatic mesh generation for 3D solid mechanics problems.

This section presents the main formulations for the T4 element.

8.1.1 Problem Setup

Consider a tetrahedral element occupying the domain Ω_e and bounded by four nodes, as illustrated in Fig.3.2. The first three nodes (1, 2, and 3) are ordered counterclockwise on a chosen base face, while node 4 is positioned such that the node ordering follows the right-hand rule. This convention ensures a positive element volume (see Section 5.7.11) [6].

The spatial coordinates of the four nodes are defined as:

- Node 1: (x_1, y_1, z_1)
- Node 2: (x_2, y_2, z_2)
- Node 3: (x_3, y_3, z_3)
- Node 4: (x_4, y_4, z_4)

The corresponding nodal displacement components are given by:

- Node 1: (u_1, v_1, w_1)
- Node 2: (u_2, v_2, w_2)
- Node 3: (u_3, v_3, w_3)
- Node 4: (u_4, v_4, w_4)

The material properties are assumed to be homogeneous and isotropic within the element domain:

- Young's modulus: E
- Poisson's ratio: ν
- aterial density: ρ

Because each node has three translational degrees of freedom, (u, v, w), a single T4 element contains a total of twelve degrees of freedom (DOFs). Consequently, the element stiffness and mass matrices are both of size 12×12, while the equivalent nodal force vector has size 12×1.

8.1.2 Element Nodal Displacement Vector

The element nodal displacement vector $\mathbf{d}_e$ is formed by vertically stacking the displacement vectors of the four nodes:

$$\mathbf{d}_e = \begin{bmatrix} \mathbf{d}_{e1} \\ \mathbf{d}_{e2} \\ \mathbf{d}_{e3} \\ \mathbf{d}_{e4} \end{bmatrix}_{12\times 1} \tag{8.1}$$

where $\mathbf{d}_{ei}$ contains the local translational components for node i of the element:

$$\mathbf{d}_{ei} = \begin{bmatrix} u_i \\ v_i \\ w_i \end{bmatrix}_{3\times 1} \tag{8.2}$$

8.1.3 Displacement Approximation

Within the domain of the tetrahedral element, the continuous displacement field is approximated using a linear interpolation of the nodal values:

$$u(x, y, z) = \sum_{i=1}^{4} N_i(x, y, z)u_i \tag{8.3}$$

$$v(x, y, z) = \sum_{i=1}^{4} N_i(x, y, z)v_i \tag{8.4}$$

$$w(x, y, z) = \sum_{i=1}^{4} N_i(x, y, z)w_i \tag{8.5}$$

By introducing the continuous displacement vector field

$$\mathbf{u}(x, y, z) = \begin{bmatrix} u \\ v \\ w \end{bmatrix}_{3\times 1} \tag{8.6}$$

the interpolation equations can be written compactly in matrix form as

$$\underbrace{\mathbf{u}(x, y, z)}_{3\times 1} = \underbrace{\mathbf{N}_e}_{3\times 12} \underbrace{\mathbf{d}_e}_{12\times 1} \tag{8.7}$$

where the comprehensive element shape function matrix $\mathbf{N}_e$ is horizontally partitioned into **nodal submatrices**:

$$\mathbf{N}_e = \begin{bmatrix} \mathbf{N}_{e1} & \mathbf{N}_{e2} & \mathbf{N}_{e3} & \mathbf{N}_{e4} \end{bmatrix}_{3\times 12} \tag{8.8}$$

The structural nodal submatrix of shape functions $\mathbf{N}_{ei}$ mapping the scalar shape functions to the corresponding translational components of node i is defined as:

$$\mathbf{N}_{ei} = \begin{bmatrix} N_i & 0 & 0 \\ 0 & N_i & 0 \\ 0 & 0 & N_i \end{bmatrix}_{3\times 3} \tag{8.9}$$

These scalar shape functions were introduced in Eq.(4.121) and are expressed as linear polynomials in the spatial coordinates:

$$N_i(x, y, z) = a_i + b_i x + c_i y + d_i z, \quad i = 1, 2, 3, 4 \tag{8.10}$$

The coefficients (a_i, b_i, c_i, d_i) depend only on the nodal coordinates of the element. The four shape functions satisfy both the Kronecker delta property, $N_i(x_j, y_j, z_j) = \delta_{ij}$, and the partition of unity condition, $\sum_{i=1}^{4} N_i = 1$, as discussed in Chapter 4.

Since the shape functions vary linearly throughout the element, their spatial derivatives remain constant within the entire element domain.

8.1.4 Strain–Displacement Relation

Under the assumption of infinitesimal deformation, the three-dimensional engineering strain state is represented by the strain vector ε:

$$\varepsilon = \begin{bmatrix} \varepsilon_x & \varepsilon_y & \varepsilon_z & \gamma_{xy} & \gamma_{yz} & \gamma_{zx} \end{bmatrix}^{\mathsf{T}} \tag{8.11}$$

This strain vector is obtained by applying the differential operator in Eq.(2.3) to the displacement field.

Substituting the finite element displacement interpolation of Eq.(8.7) into the strain-displacement relations gives:

$$\underbrace{\varepsilon}_{6\times 1} = \underbrace{\mathbf{B}_e}_{6\times 12} \underbrace{\mathbf{d}_e}_{12\times 1} \tag{8.12}$$

Here, $\mathbf{B}_e$ is the element strain–displacement matrix, assembled from **nodal submatrices** as:

$$\mathbf{B}_e = \begin{bmatrix} \mathbf{B}_{e1} & \mathbf{B}_{e2} & \mathbf{B}_{e3} & \mathbf{B}_{e4} \end{bmatrix}_{6\times 12} \tag{8.13}$$

where $\mathbf{B}_{ei}$ represents the strain–displacement contribution associated with node i.

By identifying the constants b_i, c_i, and d_i with the spatial derivatives of the shape functions,

$$\frac{\partial N_i}{\partial x}, \quad \frac{\partial N_i}{\partial y}, \quad \frac{\partial N_i}{\partial z}, \tag{8.14}$$

the nodal strain–displacement matrix for node i becomes:

$$\mathbf{B}_{ei} = \begin{bmatrix} b_i & 0 & 0 \\ 0 & c_i & 0 \\ 0 & 0 & d_i \\ c_i & b_i & 0 \\ 0 & d_i & c_i \\ d_i & 0 & b_i \end{bmatrix}_{6\times3} \tag{8.15}$$

Because these derivatives are constant throughout the element, the T4 element produces a uniform strain field over its entire volume. For this reason, the linear tetrahedral element is the three-dimensional counterpart of the two-dimensional Constant Strain Triangle (CST).

8.1.5 Constitutive Relation

For a homogeneous, isotropic, linear elastic material, the stress–strain relationship is governed by the generalized Hooke's law:

$$\sigma = \mathbf{C}\varepsilon \tag{8.16}$$

where the engineering stress vector σ is defined as:

$$\sigma = \begin{bmatrix} \sigma_x & \sigma_y & \sigma_z & \tau_{xy} & \tau_{yz} & \tau_{zx} \end{bmatrix}^{\mathrm{T}} \tag{8.17}$$

The three-dimensional constitutive matrix $\mathbf{C}$ may be written in terms of the Lamé constants or the elastic constants given in Eq.(2.11) and Eq.(2.12). Its general isotropic form is:

$$\mathbf{C} = \begin{bmatrix} C_{11} & C_{12} & C_{12} & 0 & 0 & 0 \\ C_{12} & C_{11} & C_{12} & 0 & 0 & 0 \\ C_{12} & C_{12} & C_{11} & 0 & 0 & 0 \\ 0 & 0 & 0 & C_{44} & 0 & 0 \\ 0 & 0 & 0 & 0 & C_{44} & 0 \\ 0 & 0 & 0 & 0 & 0 & C_{44} \end{bmatrix} \tag{8.18}$$

where the shear parameter satisfies

$$C_{44} = \frac{1}{2}(C_{11} - C_{12}) \tag{8.19}$$

The following function computes the coefficients C_{ij} from Young's modulus E and Poisson's ratio ν.

```
# Forming Material Matrix C
def elasticity_matrix_3D(E, nu):
```

```
    """
    Form 3D isotropic elasticity matrix C using Young's modules `E`
    and Poisson's ratio `nu`.
    """
    lam = E * nu / ((1 + nu) * (1 - 2 * nu))
    mu = E / (2 * (1 + nu))

    C = np.array([[lam+2*mu, lam, lam, 0, 0,  0],
                  [lam, lam+2*mu, lam, 0, 0,  0],
                  [lam, lam, lam+2*mu, 0, 0,  0],
                  [ 0,    0,    0,    mu,    0,  0],
                  [ 0,    0,    0,     0,   mu,  0],
                  [ 0,    0,    0,     0,    0, mu] ])
    return C
```

8.1.6 Element Stiffness Matrix

By invoking the principle of virtual work over the spatial domain of the element Ω_e, the definition of the element stiffness matrix $\mathbf{k}_e$ is stated as (see Chapter 5):

$$\underbrace{\mathbf{k}_e}_{12\times 12} = \int_{\Omega_e} \mathbf{B}_e^\top \mathbf{C} \mathbf{B}_e \, d\Omega \tag{8.20}$$

The resulting element stiffness matrix is symmetric and possesses a dimension of 12×12. Because both the kinematics matrix $\mathbf{B}_e$ and the material matrix $\mathbf{C}$ comprise constant terms independent of coordinates (x, y, z), they can be pulled outside of the volume integration. The integral then simplifies down to the exact total volume of the tetrahedron, $V_e = \int_{\Omega_e} d\Omega$:

$$\boxed{\mathbf{k}_e = V_e \mathbf{B}_e^\top \mathbf{C} \mathbf{B}_e} \tag{8.21}$$

This is the widely used standard formula for the stiffness matrix of T4 elements.

8.1.7 Stiffness Matrix in Nodal Submatrix form

Expanding this global matrix product using its **nodal submatrix** blocks yields:

$$\underbrace{\mathbf{k}_e}_{12\times 12} = V_e \begin{bmatrix} \mathbf{B}_{e1}^\top \mathbf{C} \mathbf{B}_{e1} & \mathbf{B}_{e1}^\top \mathbf{C} \mathbf{B}_{e2} & \mathbf{B}_{e1}^\top \mathbf{C} \mathbf{B}_{e3} & \mathbf{B}_{e1}^\top \mathbf{C} \mathbf{B}_{e4} \\ \mathbf{B}_{e2}^\top \mathbf{C} \mathbf{B}_{e1} & \mathbf{B}_{e2}^\top \mathbf{C} \mathbf{B}_{e2} & \mathbf{B}_{e2}^\top \mathbf{C} \mathbf{B}_{e3} & \mathbf{B}_{e2}^\top \mathbf{C} \mathbf{B}_{e4} \\ \mathbf{B}_{e3}^\top \mathbf{C} \mathbf{B}_{e1} & \mathbf{B}_{e3}^\top \mathbf{C} \mathbf{B}_{e2} & \mathbf{B}_{e3}^\top \mathbf{C} \mathbf{B}_{e3} & \mathbf{B}_{e3}^\top \mathbf{C} \mathbf{B}_{e4} \\ \mathbf{B}_{e4}^\top \mathbf{C} \mathbf{B}_{e1} & \mathbf{B}_{e4}^\top \mathbf{C} \mathbf{B}_{e2} & \mathbf{B}_{e4}^\top \mathbf{C} \mathbf{B}_{e3} & \mathbf{B}_{e4}^\top \mathbf{C} \mathbf{B}_{e4} \end{bmatrix} \tag{8.22}$$

This can be cleanly written as:

$$\boxed{\mathbf{k}_e = \begin{bmatrix} \mathbf{k}_{11} & \mathbf{k}_{12} & \mathbf{k}_{13} & \mathbf{k}_{14} \\ \mathbf{k}_{21} & \mathbf{k}_{22} & \mathbf{k}_{23} & \mathbf{k}_{24} \\ \mathbf{k}_{31} & \mathbf{k}_{32} & \mathbf{k}_{33} & \mathbf{k}_{34} \\ \mathbf{k}_{41} & \mathbf{k}_{42} & \mathbf{k}_{43} & \mathbf{k}_{44} \end{bmatrix}} \tag{8.23}$$

where each individual nodal submatrix $\mathbf{k}_{ij}$ has dimensions of 3×3 and evaluates to:

$$\mathbf{k}_{ij} = V_e \bar{\mathbf{k}}_{ij} \tag{8.24}$$

Here, $\bar{\mathbf{k}}_{ij}$ defines the fundamental stiffness density matrix associated with the node couple (i, j) within the element:

$$\boxed{\underset{3\times 3}{\bar{\mathbf{k}}_{ij}} = \underset{3\times 6}{\mathbf{B}_{ei}^\top} \underset{6\times 6}{\mathbf{C}} \underset{6\times 3}{\mathbf{B}_{ej}}} \tag{8.25}$$

Note that the nodal submatrices $\mathbf{k}_{ij}$ are generally **not** symmetric unless they lie on the main diagonal $(i = j)$.

8.1.8 Code for T4 Element Stiffness Matrix

The following is Python function that computes the element stiffness matrix for T4 elements, using the standard formula Eq.(8.21):

```
# ===================================================
# T4 Element Stiffness Matrix Construction (standard)
# ===================================================
def element_ke_T4_BCB(coords, C):
    """
    T4 element stiffness matrix using the standard matrix product
    Parameters: coords, (4,3), Element nodal coordinates
                C, (6,6), Elasticity matrix
    """
    x1, y1, z1 = coords[0]
    x2, y2, z2 = coords[1]
    x3, y3, z3 = coords[2]
    x4, y4, z4 = coords[3]

    A = np.array([[1, x1, y1, z1],    # moment matrix
                  [1, x2, y2, z2],
                  [1, x3, y3, z3],
                  [1, x4, y4, z4]])
```

```
    detA = np.linalg.det(A)
    V = abs(detA) / 6.0
    invA = np.linalg.inv(A)

    grads = invA[1:, :]
    B = np.zeros((6, 12))
    for i in range(4):
        bi = grads[0, i]
        ci = grads[1, i]
        di = grads[2, i]
        col = 3 * i
        B[:, col:col+3] = np.array([[bi, 0, 0],
                                    [0, ci, 0],
                                    [0, 0, di],
                                    [ci, bi, 0],
                                    [0, di, ci],
                                    [di, 0, bi] ])
    ke = V * (B.T @ C @ B)
    return ke, B, V
```

The following snippet produces the stiffness matrix of an example T4 element computed using the standard formulation coded in `element_ke_T4_BCB`:

```
E = 1.0; nu = 0.3
C = elasticity_matrix_3D(E, nu)

# Define an arbitrary T4 element:
T4_ele = np.array([[0.0, 0., 0.1], [1., 0.2, 0.],
                   [0.5, 1., 0.2], [0.3,0.3,1.0] ])

ke_standard = element_ke_T4_BCB(T4_ele, C)[0]
print("T4 Stiffness Matrix (Partial, Top 6x6):")
print(ke_standard[:6, :6])
```

```
T4 Stiffness Matrix (Partial, Top 6x6):
[[ 0.2016    0.05618  0.07959 -0.17185  0.01426 -0.02146]
 [ 0.05618  0.10581  0.03673 -0.01138 -0.00583  0.02362]
 [ 0.07959  0.03673  0.13191 -0.0439   0.03003 -0.02078]
 [-0.17185 -0.01138 -0.0439   0.22792 -0.07311 -0.02611]
 [ 0.01426 -0.00583  0.03003 -0.07311  0.11178  0.0126 ]
 [-0.02146  0.02362 -0.02078 -0.02611  0.0126   0.08099]]
```

This matrix corresponds to the top-left block of the stiffness matrix for the example T4 element.

Next, we use Gemini to vectorize `batch_ke_T4_BCB`, improving its computational efficiency while extending the function to process multiple elements simultaneously:

```
def batch_ke_T4_BCB(all_coords, C):
    """
    Vectorized computation of T4 element stiffness matrices for a
    batch of elements or a single element.

    Parameters::
    all_coords : ndarray of shape (4, 3) or (N, 4, 3)
        Nodal coordinates for a single element or all N elements.
    C : ndarray of shape (6, 6), Elasticity matrix.

    Returns::
    ke: ndarray of shape (12, 12) or (N, 12, 12), Stiffness matrices
    B : ndarray of shape (6, 12) or (N, 6, 12),
        Strain-displacement matrices.
    V : float or ndarray of shape (N,), Volumes for the element(s).
    """
    # 1. Manage input dimensionality dynamically:
    is_single_element = (all_coords.ndim == 2)
    if is_single_element:
        coords = all_coords[np.newaxis, :, :]  # (4, 3) -> (1, 4, 3)
    else:
        coords = all_coords

    N = coords.shape[0]

    # 2. Compute edge vectors for all elements simultaneously
    v14 = coords[:, 0, :] - coords[:, 3, :]  # (N, 3)
    v24 = coords[:, 1, :] - coords[:, 3, :]  # (N, 3)
    v34 = coords[:, 2, :] - coords[:, 3, :]  # (N, 3)

    # 3. Vectorized cross product and volume computation
    cross_24_34 = np.cross(v24, v34)                # (N, 3)
    # Row-wise dot product:
    six_V = np.einsum('ij,ij->i', v14, cross_24_34)   # (N,)
    V_all = np.abs(six_V) / 6.0

    # 4. Vectorized shape function gradients
    g1 = cross_24_34                               # (N, 3)
    g2 = np.cross(v34, v14)                        # (N, 3)
    g3 = np.cross(v14, v24)                        # (N, 3)
    g4 = -(g1 + g2 + g3)                           # (N, 3)
```

```
    # Combine into a (N, 3, 4) tensor
    grads = np.stack((g1, g2, g3, g4), axis=2)        # (N, 3, 4)

    # Divide tensor by 6V (signed volume to preserve direction)
    grads /= six_V[:, np.newaxis, np.newaxis]

    # 5. Allocate and populate the (N, 6, 12) B-matrix tensor
    B_all = np.zeros((N, 6, 12), dtype=float)

    bi = grads[:, 0, :]                               # (N, 4)
    ci = grads[:, 1, :]                               # (N, 4)
    di = grads[:, 2, :]                               # (N, 4)

    # Use slice stride broadcasting across the first axis (3 DOFs)
    B_all[:, 0, 0::3]=bi; B_all[:, 1, 1::3]=ci; B_all[:, 2, 2::3]=di
    B_all[:, 3, 0::3] = ci; B_all[:, 3, 1::3] = bi
    B_all[:, 4, 1::3] = di; B_all[:, 4, 2::3] = ci
    B_all[:, 5, 0::3] = di; B_all[:, 5, 2::3] = bi

    # 6. Batched matrix multiplication: V * (B^T @ C @ B)
    BtC = np.einsum('nji,jk->nik', B_all, C)
    ke_all = np.einsum('nik,nkl->nil', BtC, B_all)

    # Apply volumes to each element matrix
    ke_all *= V_all[:, np.newaxis, np.newaxis]

    # 7. Unpack and strip the batch dimension, if single element
    if is_single_element:
        return ke_all[0], B_all[0], float(V_all[0])

    return ke_all, B_all, V_all
```

Let us verify the vectorized code against the original standard one:

```
ke_standard = element_ke_T4_BCB(T4_ele, C)[0]         # check only ke
ke_standard_fast = batch_ke_T4_BCB(T4_ele, C)[0]

print(np.isclose(ke_standard, ke_standard_fast)[:3, :10])
print(np.allclose(ke_standard, ke_standard_fast))
```

```
[[ True  True  True  True  True  True  True  True  True  True]
 [ True  True  True  True  True  True  True  True  True  True]
 [ True  True  True  True  True  True  True  True  True  True]]
True
```

The two implementations agree to machine precision. We can now evaluate their computational performance by generating a large collection of elements using the following code snippet:

```
Ne = 1000
start = time.time()
for i in range(Ne):     # run a number of times for accurate timing
    ke_standard = element_ke_T4_BCB(T4_ele, C)[0]

time_standard = time.time() - start
print(f"Time, the standard approach: {time_standard:.4f}s")

all_coords = np.repeat(T4_ele[np.newaxis, :, :], Ne, axis=0)
start = time.time()
ke_fast = batch_ke_T4_BCB(all_coords, C)[0]

time_standard_fast = (time.time() - start)
print(f"Time, standard fast: {time_standard_fast:.4f}s")
print(f"Time ratio: {time_standard/time_standard_fast:.4f}")
```

```
Time, the standard approach: 0.1169s
Time, standard fast: 0.0046s
Time ratio: 25.2109
```

For a batch of $N_e = 1000$ T3 elements, the vectorized implementation is approximately 20–30 times faster than the standard element-by-element loop. This substantial speedup is consistent with the predictions of Amdahl's Law, which typically estimates gains in the range of $10\times$ to $50\times$ for this type of interpreter-level optimization.

8.1.9 Nodal Submatrix Analytical Formula

This section introduces an analytical approach that significantly reduces the number of required mathematical operations to the minimum, leading to faster stiffness matrix computations.

The following SymPy script symbolically computes the nodal stiffness density submatrix $\bar{\mathbf{k}}_{ij}$ defined in Eq.(8.25). The resulting expressions form the Nodal Submatrix Analytical Formulas (NSAF):

```
# ----------------------------------------------------------------
# SymPy code for deriving the analytical stiffness submatrices
# for T4 elements for 3D solids
# ----------------------------------------------------------------
bi, ci, di = sp.symbols('b_i c_i d_i')  # Symbols
bj, cj, dj = sp.symbols('b_j c_j d_j')
```

```
# Elasticity matrix constants
C11, C12, C44 = sp.symbols('C_11 C_12 C_44')

Bi = sp.Matrix([[bi,  0,   0 ],       # strain-displacement matrix
                [0,   ci,  0 ],       # for node i
                [0,   0,   di],
                [ci,  bi,  0 ],
                [0,   di,  ci],
                [di,  0,   bi] ])

Bj = sp.Matrix([[bj,  0,   0 ],       # strain-displacement matrix
                [0,   cj,  0 ],       # for node j
                [0,   0,   dj],
                [cj,  bj,  0 ],
                [0,   dj,  cj],
                [dj,  0,   bj] ])

# General 3D isotropic constitutive matrix
C = sp.Matrix([[C11, C12, C12,  0,    0,    0],
               [C12, C11, C12,  0,    0,    0],
               [C12, C12, C11,  0,    0,    0],
               [0,   0,   0,   C44,   0,    0],
               [0,   0,   0,    0,   C44,   0],
               [0,   0,   0,    0,    0,  C44] ])

# kij_bar = B_i^T C B_j
kij_bar = sp.simplify(Bi.T * C * Bj)
print(f"Analytical stiffness submatrix (density:)")
display(Math(f"\\scriptsize{latex(kij_bar)}"))
```

```
Analytical stiffness submatrix (density:)
```

$$\begin{bmatrix} C_{11}b_ib_j + C_{44}c_ic_j + C_{44}d_id_j & C_{12}b_ic_j + C_{44}b_jc_i & C_{12}b_id_j + C_{44}b_jd_i \\ C_{12}b_jc_i + C_{44}b_ic_j & C_{11}c_ic_j + C_{44}b_ib_j + C_{44}d_id_j & C_{12}c_id_j + C_{44}c_jd_i \\ C_{12}b_jd_i + C_{44}b_id_j & C_{12}c_jd_i + C_{44}c_id_j & C_{11}d_id_j + C_{44}b_ib_j + C_{44}c_ic_j \end{bmatrix}$$

We have derived an analytical formula for the nodal **submatrix** of the element stiffness matrix, eliminating all operations involving zero-valued terms. The computational cost can be reduced even further by introducing shared geometric products and rewriting the matrix entries directly in terms of these common quantities. Let

$$b_{ij} = b_ib_j, \qquad c_{ij} = c_ic_j, \qquad d_{ij} = d_id_j \tag{8.26}$$

and define the mixed spatial derivative products as

$$p_{ij} = b_ic_j, \qquad q_{ij} = c_ib_j \tag{8.27}$$

$$r_{ij} = b_i d_j, \qquad s_{ij} = d_i b_j \tag{8.28}$$

$$t_{ij} = c_i d_j, \qquad u_{ij} = d_i c_j \tag{8.29}$$

With these definitions, expanding the matrix multiplication $\mathbf{B}_i^\top \mathbf{C}\mathbf{B}_j$ yields the explicit stiffness density submatrix $\bar{\mathbf{k}}_{ij}$:

$$\bar{\mathbf{k}}_{ij} = \begin{bmatrix} C_{11}b_{ij} + C_{44}(c_{ij} + d_{ij}) & C_{12}p_{ij} + C_{44}q_{ij} & C_{12}r_{ij} + C_{44}s_{ij} \\ C_{12}q_{ij} + C_{44}p_{ij} & C_{11}c_{ij} + C_{44}(b_{ij} + d_{ij}) & C_{12}t_{ij} + C_{44}u_{ij} \\ C_{12}s_{ij} + C_{44}r_{ij} & C_{12}u_{ij} + C_{44}t_{ij} & C_{11}d_{ij} + C_{44}(b_{ij} + c_{ij}) \end{bmatrix} \tag{8.30}$$

Using this explicit analytical form, the number of operations required to evaluate the element stiffness matrix is reduced by approximately a factor of 7 compared to the direct matrix multiplication in Eq.(8.25).

This reduction can be verified through a floating-point operation (FLOP) count. The standard matrix multiplication in Eq.(8.25) requires 297 FLOPs, whereas the analytical expression in Eq.(8.30) reduces the cost to only 42 FLOPs per submatrix.

8.1.10 NSAF Code for T4 Elements

The following function computes the stiffness matrices of T4 elements using the derived NSAF expressions together with Gemini-based vectorization. The implementation supports batch processing, allowing multiple elements to be evaluated simultaneously, while also remaining applicable to a single element.

```python
def batch_ke_T4_NSAF(all_coords, C):
    """
    Vectorized T4 stiffness matrix computation using the
    Nodal Submatrix Analytical Formulation (NSAF).

    Parameters::
        all_coords : (4,3) or (N,4,3) array
            Nodal coordinates of one or many T4 elements.
        C : (6,6) array, Elastic constitutive matrix.

    Returns::
        ke_all : (12,12) or (N,12,12) Element stiffness matrices.
        B_all : (6,12) or (N,6,12), Strain-displacement matrices.
        V_all : float or (N,), Element volume(s).
    """
    # Convert a single element into a batch of size 1
    coords = all_coords[np.newaxis, :, :] if all_coords.ndim == 2 \
                                          else all_coords
    N = coords.shape[0]
```

```
# 1. Element geometry
# Edge vectors referenced to node 4
v14 = coords[:, 0] - coords[:, 3]
v24 = coords[:, 1] - coords[:, 3]
v34 = coords[:, 2] - coords[:, 3]

# Tetrahedron volume from the scalar triple product
vol6 = np.einsum('ij,ij->i', v14, np.cross(v24, v34))
V_all = np.abs(vol6) / 6.0

# Shape-function gradients (constant within a T4 element)
g1 = np.cross(v24, v34)
g2 = np.cross(v34, v14)
g3 = np.cross(v14, v24)
g4 = -(g1 + g2 + g3)

all_grads = np.stack([g1, g2, g3, g4], axis=2) \
               / vol6[:, None, None]

bn = all_grads[:, 0, :]
cn = all_grads[:, 1, :]
dn = all_grads[:, 2, :]

# 2. NSAF stiffness matrix assembly
# Only three material constants are needed for isotropic
# elasticity when using the NSAF formulation.
c11, c12, c44 = C[0, 0], C[0, 1], C[3, 3]

V_scale = V_all[:, None, None]

# Gradient outer-product matrices
BB, CC, DD = [np.einsum('ni,nj->nij', g, g) * V_scale
                                for g in (bn, cn, dn) ]

BC = np.einsum('ni,nj->nij', bn, cn) * V_scale
BD = np.einsum('ni,nj->nij', bn, dn) * V_scale
CD = np.einsum('ni,nj->nij', cn, dn) * V_scale

ke_all = np.zeros((N, 12, 12))

# Normal-stress contributions
ke_all[:, 0::3, 0::3] = c11 * BB + c44 * (CC + DD)
ke_all[:, 1::3, 1::3] = c11 * CC + c44 * (BB + DD)
ke_all[:, 2::3, 2::3] = c11 * DD + c44 * (BB + CC)
```

```
    # Shear-coupling contributions
    def fill_off(i, j, M):
        block = c12 * M + c44 * M.transpose(0, 2, 1)
        ke_all[:, i::3, j::3] = block
        ke_all[:, j::3, i::3] = block.transpose(0, 2, 1)

    fill_off(0, 1, BC)   # u-v coupling
    fill_off(0, 2, BD)   # u-w coupling
    fill_off(1, 2, CD)   # v-w coupling

    # ---------------------------------------------------------
    # 3. Strain-displacement matrix (for stress recovery)
    # ---------------------------------------------------------
    B_all = np.zeros((N, 6, 12))

    B_all[:, 0, 0::3] = bn        # εxx
    B_all[:, 1, 1::3] = cn        # εyy
    B_all[:, 2, 2::3] = dn        # εzz

    B_all[:, 3, 0::3] = cn        #  xy
    B_all[:, 3, 1::3] = bn

    B_all[:, 4, 1::3] = dn        #  yz
    B_all[:, 4, 2::3] = cn

    B_all[:, 5, 0::3] = dn        #  xz
    B_all[:, 5, 2::3] = bn

    # Return a single matrix if a single element was supplied
    if all_coords.ndim == 2:
        return ke_all[0], B_all[0], float(V_all[0])

    return ke_all, B_all, V_all
```

8.1.11 Performance of NSAF for T4 Elements

First, Let us verify these codes for element stiffness matrix computation:

```
E = 1.0; nu = 0.3
C = elasticity_matrix_3D(E, nu)

# Define an arbitrary T4 element:
T4_ele = np.array([[0,0,0.1], [1,0.2,0], [0.5,1,0.2], [0.3,0.3,1.]])
```

```
ke_standard = element_ke_T4_BCB(T4_ele, C)[0]        # check only ke
ke_NSAF = batch_ke_T4_NSAF(T4_ele, C)[0]

print(np.isclose(ke_standard, ke_NSAF)[:3, :10])
print(np.allclose(ke_standard, ke_NSAF))
```

```
[[ True  True  True  True  True  True  True  True  True  True]
 [ True  True  True  True  True  True  True  True  True  True]
 [ True  True  True  True  True  True  True  True  True  True]]
True
```

The outputs of the two implementations—the standard formulation and the NSAF formulation—agree to machine precision.

The following snippet evaluates the performance of the NSAF formulation over a large set of elements.

```
Ne = 100_000
all_coords = np.repeat(T4_ele[np.newaxis, :, :], Ne, axis=0)

start = time.time()
ke_batch = batch_ke_T4_BCB(all_coords, C)[0]
time_BCBbatch = (time.time() - start)
print(f"Time, time_BCBbatch: {time_BCBbatch:.4f}s")

start = time.time()
ke_batch = batch_ke_T4_NSAF(all_coords, C)#[0]
time_NSAFbatch = (time.time() - start)
print(f"Time, NSAF batch: {time_NSAFbatch:.4f}s")
print(f"Time ratio: {time_BCBbatch/time_NSAFbatch:.4f}")
```

```
Time, time_BCBbatch: 0.4493s
Time, NSAF batch: 0.2594s
Time ratio: 1.7322
```

For a batch of $N_e = 100{,}000$ T4 elements, the vectorized NSAF implementation based on Eq.(8.30) is approximately 1.5–2 times faster than the vectorized implementation based on the standard formulation in Eq.(8.21). This result is consistent with the discussion in the previous chapter on the key principles of computational efficiency for T4 elements, although the full reduction in operation counts is not fully realized in the current implementations.

8.1.12 Consistent Mass Matrix

The consistent element mass matrix is obtained by integrating the products of shape functions over the tetrahedral element domain Ω_e, as established in the foundational

FEM theory (Chapter 5). The derivation closely follows that of the T3 element in the previous chapter, with the primary difference being the use of volume (barycentric) coordinates in the present 3D formulation.

$$\underbrace{\mathbf{m}_e}_{12\times12} = \int_{\Omega_e} \rho \mathbf{N}_e^\top \mathbf{N}_e \, d\Omega \tag{8.31}$$

By transforming to natural tetrahedral volume coordinates ($N_i = L_i$) and executing the space integration via the Eisenberg–Malvern integration formula (Eq.??), the analytical mass matrix is evaluated exactly as:

$$\boxed{\mathbf{m}_e = \frac{\rho V_e}{20}\begin{bmatrix} 2\mathbf{I} & \mathbf{I} & \mathbf{I} & \mathbf{I} \\ \mathbf{I} & 2\mathbf{I} & \mathbf{I} & \mathbf{I} \\ \mathbf{I} & \mathbf{I} & 2\mathbf{I} & \mathbf{I} \\ \mathbf{I} & \mathbf{I} & \mathbf{I} & 2\mathbf{I} \end{bmatrix}} \tag{8.32}$$

where $\mathbf{I}$ denotes the 3×3 identity matrix and V_e is the element volume.

8.1.13 Nodal Force Vector

For a distributed body force vector $\mathbf{b}$ acting through the element volume, defined as

$$\mathbf{b} = \begin{bmatrix} b_x \\ b_y \\ b_z \end{bmatrix} \tag{8.33}$$

the energetically equivalent nodal force vector is computed using the standard virtual work expression:

$$\underbrace{\mathbf{f}_e^{(b)}}_{12\times1} = \int_{\Omega_e} \mathbf{N}_e^\top \mathbf{b} \, d\Omega \tag{8.34}$$

For a uniform (constant) body force distribution, converting the shape functions to volume coordinates ($N_i = L_i$) and applying the Eisenberg–Malvern integration rule evaluates this volume integral to:

$$\boxed{\mathbf{f}_e^{(b)} = \begin{bmatrix} \mathbf{f}_{e1}^{(b)} \\ \mathbf{f}_{e2}^{(b)} \\ \mathbf{f}_{e3}^{(b)} \\ \mathbf{f}_{e4}^{(b)} \end{bmatrix}_{12\times1}} \tag{8.35}$$

where the force contribution apportioned to an individual node i is:

$$\mathbf{f}_{ei}^{(b)} = \frac{V_e}{4}\begin{bmatrix} b_x \\ b_y \\ b_z \end{bmatrix}_{3\times1} \tag{8.36}$$

As shown, the total integrated body force over the tetrahedral volume, $(V_e b_x, V_e b_y, V_e b_z)$, is distributed equally among the four nodes. This is a direct consequence of the linear interpolation used within the element.

8.1.14 Properties of the T4 Element

1. **Geometric Adaptability:** As the simplest 3D solid element, the T4 element provides excellent flexibility for meshing complex geometries and supports fully automated mesh generation.
2. **No Numerical Integration:** Because the shape function derivatives are constant, all element-level quantities ($\mathbf{k}_e$, $\mathbf{m}_e$) and force vector ($\mathbf{f}_e$) can be evaluated using closed-form expressions. This improves computational efficiency and avoids numerical integration errors.
3. **Constant Fields:** The element enforces uniform strain and stress states throughout its volume, since the displacement gradients are constant within the element.
4. **Linear Displacement Kinematics:** The displacement field varies linearly over the tetrahedral domain, ensuring displacement compatibility across shared faces.
5. **Completeness & Patch Test:** The formulation exactly reproduces rigid-body motions and constant strain states, and therefore satisfies the patch test.
6. **Rigid-Body Modes:** At the element level, there are six rigid-body modes: three translations and three rotations.
7. **Limitations in High-Gradient Regions:** Due to its constant strain assumption, fine meshes are required to accurately capture bending behavior, steep gradients, or localized stress concentrations.

8.2 H8 Element Matrices and Codes

Building on the two-dimensional Q4 formulation, we now introduce the eight-node trilinear hexahedral element (H8) for three-dimensional solid mechanics. This is one of the most commonly used solid elements due to its simplicity, computational efficiency, and compatibility with structured meshing.

The H8 element uses trilinear interpolation within an isoparametric framework and is well suited for modeling general 3D continua.

This section presents the matrix formulation for the H8 element.

8.2.1 Problem Setup

Consider a hexahedral element with eight nodes (see Fig.3.2):

- Node i has coordinates (x_i, y_i, z_i),
- and displacement components (u_i, v_i, w_i),

for $i = 1, \dots, 8$.

The element contains: * 8 nodes, * 3 displacement degrees of freedom per node, * 24 total element degrees of freedom.

The element nodal displacement vector is constructed by stacking vertically the displacement components of all nodes sequentially:

$$\underbrace{\mathbf{d}_e}_{24\times 1} = \begin{bmatrix} u_1 & v_1 & w_1 & \cdots & u_8 & v_8 & w_8 \end{bmatrix}^\top \tag{8.37}$$

8.2.2 Isoparametric Formulation

The H8 element is formulated using the natural (or parent) coordinates

$$(\xi, \eta, \zeta) \in [-1, 1]^3 \tag{8.38}$$

defined over a bi-unit reference cube.

In accordance with the isoparametric concept, the physical coordinates within the element are mapped from the natural coordinates using the same shape functions employed for the displacement fields:

$$x(\xi, \eta, \zeta) = \sum_{i=1}^{8} N_i(\xi, \eta, \zeta) x_i \tag{8.39}$$

$$y(\xi, \eta, \zeta) = \sum_{i=1}^{8} N_i(\xi, \eta, \zeta) y_i \tag{8.40}$$

$$z(\xi, \eta, \zeta) = \sum_{i=1}^{8} N_i(\xi, \eta, \zeta) z_i \tag{8.41}$$

where N_i are the trilinear shape functions derived in Eq.(4.156):

$$N_i(\xi, \eta, \zeta) = \frac{1}{8}(1 + \xi\xi_i)(1 + \eta\eta_i)(1 + \zeta\zeta_i) \tag{8.42}$$

Here, $(\xi_i, \eta_i, \zeta_i) \in \{-1, +1\}^3$ represent the localized natural coordinates of node i in the reference coordinate system.

8.2.3 Displacement Approximation

The continuous displacement field within the element is approximated in terms of the nodal displacement vector by

$$\underbrace{\mathbf{u}(x,y,z)}_{3\times 1} = \begin{bmatrix} u \\ v \\ w \end{bmatrix} = \underbrace{\mathbf{N}_e}_{3\times 24} \underbrace{\mathbf{d}_e}_{24\times 1} \tag{8.43}$$

where the comprehensive element shape function matrix $\mathbf{N}_e$ is constructed in the **nodal submatrix** blocks as

$$\begin{aligned} \mathbf{N}_e &= \begin{bmatrix} N_1 & 0 & 0 & \cdots & N_8 & 0 & 0 \\ 0 & N_1 & 0 & \cdots & 0 & N_8 & 0 \\ 0 & 0 & N_1 & \cdots & 0 & 0 & N_8 \end{bmatrix}_{3\times 24} \\ &= \begin{bmatrix} \mathbf{N}_1 & \mathbf{N}_2 & \cdots & \mathbf{N}_8 \end{bmatrix} \end{aligned} \tag{8.44}$$

For each individual node i, the nodal submatrix $\mathbf{N}_i$ becomes:

$$\mathbf{N}_i = \begin{bmatrix} N_i & 0 & 0 \\ 0 & N_i & 0 \\ 0 & 0 & N_i \end{bmatrix}_{3\times 3} \tag{8.45}$$

The following code computing the shape functions for H8 elements based on the formulas given above.

```
def shape_functions_H8(xi, eta, zeta):
    ''' Generate H8 Shape Functions for 3D solids.
    '''
    N = np.array([(1-xi)*(1-eta)*(1-zeta), (1+xi)*(1-eta)*(1-zeta),
                  (1+xi)*(1+eta)*(1-zeta), (1-xi)*(1+eta)*(1-zeta),
                  (1-xi)*(1-eta)*(1+zeta), (1+xi)*(1-eta)*(1+zeta),
                  (1+xi)*(1+eta)*(1+zeta), (1-xi)*(1+eta)*(1+zeta)
                  ]) / 8.0

    dN_dxi = np.array([[-(1-eta)*(1-zeta), (1-eta)*(1-zeta),
                         (1+eta)*(1-zeta),-(1+eta)*(1-zeta),
                        -(1-eta)*(1+zeta), (1-eta)*(1+zeta),
                         (1+eta)*(1+zeta),-(1+eta)*(1+zeta)],

                       [-(1-xi)*(1-zeta), -(1+xi)*(1-zeta),
                         (1+xi)*(1-zeta),  (1-xi)*(1-zeta),
                        -(1-xi)*(1+zeta), -(1+xi)*(1+zeta),
                         (1+xi)*(1+zeta),  (1-xi)*(1+zeta)],
```

```
                    [-(1-xi)*(1-eta),  -(1+xi)*(1-eta),
                     -(1+xi)*(1+eta),  -(1-xi)*(1+eta),
                      (1-xi)*(1-eta),   (1+xi)*(1-eta),
                      (1+xi)*(1+eta),   (1-xi)*(1+eta)] ]) / 8.0
    return N, dN_dxi
```

8.2.4 Jacobian Matrix

To calculate derivatives and map volume elements, the Jacobian matrix $\mathbf{J}$ is established to relate the partial derivatives between the physical and natural coordinate systems:

$$\mathbf{J} = \begin{bmatrix} \frac{\partial x}{\partial \xi} & \frac{\partial y}{\partial \xi} & \frac{\partial z}{\partial \xi} \\ \frac{\partial x}{\partial \eta} & \frac{\partial y}{\partial \eta} & \frac{\partial z}{\partial \eta} \\ \frac{\partial x}{\partial \zeta} & \frac{\partial y}{\partial \zeta} & \frac{\partial z}{\partial \zeta} \end{bmatrix}_{3\times 3} \tag{8.46}$$

The spatial gradients of the shape functions with respect to the physical coordinates (x, y, z) are required to construct the strain–displacement matrix. These derivatives are obtained by transforming the gradients in the natural coordinate system (ξ, η, ζ) through the inverse of the Jacobian matrix.

$$\begin{bmatrix} \frac{\partial N_i}{\partial x} \\ \frac{\partial N_i}{\partial y} \\ \frac{\partial N_i}{\partial z} \end{bmatrix} = \mathbf{J}^{-1} \begin{bmatrix} \frac{\partial N_i}{\partial \xi} \\ \frac{\partial N_i}{\partial \eta} \\ \frac{\partial N_i}{\partial \zeta} \end{bmatrix} \tag{8.47}$$

For the inverse Jacobian matrix $\mathbf{J}^{-1}$ to exist and remain well-conditioned, the element geometry must avoid severe distortion or self-intersection. In practice, the element should remain as close as possible to a regular hexahedron or parallelepiped to ensure $\mathbf{J}$ is invertible. This requirement is a fundamental reason for limiting element distortion, as discussed in Chapter 3.

8.2.5 Strain–Displacement Relation

The full engineering strain vector for three-dimensional elasticity is expressed as

$$\varepsilon = \begin{bmatrix} \varepsilon_x & \varepsilon_y & \varepsilon_z & \gamma_{xy} & \gamma_{yz} & \gamma_{zx} \end{bmatrix}^{\mathsf{T}} \tag{8.48}$$

The strain–displacement relationship is given in matrix form as

$$\underbrace{\varepsilon}_{6\times 1} = \underbrace{\mathbf{B}_e}_{6\times 24} \underbrace{\mathbf{d}_e}_{24\times 1} \tag{8.49}$$

where the comprehensive element strain–displacement matrix $\mathbf{B}_e$ maps the derivative contributions across all 8 nodes, which can also be written in **nodal submatrix** blocks:

$$\mathbf{B}_e = \begin{bmatrix} \frac{\partial N_1}{\partial x} & 0 & 0 & \cdots & \frac{\partial N_8}{\partial x} & 0 & 0 \\ 0 & \frac{\partial N_1}{\partial y} & 0 & \cdots & 0 & \frac{\partial N_8}{\partial y} & 0 \\ 0 & 0 & \frac{\partial N_1}{\partial z} & \cdots & 0 & 0 & \frac{\partial N_8}{\partial z} \\ \frac{\partial N_1}{\partial y} & \frac{\partial N_1}{\partial x} & 0 & \cdots & \frac{\partial N_8}{\partial y} & \frac{\partial N_8}{\partial x} & 0 \\ 0 & \frac{\partial N_1}{\partial z} & \frac{\partial N_1}{\partial y} & \cdots & 0 & \frac{\partial N_8}{\partial z} & \frac{\partial N_8}{\partial y} \\ \frac{\partial N_1}{\partial z} & 0 & \frac{\partial N_1}{\partial x} & \cdots & \frac{\partial N_8}{\partial z} & 0 & \frac{\partial N_8}{\partial x} \end{bmatrix}_{6\times 24} = \begin{bmatrix} \mathbf{B}_1 & \mathbf{B}_2 & \cdots & \mathbf{B}_8 \end{bmatrix} \tag{8.50}$$

Unlike the linear tetrahedron element, the spatial derivatives of the trilinear shape functions within the H8 element are functions of coordinates (ξ, η, ζ).

8.2.6 Element Stiffness Matrix Formulation

Following standard finite element theory, the element stiffness matrix $\mathbf{k}_e$ is obtained by integrating the triple matrix product of the kinematic and constitutive relations over the element volume Ω_e (see Chapter 5):

$$\underbrace{\mathbf{k}_e}_{24\times 24} = \int_{\Omega_e} \mathbf{B}_e^{\top} \mathbf{C} \mathbf{B}_e \, d\Omega \tag{8.51}$$

where $\mathbf{C}$ is the material constitutive matrix defined in Eq.~(8.18).

By applying a parametric mapping from the physical coordinates (x, y, z) to the natural coordinate system (ξ, η, ζ), the volume integral is transformed onto the bi-unit reference cube domain:

$$\underbrace{\mathbf{k}_e}_{24\times 24} = \int_{-1}^{1} \int_{-1}^{1} \int_{-1}^{1} \underbrace{\mathbf{B}_e^{\top} \mathbf{C} \mathbf{B}_e}_{\mathbf{k}_e} \det(\mathbf{J}) \, d\xi d\eta d\zeta \tag{8.52}$$

where:

- $\mathbf{B}_e = \mathbf{B}_e(\xi, \eta, \zeta)$ is the natural coordinate-dependent strain–displacement matrix.
- $\mathbf{C}$ is the constant material constitutive matrix.
- $\mathbf{J}$ is the coordinate Jacobian matrix.

The determinant of the Jacobian, $\det(\mathbf{J})$, physically scales the differential volume element from the natural domain to the physical domain ($d\Omega = \det(\mathbf{J}) \, d\xi d\eta d\zeta$).

This requires that the physical hexahedral element geometry remain strictly convex throughout its domain to prevent a **negative Jacobian determinant**, which would violate physical mapping conditions and cause mathematical singularities during the inversion process. This is another fundamental reason for limiting element distortion, as discussed in Chapter 3.

Conventional textbooks conclude the analytical derivation at Eq.(8.52) and pass the dense matrix multiplication directly to a numerical quadrature loop using a $2 \times 2 \times 2$ Gauss-Legendre quadrature rule:

$$\boxed{\mathbf{k}_e = \sum_{i=1}^{2}\sum_{j=1}^{2}\sum_{k=1}^{2} \bar{\mathbf{k}}_e(\xi_i, \eta_j, \zeta_k)\ \det\big(\mathbf{J}(\xi_i, \eta_j, \zeta_k)\big)\, w_i w_j w_k} \tag{8.53}$$

The Python function for generating the Gauss points and the corresponding weights for the Gauss quadrature is given as follows.

```
def gauss_points_2x2x2():
    '''Gauss quadrature using points: 2x2x2'''
    a = 1.0 / np.sqrt(3)
    gp = [(-a, -a, -a), ( a, -a, -a), ( a,  a, -a), (-a,  a, -a),
          (-a, -a,  a), ( a, -a,  a), ( a,  a,  a), (-a,  a,  a)]
    weights = np.ones(8)
    return gp, weights
```

8.2.7 Standard Code for H8 Stiffness Matrix

8.2.7.1 Single Element Version

The following code computes the stiffness matrix of a given single H8 element by implementing Eq.(8.53) exactly.

```
# ==============================
# H8 Element Stiffness Matrix
# ==============================
def element_ke_H8_BCB(coords, C):
    '''Compute H8 Element Stiffness Matrix'''
    ke = np.zeros((24, 24))
    gp, weights = gauss_points_2x2x2()
    for (xi, eta, zeta), w in zip(gp, weights):
        N, dN_dxi = shape_functions_H8(xi, eta, zeta)
        J = dN_dxi @ coords                        # Jacobian
        detJ = np.linalg.det(J)

        if detJ <= 0:
```

```
            raise ValueError("Negative Jacobian detected")
        invJ = np.linalg.inv(J)

        # Shape gradients in physical coordinates
        dN_dx = invJ @ dN_dxi

        # Strain-displacement matrix
        B = np.zeros((6, 24))

        for i in range(8):
            bi = dN_dx[0, i]
            ci = dN_dx[1, i]
            di = dN_dx[2, i]
            col = 3 * i
            B[:, col:col+3] = np.array([[bi, 0, 0],
                                        [0, ci, 0],
                                        [0, 0, di],
                                        [ci, bi, 0],
                                        [0, di, ci],
                                        [di, 0, bi] ])

        ke += B.T @ C @ B * detJ * w
    return ke, B, detJ
```

8.2.7.2 Example: Stiffness Matrix of an H8 Element

The following Python script computes and displays the stiffness matrix of a general H8 element as an illustrative example.

```
E = 1.0; nu = 0.3
C = elasticity_matrix_3D(E, nu)

# Define an arbitrary H8 element:
H8_ele = np.array([[0,0,0.1], [1,0.2,0], [1,1,0.2], [0,1.1,0],
                   [0,0,1.2], [1,0,1.1], [1.1,1,1], [0,1,1.3] ])

ke_standard = element_ke_H8_BCB(H8_ele, C)[0]
print("H8 Stiffness Matrix (Partial, Top 6x6):")
print(ke_standard[:6, :6])
```

```
H8 Stiffness Matrix (Partial, Top 6x6):
[[ 0.24477  0.08666  0.08742 -0.11132  0.01425  0.01868]
 [ 0.08666  0.25339  0.08445 -0.01941  0.05636  0.04211]
 [ 0.08742  0.08445  0.2392  -0.01497  0.04211  0.05387]
 [-0.11132 -0.01941 -0.01497  0.25035 -0.08067 -0.08014]
 [ 0.01425  0.05636  0.04211 -0.08067  0.24693  0.08956]
```

```
 [ 0.01868  0.04211  0.05387 -0.08014  0.08956  0.25242]]
```

8.2.7.3 Vectorized Batch Version

The following code presents a vectorized batch implementation for computing the stiffness matrices of a set of H8 elements. The implementation supports both single-element and multiple-element inputs.

```
def batch_ke_H8_BCB(all_coords, C):
    """
    Vectorized batch calculation of 3D H8 element stiffness matrices.
    Courtecy of Gemini.
    Parameters::
    all_coords : ndarray, shape (8,3) for single, (N,8,3) for batch
        Nodal coordinates for a single element or N elements.
    C : ndarray of shape (6, 6), Symmetric constitutive matrix.

    Returns::
    ke : ndarray of shape (24, 24) or (N, 24, 24)
    B_all : ndarray of shape (8, 6, 24) or (N, 8, 6, 24)
    detJ_all : ndarray of shape (8,) or (N, 8)
    """
    # 1. Handle single-element vs batched input configurations
    is_single_element = (all_coords.ndim == 2)
    if is_single_element:
        coords = all_coords[np.newaxis, :, :]  # (8, 3) -> (1, 8, 3)
    else:
        coords = all_coords

    N = coords.shape[0]

    # 2. Reference Gauss coordinate spacing
    g_val = 0.5773502691896257
    gps = np.array([
        [-g_val, -g_val, -g_val], [ g_val, -g_val, -g_val],
        [-g_val,  g_val, -g_val], [ g_val,  g_val, -g_val],
        [-g_val, -g_val,  g_val], [ g_val, -g_val,  g_val],
        [-g_val,  g_val,  g_val], [ g_val,  g_val,  g_val]
    ], dtype=float)

    # 3. Map natural parameters over a clean (8,8) coordinate matrix
    xi, eta, zeta = gps[:, 0, np.newaxis], gps[:, 1, np.newaxis],\
                                           gps[:, 2, np.newaxis]
    r0, r1 = 1.0 - xi, 1.0 + xi
```

```
s0, s1 = 1.0 - eta, 1.0 + eta
t0, t1 = 1.0 - zeta, 1.0 + zeta

dN_dxi_all = np.zeros((8, 3, 8), dtype=float)
# dN/dxi
dN_dxi_all[:, 0, :] = 0.125 * np.hstack([
    -s0*t0,  s0*t0,  s1*t0, -s1*t0, -s0*t1, s0*t1, s1*t1, -s1*t1
])
# dN/deta
dN_dxi_all[:, 1, :] = 0.125 * np.hstack([
    -r0*t0, -r1*t0,  r1*t0,  r0*t0, -r0*t1, -r1*t1, r1*t1, r0*t1
])
# dN/dzeta
dN_dxi_all[:, 2, :] = 0.125 * np.hstack([
    -r0*s0, -r1*s0, -r1*s1, -r0*s1,  r0*s0, r1*s0,  r1*s1, r0*s1
])

# 4. Batched evaluation of the 3x3 Jacobian matrices: (N,8,3,3)
J = np.einsum('gij,njk->ngik', dN_dxi_all, coords)

# 5. Vectorized analytical 3x3 Determinant calculation
detJ_all = (J[:,:, 0, 0] *(J[:,:, 1, 1] * J[:,:, 2, 2] \
            - J[:,:, 1, 2] * J[:,:, 2, 1])- J[:,:, 0, 1] *\
            (J[:,:, 1, 0] * J[:,:, 2, 2] - J[:,:, 1, 2] *\
             J[:,:, 2, 0])+ J[:,:, 0, 2] *(J[:,:, 1, 0] *\
             J[:,:, 2, 1] - J[:,:, 1, 1] * J[:,:, 2, 0]))

if np.any(detJ_all <= 0.0):
    raise ValueError("Numerical instability: Negative or zero\
                      Jacobian detected inside mesh elements.")

# 6. Vectorized analytical 3x3 Matrix Inversion
invJ = np.empty((N, 8, 3, 3), dtype=float)
inv_detJ = 1.0 / detJ_all

invJ[:,:, 0, 0] = (J[:,:, 1, 1] * J[:,:, 2, 2] - J[:,:, 1, 2] *\
                   J[:,:, 2, 1]) * inv_detJ
invJ[:,:, 0, 1] = (J[:,:, 0, 2] * J[:,:, 2, 1] - J[:,:, 0, 1] *\
                   J[:,:, 2, 2]) * inv_detJ
invJ[:,:, 0, 2] = (J[:,:, 0, 1] * J[:,:, 1, 2] - J[:,:, 0, 2] *\
                   J[:,:, 1, 1]) * inv_detJ
invJ[:,:, 1, 0] = (J[:,:, 1, 2] * J[:,:, 2, 0] - J[:,:, 1, 0] *\
                   J[:,:, 2, 2]) * inv_detJ
invJ[:,:, 1, 1] = (J[:,:, 0, 0] * J[:,:, 2, 2] - J[:,:, 0, 2] *\
```

```
                           J[:,:, 2, 0]) * inv_detJ
    invJ[:,:, 1, 2] = (J[:,:, 0, 2] * J[:,:, 1, 0] - J[:,:, 0, 0] *\
                           J[:,:, 1, 2]) * inv_detJ
    invJ[:,:, 2, 0] = (J[:,:, 1, 0] * J[:,:, 2, 1] - J[:,:, 1, 1] *\
                           J[:,:, 2, 0]) * inv_detJ
    invJ[:,:, 2, 1] = (J[:,:, 0, 1] * J[:,:, 2, 0] - J[:,:, 0, 0] *\
                           J[:,:, 2, 1]) * inv_detJ
    invJ[:,:, 2, 2] = (J[:,:, 0, 0] * J[:,:, 1, 1] - J[:,:, 0, 1] *\
                           J[:,:, 1, 0]) * inv_detJ

    # Transform natural shape gradients to spatial derivatives.
    # Shape: (N, 8, 3, 8)
    dN_dx = np.einsum('ngij,gjk->ngik', invJ, dN_dxi_all)

    # 7. Coordinate striding into global Strain-Displacement (B)
    B_all = np.zeros((N, 8, 6, 24), dtype=float)
    col_x = slice(0, 24, 3)
    col_y = slice(1, 24, 3)
    col_z = slice(2, 24, 3)

    B_all[:,:, 0, col_x] = dN_dx[:,:, 0, :]
    B_all[:,:, 1, col_y] = dN_dx[:,:, 1, :]
    B_all[:,:, 2, col_z] = dN_dx[:,:, 2, :]

    B_all[:,:, 3, col_x] = dN_dx[:,:, 1, :]
    B_all[:,:, 3, col_y] = dN_dx[:,:, 0, :]

    B_all[:,:, 4, col_y] = dN_dx[:,:, 2, :]
    B_all[:,:, 4, col_z] = dN_dx[:,:, 1, :]

    B_all[:,:, 5, col_x] = dN_dx[:,:, 2, :]
    B_all[:,:, 5, col_z] = dN_dx[:,:, 0, :]

    # 8. Complete Gaussian Quadrature via Einstein Summation
    BtC = np.einsum('ngji,jk->ngik', B_all, C)
    ke_g = np.einsum('ngij,ngjk->ngik', BtC, B_all)

    # Multiply by localized element volume weight patches and
    # collapse Gauss axis
    ke_all=np.sum(ke_g*detJ_all[:,:, np.newaxis, np.newaxis],axis=1)

    # 9. Strip the batch layers, if single element matrix
    if is_single_element:
        return ke_all[0], B_all[0], detJ_all[0]
```

```
    return ke_all, B_all, detJ_all
```

8.2.7.4 Performance of the Vectorized Code

Let us verify the optimized code against the original standard one:

```
ke_standard = element_ke_H8_BCB(H8_ele, C)[0]    # check only ke
ke_standard_batch = batch_ke_H8_BCB(H8_ele, C)[0]

#print(np.isclose(ke_standard[:6,:6], ke_standard_batch[:6,:6]))
print(np.allclose(ke_standard, ke_standard_batch))
```

```
True
```

The following snippet compares the performance of `batch_ke_H8_BCB` for a batch of elements with the vectorized `element_ke_H8_BCB`, which is applied element-by-element in a loop.

```
#--- element_ke_H8_BCB --- vs.--- batch_ke_H8_BCB ---

Ne = 1_000                 # test on Ne elements
start = time.time()
for i in range(Ne):        # run a number of times for accurate timing
    ke_standard = element_ke_H8_BCB(H8_ele, C)[0]

time_standard = time.time() - start
print(f"Time, the standard approach: {time_standard:.4f}s")

# Create a batch of Ne elements for testing purpose:
all_coords = np.repeat(H8_ele[np.newaxis, :, :], Ne, axis=0)
start = time.time()
ke_batch = batch_ke_H8_BCB(all_coords, C)[0]
time_batch = (time.time() - start)
print(f"Time, standard fast: {time_batch:.4f}s")
print(f"Time ratio: {time_standard/time_batch:.4f}")
```

```
Time, the standard approach: 1.1458s
Time, standard fast: 0.0939s
Time ratio: 12.1992
```

For a batch of $N_e = 1{,}000$ T3 elements, the vectorized implementation is approximately 12 times faster than the standard single-element loop-based implementation. This performance gain is consistent with the speedup trends predicted by Amdahl's Law.

Further improvements in efficiency require the use of the Nodal Submatrix Analytical Formula.

8.2.8 H8 Nodal Submatrix Analytical Formula

8.2.8.1 Nodal Submatrix of Strain–Displacement

To reveal the underlying algebraic structure of the integrand, the 6×24 strain–displacement matrix is partitioned horizontally into eight distinct **nodal submatrices** as follows:

$$\mathbf{B}_e = \begin{bmatrix} \mathbf{B}_1 & \mathbf{B}_2 & \mathbf{B}_3 & \mathbf{B}_4 & \mathbf{B}_5 & \mathbf{B}_6 & \mathbf{B}_7 & \mathbf{B}_8 \end{bmatrix}_{6\times 24} \tag{8.54}$$

where each nodal contribution $\mathbf{B}_i$ is a 6×3 matrix associated with node i:

$$\mathbf{B}_i = \begin{bmatrix} g_i & 0 & 0 \\ 0 & h_i & 0 \\ 0 & 0 & r_i \\ h_i & g_i & 0 \\ 0 & r_i & h_i \\ r_i & 0 & g_i \end{bmatrix}_{6\times 3} \tag{8.55}$$

The parameters g_i, h_i, and r_i represent the spatial derivatives of the shape functions with respect to the physical coordinate axes:

$$g_i = \frac{\partial N_i}{\partial x}, \qquad h_i = \frac{\partial N_i}{\partial y}, \qquad r_i = \frac{\partial N_i}{\partial z} \tag{8.56}$$

Because these are trilinear shape functions mapped from an arbitrary hexahedron, these derivatives vary with position and are functions of (ξ, η, ζ).

8.2.8.2 Stiffness Matrix in Nodal Submatrix Blocks

Substituting the **nodal submatrix** block decomposition into the integrand product yields the global element stiffness density matrix $\bar{\mathbf{k}}_e$:

$$\bar{\mathbf{k}}_e = \mathbf{B}_e^\top \mathbf{C} \mathbf{B}_e = \begin{bmatrix} \bar{\mathbf{k}}_{11} & \bar{\mathbf{k}}_{12} & \cdots & \bar{\mathbf{k}}_{18} \\ \bar{\mathbf{k}}_{21} & \bar{\mathbf{k}}_{22} & \cdots & \bar{\mathbf{k}}_{28} \\ \vdots & \vdots & \ddots & \vdots \\ \bar{\mathbf{k}}_{81} & \bar{\mathbf{k}}_{82} & \cdots & \bar{\mathbf{k}}_{88} \end{bmatrix}_{24\times 24} \tag{8.57}$$

where each individual block is a 3×3 submatrix tracking the interaction between node pair (i, j), defined by:

$$\underbrace{\bar{\mathbf{k}}_{ij}}_{3\times 3} = \mathbf{B}_i^\top \mathbf{C} \mathbf{B}_j \tag{8.58}$$

Using this block structure, the following section derives analytical expressions for the element stiffness matrix of the H8 element.

8.2.8.3 H8 Nodal Submatrix Analytical Formula

Following the same algebraic expansion procedure used for the T4 element in Eq.(8.30), the explicit analytical form of each 3×3 stiffness density submatrix for the H8 element is obtained as:

$$\bar{\mathbf{k}}_{ij} = \begin{bmatrix} C_{11}g_{ij} + C_{44}(h_{ij} + r_{ij}) & C_{12}s_{ij} + C_{44}t_{ij} & C_{12}o_{ij} + C_{44}p_{ij} \\ C_{12}t_{ij} + C_{44}s_{ij} & C_{11}h_{ij} + C_{44}(g_{ij} + r_{ij}) & C_{12}q_{ij} + C_{44}l_{ij} \\ C_{12}p_{ij} + C_{44}o_{ij} & C_{12}l_{ij} + C_{44}q_{ij} & C_{11}r_{ij} + C_{44}(g_{ij} + h_{ij}) \end{bmatrix}_{3\times 3} \tag{8.59}$$

where the reusable **shared geometric products** are defined as:

$$g_{ij} = g_i g_j, \qquad h_{ij} = h_i h_j, \qquad r_{ij} = r_i r_j \tag{8.60}$$

$$s_{ij} = g_i h_j, \qquad t_{ij} = h_i g_j \tag{8.61}$$

$$o_{ij} = g_i r_j, \qquad p_{ij} = r_i g_j \tag{8.62}$$

$$q_{ij} = h_i r_j, \qquad l_{ij} = r_i h_j \tag{8.63}$$

8.2.8.4 Numerical Integration

Unlike the simpler tetrahedral T4 element, the spatial gradients of the trilinear H8 shape functions depend on the natural coordinates. As a result, the derivative fields (g_i, h_i, r_i) and all associated geometric products $(g_{ij}, h_{ij}, r_{ij}, s_{ij}, t_{ij}, o_{ij}, p_{ij}, q_{ij}, l_{ij})$ vary throughout the element volume.

Therefore, the stiffness matrix must be evaluated numerically using Gaussian quadrature. For a standard, undistorted, or moderately deformed H8 element, the classical full integration scheme is the $2 \times 2 \times 2$ Gauss–Legendre quadrature rule. This scheme employs eight integration points within the reference domain located at:

$$\xi_i, \eta_j, \zeta_k = \pm\frac{1}{\sqrt{3}} \approx \pm 0.577350269 \tag{8.64}$$

Accumulating the factored submatrix densities over the quadrature points yields the final element stiffness matrix:

$$\boxed{\mathbf{k}_{ij} = \sum_{i=1}^{2}\sum_{j=1}^{2}\sum_{k=1}^{2} \bar{\mathbf{k}}_{ij}(\xi_i, \eta_j, \zeta_k)\, \det(\mathbf{J}(\xi_i, \eta_j, \zeta_k))\, w_i w_j w_k} \tag{8.65}$$

where (ξ_i, η_j, ζ_k) are the specific coordinates of the Gauss integration points, and $w_i = w_j = w_k = 1.0$ are the standard weights associated with the two-point 1D Gauss rule. Combining the pre-factored block structure of Eq.(8.59) with the integration loop in Eq.(8.65) provides an exceptionally efficient framework for high-performance finite element implementations.

8.2.8.5 NSAF Code for H8 Stiffness Matrix

Using Eq.(8.59), the following function computes the stiffness matrix for H8 elements. It works for a single element or a batch elements, and is vectorized using Gemini.

```
def batch_ke_H8_NSAF(all_coords, C):
    """
    Vectorized batched calculation of 3D H8 stiffness matrices.
    Returns:
        ke_all: Stiffness matrices, shape (N, 24, 24) or (24, 24)
        B: Strain-displacement matrices,shape (N,8,6,24) or (8,6,24)
        detJ_all: Determinants of the Jacobians, shape (N,8) or (8,)
    """
    is_single_element = (all_coords.ndim == 2)
    if is_single_element:
        coords = all_coords[np.newaxis, :, :]
    else:
        coords = all_coords

    N = coords.shape[0]

    # 1. Reference Gauss coordinate spacing
    g_val = 0.5773502691896257
    gps = np.array([
        [-g_val, -g_val, -g_val], [ g_val, -g_val, -g_val],
        [-g_val,  g_val, -g_val], [ g_val,  g_val, -g_val],
        [-g_val, -g_val,  g_val], [ g_val, -g_val,  g_val],
        [-g_val,  g_val,  g_val], [ g_val,  g_val,  g_val]
    ], dtype=float)

    xi, eta, zeta = gps[:, 0, np.newaxis], gps[:, 1, np.newaxis],\
                                            gps[:, 2, np.newaxis]
    r0, r1 = 1.0 - xi, 1.0 + xi
    s0, s1 = 1.0 - eta, 1.0 + eta
    t0, t1 = 1.0 - zeta, 1.0 + zeta

    # 2. Reference natural derivative tensor (8, 3, 8)
    dN_dxi_all = np.zeros((8, 3, 8), dtype=float)
    dN_dxi_all[:,0,:] = 0.125*np.hstack([-s0*t0,  s0*t0,  s1*t0,\
                          -s1*t0, -s0*t1,  s0*t1,  s1*t1, -s1*t1])
    dN_dxi_all[:,1,:] = 0.125*np.hstack([-r0*t0, -r1*t0,  r1*t0,\
                           r0*t0, -r0*t1, -r1*t1,  r1*t1,  r0*t1])
    dN_dxi_all[:,2,:] = 0.125*np.hstack([-r0*s0, -r1*s0, -r1*s1,\
                          -r0*s1,  r0*s0,  r1*s0,  r1*s1,  r0*s1])
```

```
# 3. Batched evaluation of 3x3 Jacobians
J = np.einsum('gij,njk->ngik', dN_dxi_all, coords)

# 4. Fast analytical Determinant
detJ_all = (J[:,:, 0, 0] * (J[:,:, 1, 1] * J[:,:, 2, 2] -\
                            J[:,:, 1, 2] * J[:,:, 2, 1]) -
            J[:,:, 0, 1] * (J[:,:, 1, 0] * J[:,:, 2, 2] - \
                            J[:,:, 1, 2] * J[:,:, 2, 0]) +
            J[:,:, 0, 2] * (J[:,:, 1, 0] * J[:,:, 2, 1] - \
                            J[:,:, 1, 1] * J[:,:, 2, 0]))

# 5. Fast analytical Inverse
invJ = np.empty((N, 8, 3, 3), dtype=float)
inv_detJ = 1.0 / detJ_all

invJ[:,:, 0, 0] = (J[:,:, 1, 1] * J[:,:, 2, 2] - J[:,:, 1, 2] *\
                  J[:,:, 2, 1]) * inv_detJ
invJ[:,:, 0, 1] = (J[:,:, 0, 2] * J[:,:, 2, 1] - J[:,:, 0, 1] *\
                  J[:,:, 2, 2]) * inv_detJ
invJ[:,:, 0, 2] = (J[:,:, 0, 1] * J[:,:, 1, 2] - J[:,:, 0, 2] *\
                  J[:,:, 1, 1]) * inv_detJ
invJ[:,:, 1, 0] = (J[:,:, 1, 2] * J[:,:, 2, 0] - J[:,:, 1, 0] *\
                  J[:,:, 2, 2]) * inv_detJ
invJ[:,:, 1, 1] = (J[:,:, 0, 0] * J[:,:, 2, 2] - J[:,:, 0, 2] *\
                  J[:,:, 2, 0]) * inv_detJ
invJ[:,:, 1, 2] = (J[:,:, 0, 2] * J[:,:, 1, 0] - J[:,:, 0, 0] *\
                  J[:,:, 1, 2]) * inv_detJ
invJ[:,:, 2, 0] = (J[:,:, 1, 0] * J[:,:, 2, 1] - J[:,:, 1, 1] *\
                  J[:,:, 2, 0]) * inv_detJ
invJ[:,:, 2, 1] = (J[:,:, 0, 1] * J[:,:, 2, 0] - J[:,:, 0, 0] *\
                  J[:,:, 2, 1]) * inv_detJ
invJ[:,:, 2, 2] = (J[:,:, 0, 0] * J[:,:, 1, 1] - J[:,:, 0, 1] *\
                  J[:,:, 1, 0]) * inv_detJ

# Transform to spatial gradients: dN_dx layout is (N, 8, 3, 8)
dN_dx = np.einsum('ngij,gjk->ngik', invJ, dN_dxi_all)

# 6. Dense Spatial Allocation
B = np.zeros((N, 8, 6, 24), dtype=float)
idx_x = np.arange(0, 24, 3)
idx_y = np.arange(1, 24, 3)
idx_z = np.arange(2, 24, 3)
```

```
    B[:,:, 0, idx_x] = dN_dx[:,:, 0, :]
    B[:,:, 1, idx_y] = dN_dx[:,:, 1, :]
    B[:,:, 2, idx_z] = dN_dx[:,:, 2, :]

    B[:,:, 3, idx_x] = dN_dx[:,:, 1, :]
    B[:,:, 3, idx_y] = dN_dx[:,:, 0, :]

    B[:,:, 4, idx_y] = dN_dx[:,:, 2, :]
    B[:,:, 4, idx_z] = dN_dx[:,:, 1, :]

    B[:,:, 5, idx_x] = dN_dx[:,:, 2, :]
    B[:,:, 5, idx_z] = dN_dx[:,:, 0, :]

    # Integrate scaling weights directly into the operator matrix
    B_weighted = B * detJ_all[:,:, np.newaxis, np.newaxis]

    # 7. HIGH-SPEED MATRIX RESHAPING FOR PARALLEL BLAS
    B_flat_rows = B.reshape(N * 8, 6, 24)
    B_w_rows = B_weighted.reshape(N * 8, 6, 24)

    Bt_C = np.matmul(B_w_rows.transpose(0, 2, 1), C)  # (N*8, 24, 6)
    ke_gps = np.matmul(Bt_C, B_flat_rows)              # (N*8, 24, 24)

    # Reshape back and sum across the 8 Gauss points down axis 1
    ke_all = ke_gps.reshape(N, 8, 24, 24).sum(axis=1)

    # Maintain single element array consistency if requested
    if is_single_element:
        return ke_all[0], B[0], detJ_all[0]

    return ke_all, B, detJ_all
```

8.2.8.6 H8 NSAF Performance Evaluation

The following snippet compares the performance of the NSAF formulation in Eq.(8.59) with the standard formulation in Eq.(8.53).

```
# --- batch_ke_H8_BCB---vs.---batch_ke_H8_NSAF---

Ne = 1_000  # create a batch of Ne elements for testing purpose:
all_coords = np.repeat(H8_ele[np.newaxis, :, :], Ne, axis=0)

start = time.time()
ke_batch = batch_ke_H8_BCB(all_coords, C)[0]
```

```
time_BCBbatch = (time.time() - start)
print(f"Time, time_BCBbatch: {time_BCBbatch:.4f}s")
#print(f"Time ratio: {time_standard/time_batch:.4f}")

start = time.time()
ke_batch = batch_ke_H8_NSAF(all_coords, C)[0]
time_NSAFbatch = (time.time() - start)
print(f"Time, NSAF batch: {time_NSAFbatch:.4f}s")
print(f"Time ratio: {time_BCBbatch/time_NSAFbatch:.4f}")
```

```
Time, time_BCBbatch: 0.0977s
Time, NSAF batch: 0.0307s
Time ratio: 3.1822
```

It is observed that, for a batch of $N_e = 1{,}000$ H8 elements, the vectorized NSAF implementation based on Eq.(8.59) is approximately three times faster than the vectorized implementation using the standard formulation in Eq.(8.53). This result further supports the earlier remarks on the core principles of computational efficiency made in the context of T3 elements.

8.2.9 Element Mass Matrix

Following standard finite element theory, the consistent element mass matrix $\mathbf{m}_e$ represents the distributed inertial properties of the continuum by integrating the product of the material density ρ and the displacement interpolation functions:

$$\underbrace{\mathbf{m}_e}_{24\times 24} = \int_{\Omega_e} \rho \mathbf{N}_e^\top \mathbf{N}_e \, d\Omega \tag{8.66}$$

Mapping the physical volume element to the bi-unit reference cube via parametric natural transformation transforms the domain boundaries into standard rectangular limits:

$$\underbrace{\mathbf{m}_e}_{24\times 24} = \int_{-1}^{1}\int_{-1}^{1}\int_{-1}^{1} \rho \mathbf{N}_e^\top \mathbf{N}_e \det(\mathbf{J})\, d\xi d\eta d\zeta \tag{8.67}$$

By matching the block-structured design used for the kinematics definitions, the global 24×24 consistent mass matrix can be partitioned into an 8×8 grid of individual nodal interaction blocks:

$$\boxed{\mathbf{m}_e = \begin{bmatrix} \mathbf{m}_{11} & \mathbf{m}_{12} & \cdots & \mathbf{m}_{18} \\ \mathbf{m}_{21} & \mathbf{m}_{22} & \cdots & \mathbf{m}_{28} \\ \vdots & \vdots & \ddots & \vdots \\ \mathbf{m}_{81} & \mathbf{m}_{82} & \cdots & \mathbf{m}_{88} \end{bmatrix}} \tag{8.68}$$

where each submatrix block $\mathbf{m}_{ij}$ is a 3×3 matrix representing the inertial coupling between node i and node j:

$$\underbrace{\mathbf{m}_{ij}}_{3\times 3} = \int_{-1}^{1}\int_{-1}^{1}\int_{-1}^{1} \rho \begin{bmatrix} N_iN_j & 0 & 0 \\ 0 & N_iN_j & 0 \\ 0 & 0 & N_iN_j \end{bmatrix} \det(\mathbf{J})\, d\xi d\eta d\zeta \tag{8.69}$$

Because the coordinate shape components are decoupled across orthogonal translational degrees of freedom, the structural block contains zero cross-coupling terms. The submatrix simplifies to a scalar diagonal topology:

$$\mathbf{m}_{ij} = \mathbf{I}_3 \cdot \int_{-1}^{1}\int_{-1}^{1}\int_{-1}^{1} \rho N_i N_j \det(\mathbf{J})\, d\xi d\eta d\zeta \tag{8.70}$$

where $\mathbf{I}_3$ represents a 3×3 identity matrix. Evaluating the independent scalar scalar integral component requires integrating the shape function product:

$$m_{ij} = \int_{-1}^{1}\int_{-1}^{1}\int_{-1}^{1} \rho N_i N_j \det(\mathbf{J})\, d\xi d\eta d\zeta \tag{8.71}$$

In conventional finite element solvers, Eq.(8.71) is typically evaluated numerically using a $2 \times 2 \times 2$ Gaussian quadrature scheme. However, since the integrand consists of polynomial terms (products of trilinear shape functions scaled by the Jacobian determinant), the expression can be integrated exactly in closed form. This can be achieved by extending the analytical techniques developed for the Q4 quadrilateral element.

Although the full algebraic expansion is omitted here due to its length, the derivation removes a key technical barrier to implementing the NSAF mass matrix in a high-performance FEM framework.

8.2.10 Lumped Mass Matrix

For explicit dynamic simulations, the lumped mass matrix may be used for efficiency:

$$\mathbf{m}_e^{\text{lumped}} = \frac{\rho V}{8}\mathbf{I}_{24} \tag{8.72}$$

where $V = \int_{\Omega_e} d\Omega$ denotes the total volume of the hexahedral element. In this formulation, the total element mass is distributed uniformly, assigning exactly one-eighth of the mass to each of the eight nodes. This removes coupling between degrees of freedom, making the global mass matrix diagonal and its inversion computationally trivial.

8.2.11 Equivalent Nodal Force Vector

8.2.11.1 Body Force

The equivalent nodal force vector $\mathbf{f}_e$ corresponding to a distributed body force field $\mathbf{b} = [b_x, b_y, b_z]^\top$ is given by:

$$\underbrace{\mathbf{f}_e}_{24\times 1} = \int_{\Omega_e} \mathbf{N}_e^\top \mathbf{b}\, d\Omega \tag{8.73}$$

Applying the parametric spatial transformation maps the configuration directly to the reference domain coordinates:

$$\underbrace{\mathbf{f}_e}_{24\times 1} = \int_{-1}^{1}\int_{-1}^{1}\int_{-1}^{1} \mathbf{N}_e^\top \mathbf{b}\ \det(\mathbf{J})\, d\xi d\eta d\zeta \tag{8.74}$$

This volume integral is evaluated numerically using standard three-dimensional Gaussian quadrature over the element domain. The required number of Gauss points also depends on the form of the body force field $\mathbf{b}(\mathbf{x})$. If $\mathbf{b}(\mathbf{x})$ is sufficiently smooth and integrable, Eq.(8.74) can also be evaluated in closed analytical form, following the same strategy used for the Q4 quadrilateral element. The full expansion is omitted due to the large number of geometric permutations involved in three-dimensional integration.

8.2.11.2 Surface Traction

When an external surface traction field $\bar{\mathbf{t}} = [t_x, t_y, t_z]^\top$ acts on a boundary face Γ_s, the equivalent nodal force vector is obtained by integrating over the surface domain:

$$\underbrace{\mathbf{f}_{\text{face}}}_{12\times 1} = \int_{\Gamma_s} \mathbf{N}_{\text{face}}^\top \bar{\mathbf{t}}\, d\Gamma \tag{8.75}$$

Here, $\mathbf{N}_{\text{face}}$ denotes the reduced two-dimensional trilinear shape function matrix defined on the loaded quadrilateral face. The surface integral is typically evaluated using a 2×2 Gaussian quadrature rule over the boundary face.

8.2.12 Features of H8 Elements

1. **Trilinear Interpolation Fields:** The H8 hexahedral element uses trilinear interpolation functions to approximate the displacement field. This leads to linear variation along each natural coordinate direction and results in linearly varying strain fields within the element.

2. **Computational Efficiency:** Compared to tetrahedral elements, the H8 element generally provides better accuracy per degree of freedom, making it an efficient choice for large-scale three-dimensional simulations.

3. **Isoparametric Mapping Flexibility:** The isoparametric formulation allows the element to conform to curved or irregular geometries. However, excessive distortion or non-convex mappings can lead to poor Jacobian conditioning and reduced numerical accuracy.

Overall, the H8 hexahedral element offers a good balance between computational complexity and accuracy, and it remains a standard element type in many industrial finite element codes.

8.3 FEM Solver for 3D Solids

This section presents a complete, modular Finite Element Method (FEM) framework for three-dimensional solid continua, developed based on the theoretical formulations derived in the preceding sections. To demonstrate the algorithmic structure, verification strategy, and convergence behavior of 3D solid elements, we consider a classical structural benchmark problem.

8.3.1 Cantilever Beam in 3D Elasticity

This benchmark problem is widely used in computational mechanics. It is simple enough to implement from scratch without relying on commercial black-box solvers, yet it captures all essential components of a full 3D FEM workflow: domain discretization, kinematic mapping, global assembly, boundary condition enforcement, and stress recovery.

8.3.1.1 Problem Description

Consider a homogeneous, isotropic rectangular cantilever beam occupying the three-dimensional domain:

$$\Omega = [0, L] \times [0, H] \times [0, W] \tag{8.76}$$

The boundary conditions and loading configuration are defined as follows:

- **Clamped boundary:** The cross-section at $x = 0$ is fully fixed, with all translational degrees of freedom constrained.
- **Transverse loading:** The free end at $x = L$ is subjected to a transverse load in the negative y-direction.

The cross-sectional dimensions are set as $H = 1$ m and $W = 1$ m. The beam length L can be varied to study different mechanical regimes. A large value of L produces a slender beam suitable for comparison with Euler–Bernoulli beam theory [4], while a smaller L reveals full 3D stress and deformation effects. In this example, we use $L = 5$ m. Readers are encouraged to experiment with other values using the same code.

All quantities are expressed in SI units, resulting in displacements in meters (m) and stresses in Pascals (Pa).

8.3.1.2 Material Properties

The material is modeled as a linear elastic, isotropic solid defined by Young's modulus E and Poisson's ratio ν:

$$E = 2.1 \times 10^{11} \text{ Pa}, \qquad \nu = 0.3 \tag{8.77}$$

8.3.1.3 Boundary Conditions

Kinematic Constraints (Clamped End) To represent a fully fixed support, all displacement components are constrained on the plane $x = 0$:

$$u_x = u_y = u_z = 0 \quad \text{at} \quad x = 0 \tag{8.78}$$

Force Boundary Conditions (Free End Loading) The loading at the free end ($x = L$) can be applied in two equivalent numerical forms, depending on the modeling objective:

1. **Distributed surface traction:** A uniform shear traction is applied over the entire end face:

$$\mathbf{t} = \begin{bmatrix} 0 \\ 0 \\ -q \end{bmatrix} \quad \text{(Pa)} \tag{8.79}$$

2. **Concentrated nodal load:** A point force is applied at selected nodes on the free end:

$$F_z = -P \quad \text{(N)} \tag{8.80}$$

The concentrated force formulation is often preferred for introductory implementations, as it avoids surface integration and direct construction of equivalent nodal traction vectors. It is used in the following codes.

8.3.1.4 Pedagogical Insights and Behavior

The cantilever beam example highlights several fundamental behaviors of low-order finite elements:

- **Piecewise-constant stress fields:** With linear tetrahedral elements (T4), strains and stresses remain constant within each element due to linear displacement interpolation.
- **Stress discontinuities:** Stresses are discontinuous across element boundaries, motivating the use of smoothing or averaging techniques in post-processing.
- **Convergence behavior:** The solution exhibits stable, monotonic convergence toward the reference energy as mesh density increases.

Mesh Refinement Study This model provides a natural framework for mesh convergence analysis. By progressively refining the mesh, one can track convergence of:

- Tip displacement at the loaded end
- Total strain energy
- Nodal stress error indicators

Analytical Validation Metric For sufficiently slender beams ($L \gg H, W$), the numerical tip deflection can be compared with the Euler–Bernoulli beam solution [4]:

$$u_y(L) = \frac{PL^3}{3EI} \tag{8.81}$$

where the second moment of area is:

$$I = \frac{WH^3}{12} \tag{8.82}$$

The Euler-Bernoulli beam theory predicts the maximum bending stress at the fixed root, where the moment has the maximum. The formula to compute the maximum normals stress (on the top or bottom surface) is

$$\sigma_{\max} = \frac{Mc}{I},$$

where

$$M = PL, \qquad c = \frac{H}{2}.$$

Substituting gives the simple formula:

$$\sigma_{\max} = \frac{6PL}{WH^2}.$$

It is usually more often to measure the magnitude of the normal stress:

$$|\sigma_{\max}| = \frac{6|P|L}{WH^2}.$$

The following Python code computes these values for a beam with its dimensions defined. These values will be used in a comparison study against FEM models.

```
# ================================================================
# Euler-Bernoulli beam bending stress for the example problem
# ================================================================

P = -1e7                        # N, load at beam tip, 10,000,000 Newtons
L = 5.0; H = 1.0; W = 1.0 # Beam eometry: 5m long, 1m high, 1m wide
E = 210e9; nu = 0.3             # Material, Steel
```

```
# Analytical Euler-Bernoulli beam tip displacement
I = W * H**3 / 12.0                              # Second moment of area
uy_ref = P * L**3 / (3 * E * I)                  # Tip displacement

print("Reference solutions from the Euler-Bernoulli beam theory:")
print("---------------------------")
print("Second moment of area I =", I)
print("Analytical tip displacement uy(L) =", uy_ref, "m")

M_root = abs(P) * L         # root bending moment
c = H / 2.0                 # distance to extreme fiber
sigma_root = M_root * c / I

print(f"Root moment M = {M_root:.6e} N·m")
print(f"Maximum root stress = {sigma_root:.6e} Pa")
print(f"Maximum root stress = {sigma_root/1e6:.3f} MPa")
```

```
Reference solutions from the Euler-Bernoulli beam theory:
---------------------------
Second moment of area I = 0.08333333333333333
Analytical tip displacement uy(L) = -0.023809523809523808 m
Root moment M = 5.000000e+07 N·m
Maximum root stress = 3.000000e+08 Pa
Maximum root stress = 300.000 MPa
```

It is important to emphasize that this expression is not the exact solution of 3D elasticity, but rather a reduced beam-theory approximation. For slender geometries, it provides an accurate reference, while the full FEM formulation remains valid for arbitrary aspect ratios and fully captures three-dimensional deformation effects without relying on beam assumptions.

8.3.2 Use of T4 Elements

The overall workflow follows the same structure presented in the previous chapter for two-dimensional solids, beginning with mesh generation. Before a 3D finite element analysis can be performed, the physical geometry must be discretized into a high-quality finite element mesh. This stage is commonly referred to as **pre-processing**.

Compared with 2D problems, mesh generation for 3D solids is significantly more challenging due to the increased geometric complexity and the need to create valid volumetric elements. For complex industrial components, dedicated CAE pre-processors and specialized mesh-generation software are typically required.

In this chapter, however, we consider a simple cantilever beam to demonstrate the development of a 3D FEM solver. Because the geometry is straightforward, the mesh can be generated directly within Python, allowing us to focus on the finite element formulation and implementation rather than on geometric modeling issues.

8.3.2.1 Mesh Generation for 3D Solids

We begin by discretizing the 3D beam using four-node linear tetrahedral (T4) elements, constructing the corresponding FEM model, and computing the numerical solution.

The following workflow illustrates how to generate a tetrahedral mesh for the cantilever beam using the `MeshPy` interface to the `TetGen` tetrahedral mesh generator. This example demonstrates the essential steps involved in creating a volumetric finite element mesh suitable for three-dimensional solid mechanics analysis.

1. **Boundary definition:** The rectangular geometry is defined by specifying its eight corner points in a `points` array.

2. **Facet construction:** The six boundary faces of the beam are defined in a `facets` list. Each facet corresponds to a quadrilateral surface described by four vertex indices. Together, these facets form a closed, watertight representation of the 3D domain.

3. **Mesh initialization:** The boundary description is loaded into a `MeshInfo` object:

```
mesh_info = mt.MeshInfo()
mesh_info.set_points(points)
mesh_info.set_facets(facets)
```

4. **Tetrahedral discretization:** The volumetric mesh is generated using the `build` function:

```
mesh = mt.build(mesh_info, max_volume=h**3)
```

The parameter `max_volume = h^3` controls the maximum allowable volume of each tetrahedral element. It acts as the primary mesh density parameter: smaller values of `h` produce finer meshes with more elements, while larger values produce coarser discretizations.

5. **Data extraction:** After mesh generation, the nodal coordinates and element connectivity are extracted into NumPy arrays for use in the solver:

```
nodes = np.array(mesh.points)
elements = np.array(mesh.elements)
```

Here, `nodes` contains the global coordinates (x, y, z) of all mesh vertices, while `elements` defines the tetrahedral connectivity, where each row lists the four node indices of a T4 element.

The Python implementation for generating the 3D cantilever beam mesh is provided below.

```
# ========================================
# Mesh Generation for 3D Cantilever Beam
# ========================================
import meshpy.tet as mt

def generate_mesh_T4(L, H, W, h):
    """
    Creates a 3D box geometry and mesh of Tetrahedral (T4) elements
    - L, H, W: Length, Height, and Width of the beam.
    - h: Target element size (smaller h = more accurate but slower)
    """
    # Define the 8 corners of the beam
    points = [(0,0,0), (L,0,0), (L,H,0), (0,H,0), \
              (0,0,W), (L,0,W), (L,H,W), (0,H,W)]
    # Define the 6 faces (facets) of the box
    facets = [[0,1,2,3], [4,5,6,7], [0,1,5,4], \
              [1,2,6,5], [2,3,7,6], [3,0,4,7]]

    mesh_info = mt.MeshInfo()
    mesh_info.set_points(points)
    mesh_info.set_facets(facets)

    # Use MeshPy to generate the T4 mesh based on max_volume
    mesh = mt.build(mesh_info, max_volume=h**3)
    return np.array(mesh.points), np.array(mesh.elements)
```

8.3.2.2 FEM Utility Codes for 3D Solids

We now develop a set of FEM utility functions for 3D solid analysis and apply them to the cantilever beam problem. The overall code structure closely follows the 2D implementation, with a straightforward extension to include the third spatial dimension. The implementation is organized in a modular format, and each component is documented with detailed comments for clarity and ease of use.

For 3D problems, the global stiffness matrix is typically highly sparse. For this reason, we use the sparse linear solvers available in `scipy.sparse.linalg`, which are well-suited for large-scale finite element systems.

```
# Modular Python FEM Code for 3D Cantilever Beam Using T4 Elements
import scipy.sparse as sps
import scipy.sparse.linalg as spla

# ============================================
# Apply Dirichlet Boundary Conditions (DBCs)
```

```
# ============================================
def apply_DBC_3D(K, F, fixed_dofs):
    """
    Applies Dirichlet BCs (Fixed supports). Modified the
    matrix K so that 'fixed' nodes cannot move (displacement = 0).
    This is done by setting the diagonal of K to 1 corresponding
    to thes fixed DOFs.
    """
    # Convert to 'List of Lists' format for easy modification
    K = K.tolil()
    for dof in fixed_dofs:
        K[dof, :] = 0       # Clear the row
        K[:, dof] = 0       # Clear the column
        K[dof, dof] = 1     # Set diagonal to 1 to enforce 1*u = 0
        F[dof] = 0          # No force is applied to fixed DOFs
    return K.tocsr(), F

def apply_tip_load(nodes, F, P=-1.0):
    """
    Applies a downward (z) force (P) at the end of the beam (at x=L).
    P is distributed equally among all nodes at the tip surface.
    """
    # Find the tip surface:
    tip_nodes=np.where(nodes[:,0] >= np.max(nodes[:,0]) - 1e-7)[0]
    for n in tip_nodes:
        # 3*n + 1 targets the Y-direction DOF of node n
        F[3*n + 2] += P / len(tip_nodes)
    return F

def solve_system(K, F):
    """Solves the linear system K * D = F to find the nodal
    displacements D at all nodes in the mesh, using sparse solver.
    """
    return spla.spsolve(K, F)

def assemble_system_T4(nodes, elements, element_ke_func, C):
    """
    Assembles Global Stiffness Matrix (K) using Vectorized Mapping.
    Generic FEM assembly routine works for T4, H8, etc.
    Parameters: nodes, array, global nodal coordinates
                elements, array, element connectivity
                element_ke_func, callable, element stiffness routine
                C : array, constitutive matrix
    """
```

```
    n_nodes = len(nodes)
    n_elems = len(elements)
    n_dofs = 3 * n_nodes

    # 1. Batch compute all 12x12 stiffness matrices at once:
    ke_all, _, _ = element_ke_func(nodes[elements], C)

    # 2. Vectorized DOF Mapping
    # We calculate the 12 DOFs for all elements simultaneously
    g_nodes = elements      # Shape (N, 4)
    g_dofs = np.zeros((n_elems, 12), dtype=np.int32)

    # Map each of the 4 nodes to its 3 DOFs (x, y, z)
    for i in range(4):
        # Result: [3*n, 3*n+1, 3*n+2] for every element's i-th node
        g_dofs[:, 3*i : 3*i+3] = 3*g_nodes[:,[i]]+np.array([0,1,2])

    # 3. Use 'repeat' and 'tile' to create the (Row, Col) pairs
    #    for the sparse matrix. This avoids the
    #    slow np.meshgrid and .extend() calls inside a loop
    rows = np.repeat(g_dofs, 12, axis=1).flatten()
    cols = np.tile(g_dofs, (1, 12)).flatten()
    vals = ke_all.flatten()

    # 4. Construct the sparse matrix in one shot
    K = sps.coo_matrix((vals, (rows, cols)), \
                        shape=(n_dofs, n_dofs)).tocsr()
    return K, np.zeros(n_dofs)
```

```
# =====================
# Main Solver Pipeline
# =====================

def FEM_Soler_3D_Beam_T4(L, H, W, E, nu, P, h=0.1):
    """
    The master function to build FEM model and solve KD=F.
       Inputs:: 3D Beam geomerty, L, H, W
                Young's modulus E, and Poission's ratio nu
                Loading at the beam tip P
                h: mesh size
       Returns::
                D:nodal displacements
                nodal_stress: averaged elemental stresses at nodes
                nodes: nodes of the model created
```

```
                 elements: elements of the model created
    """
    # Step 1: Pre-calculations
    C = elasticity_matrix_3D(E, nu)
    nodes, elements = generate_mesh_T4(L, H, W, h)

    # Step 2: Global System Assembly
    K, F = assemble_system_T4(nodes, elements, batch_ke_T4_NSAF, C)

    # Step 3: Apply Loads (Pushing down on the tip)
    F = apply_tip_load(nodes, F, P)

    # Step 4: Apply Boundary Conditions (Clamping the base at x = 0)
    fixed_nodes = np.where(nodes[:, 0] < 1e-7)[0]
    fixed_dofs = np.stack([3*fixed_nodes, 3*fixed_nodes+1,\
                           3*fixed_nodes+2], axis=1).flatten()
    K, F = apply_DBC_3D(K, F, fixed_dofs)

    # Step 5: Solve for Displacements
    D = solve_system(K, F)

    # Step 6: Post-process to find Stresses
    nodal_stress=nodal_stresses_T4_vectorized(nodes, elements, D, C)

    # Step 7: Visualization
    plot_nodal_stress(nodes, D, nodal_stress[:, 0],\
                         title='Sigma_xx (Normal Stress)')
    plot_nodal_stress(nodes, D, von_mises(nodal_stress),\
                         title='Von Mises Stress')
    plot_deformed_mesh(nodes, D, scale=100)
    # scale=100 exaggerates movement to make it visible

    print(f"Nodes: {len(nodes)} | Elements: {len(elements)}")
    print(f"Max Tip Deflection: {np.min(D):.4e} meters")

    return D, nodal_stress, nodes, elements
```

8.3.2.3 Utility Functions for Post-processing Results

```
# ============================================
#  Stress and Visualization
# ============================================
def von_mises(stress):
```

```
    """Calculates the Von Mises Equivalent Stress
    (a scalar often used in a failure criteria).
    """
    sx, sy, sz, txy, tyz, txz = stress.T
    svm = np.sqrt(0.5*((sx-sy)**2 + (sy-sz)**2 + (sz-sx)**2) +\
                  3*(txy**2 + tyz**2 + txz**2))
    return svm

def nodal_stresses_T4_vectorized(nodes, elements, D, C):
    """
    Recovers stress values at each node.
    - D: The global displacement vector (found by the solver)
    - C: The 6x6 elasticity matrix
    """
    # 1. Gather nodal displacements for every element
    elem_nodes = elements.flatten()
    dofs=np.stack([3*elem_nodes, 3*elem_nodes+1, 3*elem_nodes+2],\
                                                axis=1).flatten()

    # Make sure this name matches what we use in einsum!
    De_all = D[dofs].reshape(-1, 12)

    # 2. Get the Strain-Displacement matrix (B)
    # Ensure batch_ke_T4_NSAF is returning (Ke, B, V)
    _, B_all, _ = batch_ke_T4_NSAF(nodes[elements], C)

    # 3. Calculate Stress
    # 'nij' is B_all (N_elements, 6 components, 12 DOFs)
    # 'nj'  is De_all (N_elements, 12 DOFs)
    # 'ni'  is the result (N_elements, 6 strain components)
    stress_all = (np.einsum('nij,nj->ni', B_all, De_all)) @ C.T

    # 4. Nodal Averaging
    nodal_stress = np.zeros((len(nodes), 6))
    counts = np.zeros(len(nodes))
    for i in range(4):
        np.add.at(nodal_stress, elements[:, i], stress_all)
        np.add.at(counts, elements[:, i], 1)
    return nodal_stress / counts[:, np.newaxis]

def plot_nodal_stress(nodes, D, stress_values, scale=50,
                                              title='Nodal Stress'):
    disp = D.reshape((-1,3))
    nodes_def = nodes + scale*disp
```

```
    fig = plt.figure(figsize=(10,6), dpi=300)
    ax = fig.add_subplot(111, projection='3d')

    # Stress contour
    sc = ax.scatter(nodes_def[:,0], nodes_def[:,1], nodes_def[:,2],
        c=stress_values, cmap='jet', s=10 )

    plt.colorbar(sc, ax=ax, shrink=0.75, label=title)

    ax.set_title(f'{title} on Deformed Shape (scale={scale})' )

    ax.set_xlabel('x')
    ax.set_ylabel('y')
    ax.set_zlabel('z')
    set_axes_equal(ax)
    plt.show()
```

8.3.2.4 Utility Functions for Visualization

```
# ====================
# Visualization Tools
# ====================

def set_axes_equal(ax):
    """
    Make axes of 3D plot have equal scale.
    """
    x_limits = ax.get_xlim3d()
    y_limits = ax.get_ylim3d()
    z_limits = ax.get_zlim3d()

    x_range = abs(x_limits[1] - x_limits[0])
    y_range = abs(y_limits[1] - y_limits[0])
    z_range = abs(z_limits[1] - z_limits[0])

    x_middle = np.mean(x_limits)
    y_middle = np.mean(y_limits)
    z_middle = np.mean(z_limits)

    radius = 0.5 * max([x_range, y_range, z_range])

    ax.set_xlim3d([x_middle - radius, x_middle + radius])
```

```
    ax.set_ylim3d([y_middle - radius, y_middle + radius])
    ax.set_zlim3d([z_middle - radius, z_middle + radius])

def plot_deformed_mesh(nodes, D, scale=1.0):
    """
    Plot undeformed and deformed node locations.
    """
    disp = D.reshape((-1, 3)) # displacement vector was stored in 1D
                              # change it to 2D array with 3 columns

    nodes_def = nodes + scale * disp

    fig = plt.figure(figsize=(10, 6),dpi=300)
    ax = fig.add_subplot(111, projection='3d')

    ax.scatter(nodes[:,0], nodes[:,1], nodes[:,2],
               label='Original undeformed', s=5)

    ax.scatter(nodes_def[:,0], nodes_def[:,1], nodes_def[:,2],
               label='Deformed 3D beam', s=5)

    ax.set_xlabel('x')
    ax.set_ylabel('y')
    ax.set_zlabel('z')
    set_axes_equal(ax)
    ax.legend()
    plt.show()
```

```
def find_max_stress(nodal_stress, nodes):
    '''Post-processing: find and print the maximum stresses.
    '''
    sx = nodal_stress[:,0]
    sy = nodal_stress[:,1]
    sz = nodal_stress[:,2]

    print("\FEM maximum normal stresses:")
    print(f"max sigma_xx = {np.max(sx):.6e} Pa")
    print(f"min sigma_xx = {np.min(sx):.6e} Pa")

    print(f"max sigma_yy = {np.max(sy):.6e} Pa")
    print(f"min sigma_yy = {np.min(sy):.6e} Pa")

    print(f"max sigma_zz = {np.max(sz):.6e} Pa")
    print(f"min sigma_zz = {np.min(sz):.6e} Pa")
```

```
    imax = np.argmax(np.abs(sx))

    print("\nFEM maximum |sigma_xx|:")
    print(f"Node = {imax}")
    print(f"Location = {nodes[imax]}")
    print(f"sigma_xx = {sx[imax]:.6e} Pa")

    vm = von_mises(nodal_stress)

    print("\nFEM von Mises stress:")
    print(f"max von Mises = {np.max(vm):.6e} Pa")
    print(f"min von Mises = {np.min(vm):.6e} Pa")

    imax = np.argmax(vm)

    print("\nFEM maximum von Mises stress:")
    print(f"Node = {imax}")
    print(f"Location = {nodes[imax]}")
    print(f"von Mises = {vm[imax]:.6e} Pa")
```

Using these utility functions, together with the previously introduced stiffness matrix implementation, the main program of the 3D FEM solver is presented as follows.

8.3.2.5 Solution Obtained Using T4 Mesh

We now execute the main program and compute the cantilever beam response. The results include the tip displacement of the beam as well as the full 3D displacement and stress field visualization.

```
# Execution:
# re-set the parameters, if needed.
# P = -1e7 # N, load at beam tip, 10,000,000 Newtons
# L = 5.0; H = 1.0; W = 1.0 # Beam eometry: 5m long, 1m high, 1m wide
# E = 210e9; nu = 0.3 # Material, Steel

D, nodal_stress, nodes, elements = FEM_Soler_3D_Beam_T4\
                                   (L, H, W, E, nu, P, h=0.1)
```

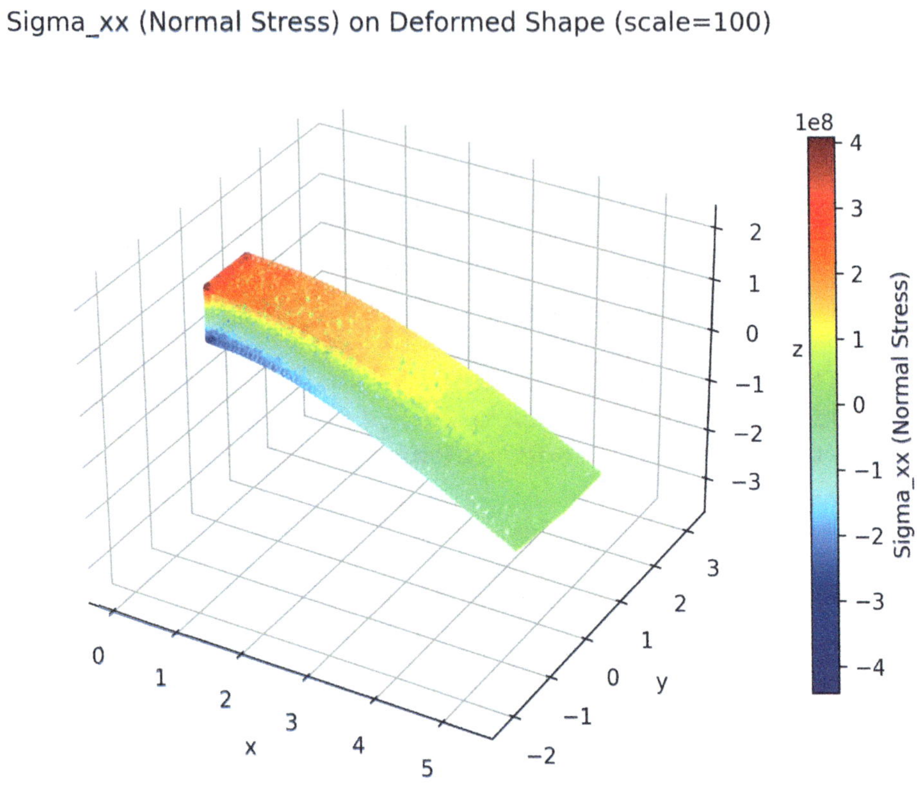

This model, consisting of 2,801 nodes and 11,305 T4 elements, was solved in approximately 4 seconds on the author's Dell laptop.

On Normal Stress:

In the foregoing plot, the 3D stress contour for the normal stress component clearly shows the characteristic bending behavior of the cantilever beam. The upper surface of the beam is subjected to tensile stress, while the lower surface experiences compressive stress, which is consistent with a downward load applied at the free end.

The stress distribution indicates that the largest normal stresses occur near the fixed root of the beam, where the bending moment attains its maximum value. As expected, the normal stress gradually decreases along the beam length and approaches zero near the free end, where the bending moment vanishes.

Furthermore, the stress contours reveal a nearly linear variation of normal stress through the beam depth. The stress is tensile on the upper surface, compressive on the lower surface, and approximately zero along the neutral plane located near the mid-depth of the beam. These observations are largely in agreement with classical Euler–Bernoulli beam theory and demonstrate that the finite element model successfully captures the dominant bending behavior of the cantilever beam.

On von Mises Stress

The 3D contour plot of the von Mises stress, as shown in the following plot, reveals a different aspect of the cantilever beam's response to a downward tip load. In addition to the normal bending stresses, the applied load generates transverse shear stresses within the beam cross-sections. According to classical beam theory, these shear stresses follow a parabolic distribution through the beam depth, reaching their maximum value along the mid-depth plane and decreasing to zero at the top and bottom surfaces.

Because the von Mises stress combines the effects of both normal and shear stress components, its distribution differs from that of the normal stress alone. Although the bending stress approaches zero near the free end where the bending moment vanishes, the shear stress remains significant because the transverse shear force is essentially constant along the beam. Consequently, the von Mises stress remains nonzero over much of the free-end cross-section and becomes zero only at locations where both the normal and shear stresses vanish, such as along the top and bottom edges of the free end.

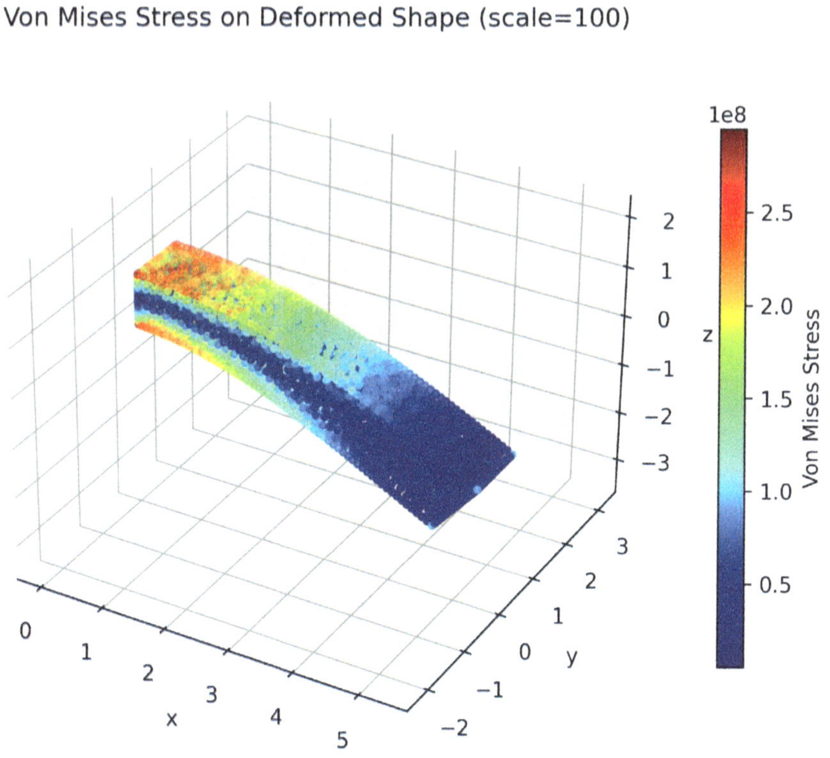

On Deflection

The 3D deformation plot shows the cantilever beam bending downward under the applied tip load. According to Euler–Bernoulli beam theory, the exact deflection curve is a cubic polynomial along the beam length. In the present finite element model, the displacement field is linear within each T4 element. However, because many elements are employed along the longitudinal direction, the piecewise-linear approximation reproduces the overall cubic deflection profile reasonably well.

The computed tip displacement, $u_y = -0.02275, \mathrm{m}$, is slightly smaller in magnitude than the reference solution, $u_y = -0.02381, \mathrm{m}$. This behavior is consistent with the well-known tendency of linear T4 elements to exhibit excessive stiffness, particularly in bending-dominated problems. As the mesh is refined—for example, by reducing the characteristic element size to $h = 0.05$—the predicted tip displacement increases in magnitude and approaches the reference solution. In other words, the numerical model becomes less stiff and more accurate as the mesh is refined. Readers are encouraged to experiment with different mesh sizes using the code presented above and observe the convergence behavior of the finite element solution.

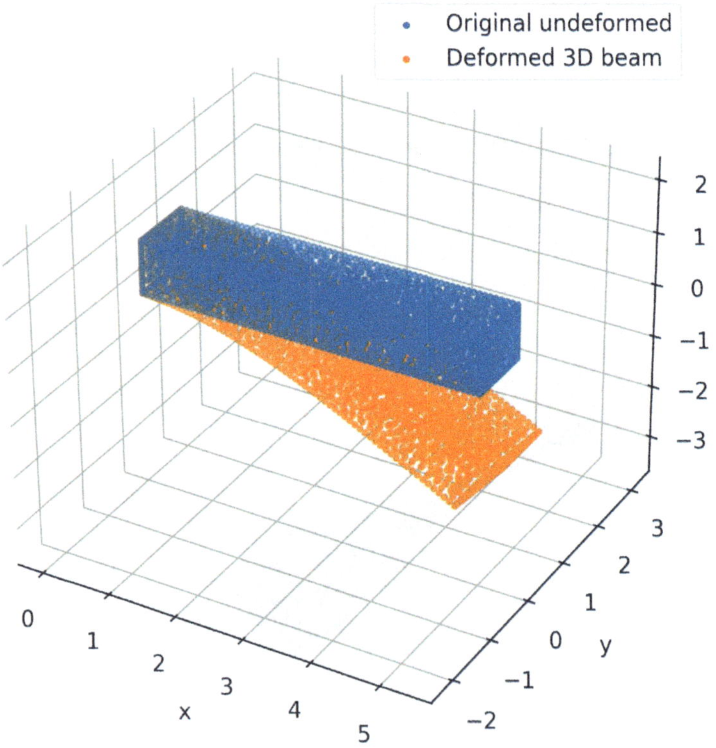


```
Nodes: 2801 | Elements: 11305
Max Tip Deflection: -2.2761e-02 meters
```

The computed tip displacement, $u_y = -0.02275, \mathrm{m}$, is slightly smaller in magnitude than the reference solution, $u_y = -0.02381, \mathrm{m}$. This behavior is consistent with the well-known tendency of linear T4 elements to exhibit excessive stiffness. As the mesh is refined—for example, by reducing the characteristic element size to $h = 0.05$—the predicted tip displacement increases in magnitude. In other words, the numerical model becomes less stiff and more accurate as the mesh is refined. Readers are encouraged to experiment with different mesh sizes using the code presented above and observe the convergence behavior.

The following code snippet computes the maximum stresses within the domain and compares the numerical results with the corresponding reference solution.

8.3.2.6 Comparison Study

```
# Comparison Study
find_max_stress(nodal_stress, nodes)

print("\nEuler-Bernoulli Beam Theory")
print(f"Maximum root stress = {sigma_root:.6e} Pa")

### Compare with FEM (take the absolute value)
sigma_fem = np.max(np.abs(nodal_stress[:,0]))

print("\nComparison")
print("----------")
print(f"Beam theory stress = {sigma_root/1e6:.3f} MPa")
print(f"FEM max sigma_xx   = {sigma_fem/1e6:.3f} MPa")
print(f"Ratio FEM/Theory   = {sigma_fem/sigma_root:.4f}")
```

```
\FEM maximum normal stresses:
max sigma_xx = 4.281191e+08 Pa
min sigma_xx = -4.104090e+08 Pa
max sigma_yy = 1.834796e+08 Pa
min sigma_yy = -1.758896e+08 Pa
max sigma_zz = 1.834796e+08 Pa
min sigma_zz = -1.758896e+08 Pa

FEM maximum |sigma_xx|:
Node = 3
Location = [0. 1. 0.]
sigma_xx = 4.281191e+08 Pa

FEM von Mises stress:
max von Mises = 2.921653e+08 Pa
min von Mises = 4.427110e+06 Pa

FEM maximum von Mises stress:
Node = 3
Location = [0. 1. 0.]
von Mises = 2.921653e+08 Pa

Euler-Bernoulli Beam Theory
```

```
Maximum root stress = 3.000000e+08 Pa

Comparison
----------
Beam theory stress = 300.000 MPa
FEM max sigma_xx   = 428.119 MPa
Ratio FEM/Theory   = 1.4271
```

It is observed that the maximum normal stress occurs at a corner point on the fixed-end surface of the beam, where a stress singularity is expected. The maximum normal stress predicted by the T4 finite element model is approximately 47% higher than the reference solution obtained from Euler–Bernoulli beam theory.

This discrepancy is not surprising. The Euler–Bernoulli solution is derived under the assumption of a one-dimensional beam and predicts a smooth stress distribution across the cross-section. In contrast, the finite element model captures the local three-dimensional stress concentration introduced by the fully fixed boundary condition. At the intersection of the constrained surface and the free surfaces of the beam, the stress field becomes singular, and the theoretical stress tends to infinity as the observation point approaches the corner.

Consequently, the maximum stress reported by the FEM model is strongly influenced by mesh resolution and typically increases with mesh refinement. The Euler–Bernoulli beam theory cannot capture this stress singularity and therefore provides only the nominal bending stress away from the fixed-end corner region.

Readers are encouraged to experiment with progressively finer meshes using the code presented above and observe the corresponding increase in the computed peak stress. This behavior is characteristic of stress singularities: as the mesh is refined, the finite element model resolves the highly localized stress field more accurately, causing the maximum stress to increase rather than converge to a finite value. The observed trend provides evidence that the model is capturing the stress singularity at the fixed-end corner of the beam.

8.3.3 Use of H8 Elements

This section presents the same solver framework using eight-node hexahedral elements (H8). Many of the utility functions developed for the T4 solver remain directly reusable, since they are independent of the specific element type and belong to the general FEM infrastructure.

These reusable components include:

- Material model: `elasticity_matrix_3D()`, which depends only on the material parameters (E, ν).
- Boundary condition enforcement: `apply_DBC_3D()`, which applies Dirichlet boundary conditions and is identical for both T4 and H8 elements.

- External loading: `apply_tip_load()`, which remains unchanged since nodal forces are independent of the interpolation scheme.
- Linear system solver: `solve_system()`, which uses a sparse iterative solver to solve the global system $\mathbf{KD} = \mathbf{F}$.
- Visualization routines: `plot_deformed_mesh()`, `find_max_stress()`, `plot_nodal_stress()`, and `set_axes_equal()`, which operate purely on nodal coordinates and displacements, and are therefore independent of element type.

Due to the change in element formulation, several components must be reimplemented because they are element-specific:

- **Mesh generation:** `generate_H8_mesh()`, which constructs a structured hexahedral (brick) mesh.
- **Element stiffness computation:** `batch_ke_H8_NSAF()` or `batch_ke_H8_BCB()`. This is the most significant modification, arising from the use of trilinear shape functions, 24 element degrees of freedom, isoparametric mapping, and numerical Gauss integration.
- **Shape functions:** `shape_functions_H8()`, which define the trilinear interpolation of geometry and displacement fields.
- **Gaussian quadrature:** `gauss_points_2x2x2()`, required because H8 elements employ isoparametric formulations and therefore require numerical integration over the parent element.
- **System assembly:** `assemble_system_H8()`, required because each H8 element contributes a 24×24 stiffness matrix and has a different connectivity structure from the T4 element.
- **Nodal stress computation:** `nodal_stresses_H8_vectorized()`, required because stresses are evaluated at multiple Gauss points within each H8 element and must be appropriately extrapolated or averaged to obtain nodal stress values.

8.3.3.1 On FEM Architecture

In a modular FEM design, only the element-level routines change when switching element types. Typically:

Reusable Infrastructure	Element-Specific Components
Material model	Shape functions
Assembly framework	Element stiffness
Solver	Numerical integration
Boundary conditions	Element DOF layout
Postprocessing	Gradient evaluation
Visualization	Reference mapping

This highlights one of the major strengths of a modular FEM implementation: only the element-dependent components need to be modified when introducing a new element type.

8.3.3.2 Utility Codes for H8 Mesh

The following section presents an FEM implementation using H8 elements.

```
# ====================
# Structured H8 Mesh
# ====================
def generate_H8_mesh(L, H, W, nx, ny, nz):
    x = np.linspace(0, L, nx + 1)
    y = np.linspace(0, H, ny + 1)
    z = np.linspace(0, W, nz + 1)

    nodes = []
    for k in range(nz + 1):
        for j in range(ny + 1):
            for i in range(nx + 1):
                nodes.append([x[i], y[j], z[k]])

    nodes = np.array(nodes)
    # Connectivity
    elements = []

    def node_id(i, j, k):
        return i + (nx + 1) * j + (nx + 1) * (ny + 1) * k

    for k in range(nz):
        for j in range(ny):
            for i in range(nx):
                n0 = node_id(i,   j,   k)
                n1 = node_id(i+1, j,   k)
                n2 = node_id(i+1, j+1, k)
                n3 = node_id(i,   j+1, k)

                n4 = node_id(i,   j,   k+1)
                n5 = node_id(i+1, j,   k+1)
                n6 = node_id(i+1, j+1, k+1)
                n7 = node_id(i,   j+1, k+1)
                elements.append([n0, n1, n2, n3, n4, n5, n6, n7])
    elements = np.array(elements)
```

```
    return nodes, elements
```

```
def nodal_stresses_H8_vectorized(nodes, elements, D, C):
    """
    Recovers stresses for H8 elements.
    Averages stresses from the 8 Gauss points to the 8 nodes.
    """
    n_elems = len(elements)
    # 1. Map global displacements to local element DOFs (N, 24)
    elem_nodes = elements.flatten()
    dofs = np.stack([3*elem_nodes, 3*elem_nodes+1, 3*elem_nodes+2],\
                                                   axis=1).flatten()
    De_all = D[dofs].reshape(n_elems, 24)

    # 2. Get the B-matrix (N, 8, 6, 24)
    _, B_all, _ = batch_ke_H8_NSAF(nodes[elements], C)

    # 3. Calculate stress at every Gauss point (N, 8, 6)
    #    Einstein sum: elements (n), gauss points (g),
    #    stress components (i), DOFs (j)
    stress_gps = np.einsum('ngij,nj->ngi', B_all, De_all) @ C.T

    # 4. Extrapolate/Average to nodes
    nodal_stress = np.zeros((len(nodes), 6))
    counts = np.zeros(len(nodes))
    for i in range(8):
        # Assume GPi is closest to Node i (Simplified extrapolation)
        np.add.at(nodal_stress, elements[:, i], stress_gps[:, i, :])
        np.add.at(counts, elements[:, i], 1)

    return nodal_stress / counts[:, np.newaxis]
```

```
def assemble_system_H8(nodes, elements, element_ke_func, C):
    n_nodes = len(nodes)
    n_elems = len(elements)
    n_dofs = 3 * n_nodes

    # UNPACK the tuple here.
    ke_all, B_all, detJ_all = element_ke_func(nodes[elements], C)

    # Vectorized DOF Mapping for 8 nodes (Hexahedron)
    g_nodes = elements   # Shape (N, 8)
    g_dofs = np.zeros((n_elems, 24), dtype=np.int32)
```

```
    # Map each of the 8 nodes to its 3 DOFs (x, y, z)
    for i in range(8):
        g_dofs[:, 3*i : 3*i+3] = 3*g_nodes[:,[i]]+np.array([0,1,2])

    # Each H8 element has 24*24 = 576 entries
    rows = np.repeat(g_dofs, 24, axis=1).flatten()
    cols = np.tile(g_dofs, (1, 24)).flatten()

    # Now ke_all is a numpy array (N, 24, 24), and flatten it.
    vals = ke_all.flatten()

    K = sps.coo_matrix((vals, (rows, cols)), \
                        shape=(n_dofs, n_dofs)).tocsr()
    return K, np.zeros(n_dofs)
```

```
# Main Program
def FEM_Soler_3D_Beam_H8(L, H, W, E, nu, P, nx=35, ny=7, nz=7):

    nodes, elements = generate_H8_mesh(L, H, W, nx, ny, nz)
    print("Nodes    :", len(nodes))
    print("Elements:", len(elements))
    print("DOFs     :", 3*len(nodes))

    C = elasticity_matrix_3D(E, nu)

    # Step 2: Global System Assembly
    K, F = assemble_system_H8(nodes, elements, batch_ke_H8_NSAF, C)

    # Step 3: Apply Loads (Pushing down on the tip)
    F = apply_tip_load(nodes, F, P)

    # Step 4: Apply Boundary Conditions (Clamping the base at x = 0)
    fixed_nodes = np.where(nodes[:, 0] < 1e-7)[0]
    fixed_dofs = np.stack([3*fixed_nodes, 3*fixed_nodes+1,\
                           3*fixed_nodes+2], axis=1).flatten()
    K, F = apply_DBC_3D(K, F, fixed_dofs)

    # Step 5: Solve for Displacements
    D = solve_system(K, F)

    # Step 6: Post-process to find Stresses
    nodal_stress=nodal_stresses_H8_vectorized(nodes, elements, D, C)
```

```
    # Step 7: Visualization (uncomment to plot):
    #plot_nodal_stress(nodes, D, nodal_stress[:, 0],\
    #                           title='Sigma_xx (Normal Stress)')
    #plot_nodal_stress(nodes, D, von_mises(nodal_stress),\
    #                           title='Von Mises Stress')
    #plot_deformed_mesh(nodes, D, scale=100)
    # scale=100 exaggerates movement to make it visible

    print(f"Nodes: {len(nodes)} | Elements: {len(elements)}")
    print(f"Max Tip Deflection: {np.min(D):.4e} meters")

    return D, nodal_stress, nodes, elements
```

8.3.3.3 Solution Obtained Using H8 Mesh

We now execute the main program and compute the cantilever beam response. The results include the tip displacement of the beam as well as the full 3D displacement field visualization.

```
# Execution:
# re-set the parameters, if needed.
# P = -1e7 # N, load at beam tip, 10,000,000 Newtons
# L = 5.0; H = 1.0; W = 1.0 # Beam eometry: 5m long, 1m high, 1m␣
↪wide
# E = 210e9; nu = 0.3 # Material, Steel

D, nodal_stress, nodes, elements = FEM_Soler_3D_Beam_H8\
                            (L, H, W, E, nu, P, nx=35, ny=7, nz=7)

print("Analytical tip displacement uy(L) =", uy_ref, "m")
```

```
Nodes    : 2304
Elements: 1715
DOFs     : 6912
Nodes: 2304 | Elements: 1715
Max Tip Deflection: -2.3910e-02 meters
Analytical tip displacement uy(L) = -0.023809523809523808 m
```

It was found earlier that the displacement $u_y = -0.02275, \mathrm{m}$ obtained from the T4 FEM model is smaller in magnitude than the reference solution, $u_y = -0.02391, \mathrm{m}$, reflecting the relatively stiff behavior of the linear tetrahedral element.

In contrast, the H8 element produces a displacement of approximately $u_y = -0.02391, \mathrm{m}$, which is much closer to the reference value and, in this case, slightly exceeds it in magnitude. The H8 model therefore exhibits a more flexible response for this benchmark problem. This improved performance can be attributed to the trilinear interpolation employed by the H8 element, which generally provides a more accurate representation of bending deformation than the linear displacement field of the T4 element.

The following code snippet examines the stress field predicted by the H8 model and compares it with the corresponding reference solution.

```
# Comparison Study
find_max_stress(nodal_stress, nodes)

print("\nEuler-Bernoulli Beam Theory")
print(f"Maximum root stress = {sigma_root:.6e} Pa")

### Compare with FEM (take the absolute value)
sigma_fem = np.max(np.abs(nodal_stress[:,0]))

print("\nComparison")
print("----------")
print(f"Beam theory stress = {sigma_root/1e6:.3f} MPa")
print(f"FEM max sigma_xx   = {sigma_fem/1e6:.3f} MPa")
print(f"Ratio FEM/Theory   = {sigma_fem/sigma_root:.4f}")
```

```
\FEM maximum normal stresses:
max sigma_xx = 3.025878e+08 Pa
min sigma_xx = -3.497519e+08 Pa
max sigma_yy = 6.637819e+07 Pa
min sigma_yy = -1.253811e+08 Pa
max sigma_zz = 7.703798e+07 Pa
min sigma_zz = -1.260821e+08 Pa

FEM maximum |sigma_xx|:
Node = 2016
Location = [0. 0. 1.]
sigma_xx = -3.497519e+08 Pa

FEM von Mises stress:
max von Mises = 2.733070e+08 Pa
min von Mises = 1.269741e+07 Pa

FEM maximum von Mises stress:
Node = 252
Location = [0. 1. 0.]
von Mises = 2.733070e+08 Pa

Euler-Bernoulli Beam Theory
```

```
Maximum root stress = 3.000000e+08 Pa

Comparison
----------
Beam theory stress = 300.000 MPa
FEM max sigma_xx   = 349.752 MPa
Ratio FEM/Theory   = 1.1658
```

It was found earlier that the maximum normal stress predicted by the T4 finite element model is approximately 43% higher than the reference solution obtained from Euler–Bernoulli beam theory. In contrast, the maximum normal stress predicted by the H8 model is only about 17% higher than the corresponding reference value. This result indicates that the H8 element provides a significantly more accurate representation of the stress field, particularly in bending-dominated problems.

It should be emphasized, however, that both models predict peak stresses at the fixed-end corner of the beam, where a stress singularity is expected. Consequently, the reported maximum stresses depend on mesh resolution and should not be interpreted as converged physical values. The Euler–Bernoulli beam solution does not capture this local three-dimensional stress concentration and therefore serves only as a reference for the nominal bending stress.

Readers are encouraged to vary the beam length to $L = 10, \text{m}$ and repeat the analysis using the same code base. This simple modification provides an opportunity to further investigate the displacement and stress behavior of both T4 and H8 elements, as well as their convergence characteristics under increasingly slender beam geometries.

8.4 Final Remarks

This chapter extended the Finite Element Method (FEM) from two-dimensional elasticity to three-dimensional solid mechanics, focusing on the formulation, implementation, and computational optimization of four-node linear tetrahedral (T4) and eight-node trilinear hexahedral (H8) elements. A modular set of FEM utility functions was developed and applied to the classical three-dimensional cantilever beam benchmark problem, demonstrating both the theoretical formulation and practical implementation of 3D solid mechanics solvers.

1. **Four-Node Linear Tetrahedron (T4):**

   The T4 element employs linear shape functions, resulting in a constant strain field within each element. This property allows the element stiffness matrix to be integrated exactly in closed form. Owing to its simple geometry and compatibility with automatic mesh generators, the T4 element is highly robust and well suited for meshing complex geometries. However, it is generally more prone to excessive stiffness, particularly in bending-dominated problems, and often requires finer meshes to achieve high accuracy.

2. **Eight-Node Trilinear Hexahedron (H8):**

   The H8 element utilizes trilinear shape functions, producing strain fields that vary throughout the element domain. Consequently, numerical integration is required to evaluate the element matrices. Compared with the T4 element, the H8 element generally provides higher accuracy per degree of freedom and exhibits superior performance in bending-dominated problems. The benchmark examples presented in this chapter demonstrated that the H8 element yields displacement and stress predictions that are closer to the reference solutions.

3. **Nodal Submatrix-Based Analytical Formulation:**

   Rather than directly evaluating the full element stiffness expression

   $$\mathbf{k}_e = \int_{\Omega_e} \mathbf{B}_e^\top \mathbf{C} \mathbf{B}_e \, d\Omega, \tag{8.83}$$

   the formulation was reorganized into a structured collection of **nodal submatrices**,

   $$\mathbf{k}_{ij} = \int_{\Omega_e} \mathbf{B}_i^\top \mathbf{C} \mathbf{B}_j \, d\Omega. \tag{8.84}$$

   This block-wise representation exposes the inherent structure of the stiffness matrix, facilitates analytical integration, and reduces the number of floating-point operations required during assembly. The resulting implementations achieved approximately a 1.5×–3× speedup in element stiffness computations compared with conventional dense formulations.

4. **Fundamental Reasons Why FEM Works**

   Throughout this book, we have seen that the finite element method not only works in practice but also rests on a solid mathematical foundation. The key reasons for its success are:

   (a) The Galerkin formulation (Chapter 5) performs an orthogonal projection of the exact solution onto the finite-dimensional approximation space, leading to a symmetric stiffness matrix $\mathbf{K}$;

   (b) The use of linearly independent and compatible shape functions (Chapters 3 and 4) establishes a finite-dimensional approximation space capable of representing admissible solutions with increasing accuracy as the mesh is refined;

   (c) The enforcement of sufficient Dirichlet boundary conditions (Chapters 6-8) removes rigid-body modes and ensures that $\mathbf{K}$ is positive definite;

   (d) Classical results from linear algebra [10] guarantee that the resulting symmetric positive-definite (SPD) system admits a unique solution, which can be obtained efficiently using direct or iterative solvers.

5. **Advanced Models Beyond the Standard FEM:**

   The T4 and H8 elements share many of the limitations already encountered in their two-dimensional counterparts, the T3 and Q4 elements. These include excessive stiffness, sensitivity to mesh distortion, dependence on numerical integration, and difficulties associated with Jacobian evaluation.

   To overcome some of these shortcomings, a number of enhanced finite element formulations have been developed. Among them, the Smoothed Finite Element Method (S-FEM) [11] offers a particularly attractive alternative framework. By smoothing strains over local smoothing domains, S-FEM can improve accuracy, enhance robustness, reduce mesh sensitivity, and alleviate many of the numerical difficulties associated with conventional finite elements. In addition to the Galerkin orthogonal projection applied to the displacement field, S-FEM employs a smoothing-based orthogonal projection of the strain field, thereby achieving improved numerical performance and additional computational advantages.

This book has focused on the principal classes of finite elements and has employed a relatively small set of element types in the numerical examples. Since the introduction of the FEM in the 1950s, however, a vast number of specialized elements and advanced formulations have been developed for different physical applications and engineering requirements. Furthermore, the examples presented throughout this book have intentionally involved relatively simple geometries to emphasize the underlying concepts and implementation details.

In **practical engineering applications involving complex geometries and large-scale solid mechanics problems**, engineers and researchers typically rely on commercial FEM software packages for geometry modeling, mesh generation using a wide variety of element types, solution analysis, and post-processing. These software systems provide robust and efficient tools for handling realistic engineering models that would be difficult and time-consuming to develop from scratch.

Several widely used FEM software packages will be briefly introduced in the following chapter.

8.5 Exercises

1: 3D Strain-Displacement Matrix

Given a 4-node tetrahedral (T4) element with the physical coordinate matrix:

$$A = \begin{bmatrix} 1 & 0 & 0 & 0 \\ 1 & 1 & 0 & 0 \\ 1 & 0 & 1 & 0 \\ 1 & 0 & 0 & 1 \end{bmatrix} \tag{8.85}$$

1. Compute the volume of the element.
2. Analytically derive the spatial derivatives of the four shape functions $(\frac{\partial N_i}{\partial x}, \frac{\partial N_i}{\partial y}, \frac{\partial N_i}{\partial z})$.
3. Construct the complete constant 6×12 strain-displacement matrix $\mathbf{B}_e$.

2: Jacobian Uniqueness and Element

An 8-node hexahedral (H8) element in physical coordinates undergoes severe distortion, causing one of its corner nodes to cross the plane of its adjacent faces (creating a non-convex geometry).

1. Analytically explain how a non-convex element shape impacts the Jacobian matrix $\mathbf{J}$ and its determinant $\det(\mathbf{J})$.
2. What are the numerical consequences of a zero or negative Jacobian determinant during the calculation of the element stiffness matrix $\mathbf{k}_e$?

3: Isotropic Constitutive

Using the 6×6 isotropic linear elastic constitutive matrix $\mathbf{C}$ parameterized by E and ν:

1. Prove that when Poisson's ratio $\nu \to 0.5$, the material matrix terms associated with normal stresses approach infinity.
2. Relate this mathematical trend to the physical phenomenon of volumetric locking in standard H8 elements under nearly incompressible conditions.

4: Nodal Block-Submatrix

Using the analytical form of the 3×3 stiffness submatrix $\bar{\mathbf{k}}_{ij}$ given in Eq. (8.59):

1. Prove that the individual submatrix block tracking a node interacting with itself ($\bar{\mathbf{k}}_{ii}$) is inherently symmetric.
2. Prove that $\bar{\mathbf{k}}_{ji} = \bar{\mathbf{k}}_{ij}^\top$, demonstrating how global matrix symmetry is preserved at the block-submatrix level.

5: Trace and Geometric

Refer to the shared geometric products defined for the analytical H8 submatrix (e.g., $g_{ij} = g_i g_j$, $s_{ij} = g_i h_j$).

1. Identify which geometric variables are invariant under a node index swap (i.e., when switching from pair (i, j) to (j, i)).
2. Rewrite the submatrix components of $\bar{\mathbf{k}}_{ji}$ explicitly using the terms defined for $\bar{\mathbf{k}}_{ij}$ to minimize unique variable storage requirements in code.

6: Complete Mass Matrix

For a 3D continuum element, the submatrix block tracking inertial coupling between nodes i and j is given by:

$$\mathbf{m}_{ij} = \mathbf{I}_3 \cdot \int_{-1}^{1}\int_{-1}^{1}\int_{-1}^{1} \rho N_i N_j \det(\mathbf{J})\, d\xi d\eta d\zeta$$

1. Explain why the off-diagonal terms within this specific 3×3 submatrix are zero, whereas the off-diagonal submatrices in the global 24×24 mass matrix are non-zero.
2. For an undistorted, regular cube H8 element centered at the origin with a constant Jacobian determinant $\det(\mathbf{J}) = \frac{V}{8}$, analytically evaluate the scalar integral component m_{11} (the diagonal mass coupling for Node 1 with itself).

7: Mass Lumping and Conservation

The consistent mass matrix distributes inertia via continuous interpolation, while the lumped mass matrix approximation simplifies this to $\mathbf{m}_e^{\text{lumped}} = \frac{\rho V}{8}\mathbf{I}_{24}$.

1. Prove that both the consistent mass matrix and the lumped mass matrix preserve the exact total mass of the element.
2. Discuss the computational advantages of using a lumped mass matrix over a consistent mass matrix in explicit dynamic transient analysis solvers.

8: Operation Count Analysis (FLOPS)

Compare the computational efficiency of the standard $\mathbf{B}^\top\mathbf{C}\mathbf{B}$ approach against the optimized nodal submatrix analytical approach for a single Gauss integration point of an H8 element.

1. Calculate the total number of floating-point multiplications required to compute $\mathbf{B}_e^\top\mathbf{C}\mathbf{B}_e$ using dense matrix operations (where $\mathbf{B}_e$ is 6×24 and $\mathbf{C}$ is 6×6).
2. Calculate the number of multiplications required when evaluating the 64 separate 3×3 submatrices using the expanded analytical block form in Eq. (8.59), counting all shared geometric products. Compare the two counts.

9: Stride Mapping and Memory Locality

In the vectorized Python implementation, loop-less assembly was achieved by assigning submatrix components directly to the global matrix using stride slices:

```
ke[row_offset::3, col_offset::3] = k_grid
```

1. Draw a schematic map showing how a 4×4 node layout (tracking only x and y components for simplicity) fills up when using standard block allocation vs. stride mapping slices.
2. Explain how stride mapping relates to data placement in computer memory structures, and why minimizing row allocations speeds up performance in interpreted languages like Python.

10: Rigid Body Mode Verification

A fully unconstrained 3D solid element must possess exactly 6 rigid body modes corresponding to zero strain energy states.

1. List the physical transformations associated with these six modes.

2. Describe an eigenvalue-based verification procedure you can run on your computed 24×24 element stiffness matrix $\mathbf{k}_e$ to guarantee that it captures all 6 rigid body modes correctly without introducing artificial energy clean errors.

Chapter 9

On Commercial FEM Packages

The Finite Element Method (FEM) is a highly developed and powerful tool for the analysis and design of complex structural systems. When encountering sophisticated engineering problems, researchers and practitioners utilize commercially available FEM software packages. These suites offer an expansive variety of element types, advanced material models, and highly optimized numerical solvers for both static and dynamic analysis. Most importantly, they all equipped with user-friendly pre- and post processors.

The commercial packages are often like a black-box, although they alls use the same FEM engine. With our discussion in the previous chapters, we should be equipped with sufficient knowledge about what is really going on inside these packages.

9.1 Major Commercial FEM Packages

The global market for commercial FEM software is dominated by a few mature, comprehensive engineering ecosystems. These packages are typically classified by their primary solution strengths, multi-physics integration capabilities, and industry-standard applications:

1. **Ansys Mechanical:** Renowned for robust multiphysics coupling (thermal, fluids, electromagnetics) and user-friendly parametric optimization.
2. **Abaqus (Dassault Systèmes):** The industry standard for highly nonlinear problems, including complex contact mechanics, large deformations, and advanced material damage/failure models.
3. **Simcenter Portfolio (Siemens Digital Industries Software):** A unified modern ecosystem integrating premier structural solvers like Simcenter Nastran with world-class pre- and post-processing tools.
4. **MSC Apex / MSC Nastran (Hexagon):** Known for legacy aerospace validation, generative design, and interactive, geometry-direct meshing environ-

ments.

5. **COMSOL Multiphysics:** Highly specialized for strongly coupled, customized multiphysics PDE modeling using flexible mathematical formulations.

Below is a detailed examination of the Siemens Simcenter architecture, highlighting its core structural engine and its expanded capabilities following the strategic integration of Altair's pre-processing technologies.

9.2 Simcenter Nastran Architecture & Capabilities

The primary finite element method (FEM) solver within the Siemens Simcenter portfolio is **Simcenter Nastran** (formerly known as NX Nastran). It is widely recognized across the aerospace, automotive, and heavy machinery sectors as a premier, industry-standard implicit structural solver engineered for performance, accuracy, and scalability.

9.2.1 Portfolio Context & Integration

Simcenter Nastran serves as the mathematical engine ("the solver") that computes the discretized governing algebraic and differential equations. Because it operates purely as a solver without a native graphical user interface (GUI), it relies on seamless integration with primary pre- and post-processors to establish an end-to-end simulation workflow:

- **Simcenter 3D:** A multi-disciplinary CAE environment tightly coupled with CAD (such as Siemens NX). It is highly optimized for managing complex geometric parameters, associative design updates, and automated multi-physics workflows.

- **Simcenter Femap:** An advanced, mesh-centric, CAD-independent finite element modeling environment. It is preferred by analysts who require meticulous, element-level control over mesh topologies, entity definitions, and direct manual card editing.

9.2.2 Core Solution Capabilities

The solver processes a broad spectrum of physics using standardized solution sequences (denoted by SOL numbers). It operates via consistent input/output file formats (such as `.dat` input decks, `.op2` binary output files, and `.pch` punch files) to maintain structured data pipelines:

1. Structural & Dynamic Analysis

- **Linear Statics & Buckling (SOL 101/105):** The baseline sequence for evaluating structural strength, stress concentrations, and critical load factors prior to the onset of geometric instability.

- **Normal Modes & Vibrations (SOL 103):** Computes natural frequencies and mode shapes. It utilizes highly optimized eigenvalue extraction routines, including **RDMODES** (Recursive Domain Lanczos Method), to solve massive structural models containing thousands of modes efficiently.
- **Dynamic Response (SOL 111/112):** Evaluates structural performance in both the frequency and time domains under transient, harmonic, or random acoustic loadings.

Figure 9.1 is an example FEM model for a gearbox housing.

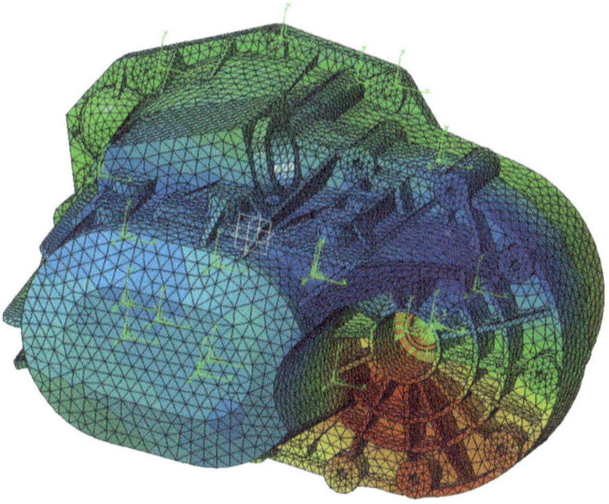

Figure 9.1: FEM model of a gearbox housing. Courtesy of Siemens DISW.

2. Advanced Nonlinear Formulations

- **Multi-step Nonlinear (SOL 401/402):** Supports large displacements, large strains, and complex constitutive material nonlinearities (e.g., plasticity, hyperelasticity, creep). It provides robust contact mechanics algorithms that handle deformable-to-deformable and rigid-body interactions with automated time-stepping control.

3. Simcenter Nastran Thermal (SOL 153 / SOL 159) For structural analysts, Simcenter Nastran includes built-in thermal solution sequences:

- **SOL 153 (Steady-State):** Solves static thermal equilibrium profiles.
- **SOL 159 (Transient):** Tracks time-varying temperature histories under fluctuating thermal loads.
- **Thermal-Structural Coupling:** This workflow is highly efficient for determining **thermal stresses**. The temperature distribution calculated by the thermal solver is directly mapped onto the structural mesh (often within the

same simulation run) to evaluate thermal expansion, distortion, and thermal fatigue.

4. Simcenter 3D Thermal & Flow Solvers For dedicated thermal engineers, Simcenter 3D provides a specialized, non-linear finite volume-based thermal engine that handles severe physics complexities:

- **Orbital Thermal Analysis:** Computes solar radiation, albedo, and earth infrared heating for satellites and spacecraft navigating transient orbital trajectories.

- **Electronic Cooling:** Provides macro-components like heat sinks, heat pipes, thermoelectric coolers (Peltier elements), and printed circuit boards (PCBs) with detailed layer-stack thermal definitions.

- **Thermal Network Modeling:** Allows users to mix 3D discretized FE/FV meshes with simplified 1D thermal resistance networks (lumped parameters) to optimize solution times for massive assemblies.

- **Advanced Contact Thermal Resistance:** Models the temperature drop across microscopic air gaps or thermal interface materials (TIMs) at structural joints, varying the thermal resistance dynamically based on localized structural contact pressures.

Figure 9.2 is an example of FEM model for an automotive headlight.

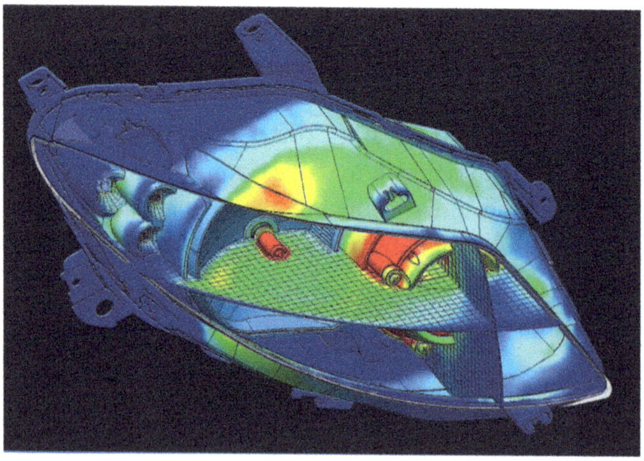

Figure 9.2: FEM model of an automotive headlight. Courtesy of Siemens DISW.

Figure 9.3 is an example of FEM model for thermal analysis in a gas turbine.

Figure 9.3: FEM model for thermal analysis in a gas turbine. Courtesy of Siemens DISW.

GasturbineThermalAnalysis

5. Specialized Disciplines

- **Aeroelasticity & Flutter (SOL 144/146):** Evaluates the complex interaction between aerodynamic loads and structural stiffness, which is a critical regulatory requirement for aircraft certification.
- **Rotor Dynamics:** Predicts the dynamic stability, Campbell diagrams, and critical speeds of rotating systems such as turbines, shafts, and propellers.

Figure 9.4 is an example of FEM model for an ATV differential.

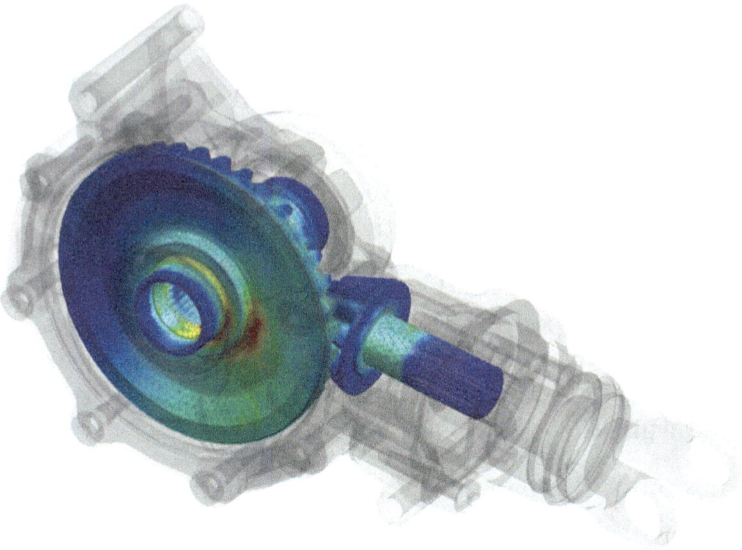

Figure 9.4: FEM model of an ATV differential. Courtesy of Siemens DISW.

9.2.3 Strategic Expansion: Altair HyperMesh Integration

Following Siemens' acquisition of Altair Engineering, the Simcenter ecosystem has significantly expanded its pre-processing capabilities by incorporating **Altair HyperMesh** into its portfolio. This integration provides analysts with an additional, world-class option for high-fidelity mesh generation alongside Simcenter 3D and Femap.

[CAD Geometry Input]

⇓

Pre-Processing & Meshing

- Simcenter 3D (CAD-Centric / Associative)
- Simcenter Femap (FEA-Centric Control)
- Altair HyperMesh (Large Assembly / AI)

Integrated Domain

(.dat / Solver Deck)

⇓

Solver

Simcenter Nastran

(SOL 101, SOL 103, SOL 401/402, etc.)

9.2.4 HyperMesh Role in the Simulation Workflow

HyperMesh is an industry-standard, solver-agnostic pre-processor designed explicitly for creating, editing, and managing massive, complex simulation models. Within the Simcenter framework, it serves several critical functions:

- **High-Fidelity Meshing for Complex Geometries:** It provides industry-leading mid-surfacing, advanced geometry cleanup tools, and fine-grained element quality controls. This ensures highly accurate mesh discretization for complex, cast, or thin-walled structures where automated tria/quad meshing typically struggles.

- **Large Assembly Management:** HyperMesh utilizes a structured hierarchy of **Parts, Subsystems, and Connectors**. This allows engineering teams to manage system-level models (such as full automotive chassis or aircraft fuselages) by leveraging automated connector controls for welds, rivets, and structural adhesives.

- **AI-Augmented Pre-Processing:** HyperMesh introduces embedded AI tools, such as automated shape and pattern recognition. It can group geometrically similar parts, automate material assignments via CAD metadata, and use **PhysicsAI** neural architectures to generate real-time structural predictions based on historical simulation data before invoking the implicit Nastran solver.

9.2.5 High-Performance Computing (HPC) Architecture

To process modern engineering models containing tens of millions of degrees of freedom (DOFs), Simcenter Nastran employs advanced parallel processing architectures:

1. **Shared Memory Parallel (SMP):** Leverages multiple CPU cores and shared memory space within a single workstation to accelerate localized element-level operations and matrix assemblies.
2. **Distributed Memory Parallel (DMP):** Distributes the global computational workload across multiple independent cluster nodes connected over a high-speed, low-latency network (e.g., InfiniBand), executing via MPI protocols.
3. **Superelements (Substructuring):** Decomposes a massive global finite element system into smaller, mathematically equivalent substructures. This allows engineering teams to solve individual components independently, store their reduced stiffness/mass matrices, and swap component variants without re-solving the entire global system.

9.2.6 Customizability via DMAP

A distinguishing technical feature of Simcenter Nastran is **DMAP (Direct Matrix Abstraction Program)**. DMAP is a proprietary, high-level structural programming language operating inside the solver environment.

It allows advanced analysts to:

- Intercept standard solution sequences at specific execution blocks.
- Manipulate global stiffness ($\mathbf{K}$), mass ($\mathbf{M}$), and damping matrices directly.
- Extract custom state variables or inject specialized constitutive algorithms.
- Export precise data structures directly to external scientific computing environments (e.g., Python using NumPy/SciPy arrays) for customized post-processing and research verification.

Bibliography

[1] Robert A. Adams and John J. F. Fournier. *Sobolev Spaces*, volume 140 of *Pure and Applied Mathematics*. Elsevier/Academic Press, Amsterdam, second edition, 2003.

[2] G.R. Liu. *Mesh free methods: moving beyond the finite element method, 2nd Ed.* CRC press, Taylor and Francis Group, New York, 2010.

[3] G.R. Liu. *Machine Learning with Python: Theory and Applications.* World Scientific, New Jersey, 2023.

[4] G.R. Liu. *Mechanics of Materials: formulations and solutions with Python.* World Scientific, New Jersey, 2024.

[5] G.R. Liu. *Numbers and Functions: Theory, Formulation, and Python codes.* World Scientific, New Jersey, 2024.

[6] G.R. Liu. *Calculus: a practical course with Python.* World Scientific, New Jersey, 2025.

[7] G.R. Liu. *PINN with Python: An Introduction.* ScienTech Publisher, Mason, USA, 2025.

[8] G.R. Liu. *Solid Mechanics with Python: Cartesian Coordinates.* World Scientific, New Jersey, 2025.

[9] G.R. Liu. *Vector Calculus: Formulations, Applications, and Python Codes.* World Scientific, New Jersey, 2025.

[10] G.R. Liu. *Linear Algebra with Python: Modern Computational Approaches.* ScienTech Publisher, Mason, USA, 2026.

[11] G.R. Liu and T.T. Nguyen. *Smoothed finite element methods.* CRC Press, New York, 2010.

[12] G.R. Liu and Siu Sin Quek. *The finite element method: a practical course, 2nd Ed.* Butterworth-Heinemann, London, 2013.

[13] G.R. Liu, J. Tani, T. Ohyoshi, and K. Watanabe. Transient waves in anisotropic laminated plates, part 1: Theory. *Journal of Vibration and Acoustics*, 113(2):230–234, 1991.

[14] G.R. Liu, J. Tani, T. Ohyoshi, and K. Watanabe. Transient waves in anisotropic laminated plates, part 2: Application. *Journal of Vibration and Acoustics*, 113(2):235–239, 1991.

[15] Yu-Qiu Long. *Finite Element Method and Its Applications.* Tsinghua University Press, Beijing, 2000.

[16] Larry J. Segerlind. *Applied Finite Element Analysis.* John Wiley & Sons, New York, 2 edition, 1984.

[17] O. C. Zienkiewicz and R. L. Taylor. *The Finite Element Method for Solid and Structural Mechanics.* Butterworth-Heinemann, Oxford, 6 edition, 2005.

Index

Made in United States
North Haven, CT
04 November 2026

21728893R00226